ENERGY, GOVERNANCE AND SECURITY IN THAILAND AND MYANMAR (BURMA)

Transforming Environmental Politics and Policy

Series Editors:

Timothy Doyle
Keele University, UK and University of Adelaide, Australia

Philip Catney
Keele University, UK

The theory and practice of environmental politics and policy are rapidly emerging as key areas of intense concern in the first, third and industrializing worlds. People of diverse nationalities, religions and cultures wrestle daily with environment and development issues central to human and non-human survival on the planet Earth. Air, Water, Earth, Fire. These central elements mix together in so many ways, spinning off new constellations of issues, ideas and actions, gathering under a multitude of banners: energy security, food sovereignty, climate change, genetic modification, environmental justice and sustainability, population growth, water quality and access, air pollution, mal-distribution and over-consumption of scarce resources, the rights of the non-human, the welfare of future citizens – the list goes on.

What is much needed in green debates is for theoretical discussions to be rooted in policy outcomes and service delivery. So, while still engaging in the theoretical realm, this series also seeks to provide a 'real world' policy-making dimension. Politics and policy making is interpreted widely here to include the territories, discourses, instruments and domains of political parties, non-governmental organizations, protest movements, corporations, international regimes, and transnational networks.

From the local to the global – and back again – this series explores environmental politics and policy within countries and cultures, researching the ways in which green issues cross North-South and East-West divides. The 'Transforming Environmental Politics and Policy' series exposes the exciting ways in which environmental politics and policy can transform political relationships, in all their forms.

Other title in the series:

Community Gardening as Social Action
Claire Nettle

Energy, Governance and Security in Thailand and Myanmar (Burma)

A Critical Approach to Environmental Politics in the South

ADAM SIMPSON
University of South Australia, Australia

Routledge
Taylor & Francis Group

LONDON AND NEW YORK

First published 2014 by Ashgate Publishing

2 Park Square, Milton Park, Abingdon, Oxfordshire OX14 4RN
52 Vanderbilt Avenue, New York, NY 10017

Routledge is an imprint of the Taylor & Francis Group, an informa business

First issued in paperback 2020

British Library Cataloguing in Publication Data
A catalogue record for this book is available from the British Library

The Library of Congress has cataloged the printed edition as follows:
Simpson, Adam.
Energy, governance and security in Thailand and Myanmar (Burma): a critical approach to environmental politics in the South / by Adam Simpson.
  pages cm — (Transforming environmental politics and policy)
Includes bibliographical references and index.
ISBN 978-1-4094-2993-7 (hardback : alk. paper) — ISBN 978-1-4094-2994-4 (ebook) — ISBN 978-1-4094-7394-7 (epub) 1. Environmental policy—Asia, South. 2. Energy policy—Asia, South. 3. Environmental protection—Asia, South. 4. Security, International—Environmental aspects. 5. Environmental justice—Asia, South. 6. South Asia—Strategic aspects. I. Title.
 GE190.A784S56 2013
 333.709591—dc23

2013009602

ISBN 13: 978-1-4094-2993-7 (hbk)
ISBN 13: 978-0-367-60543-8 (pbk)

For Ann, Graham and Guy

Contents

List of Figures and Tables

Figures

Tables

List of Acronyms

AASYC	All Arakan Students and Youth Congress
ADB	Asian Development Bank
ALD	Arakan League for Democracy
AOW	Arakan Oil Watch
ASEAN	Association of Southeast Asian Nations
ATCA	Alien Torts Claim Act
BLC	Burma Lawyers' Council
BRN	Burma Rivers Network
BSPP	Burma Socialist Programme Party
CBO	community-based organisations
CGG	compromise governance group
CNOOC	China National Offshore Oil Corporation
CNPC	China National Petroleum Corporation
CSR	corporate social responsibility
DKBA	Democratic Karen Buddhist Army
ECODEV	Economically Progressive Ecosystem Development
EG	emancipatory group
EGAT	Electricity Generating Authority of Thailand
EGG	emancipatory governance group
EGS	environmental governance state
EIA	environmental impact assessment
EITI	Extractive Industries Transparency Initiative
ERI	EarthRights International
ESCR-Net	International Network for Economic, Social and Cultural Rights
EU	European Union
FDI	foreign direct investment
FoE	Friends of the Earth
FoEI	Friends of the Earth International
GAIL	Gas Authority of India Ltd
GSP	gas separation plant
IDP	internally displaced peoples
ILO	International Labour Organization
IMF	International Monetary Fund
KCG	Kanchanaburi Conservation Group
KDRG	Karenni Development Research Group
KEG	Karenni Evergreen
KESAN	Karen Environmental and Social Action Network

KHIS	Korean House for International Solidarity
KHRG	Karen Human Rights Group
KNLA	Karen National Liberation Army (military wing of KNU)
KNPP	Karenni National Progressive Party
KNU	Karen National Union
KORD	Karen Office of Relief and Development
KRW	Karen Rivers Watch
KWO	Karen Women's Organisation
MDRI	Myanmar Development Resource Institute
MOGE	Myanmar Oil and Gas Enterprise
MoU	memorandum of understanding
MSG	Multi-Stakeholder Group
NCGUB	National Coalition Government of the Union of Burma
NGO	non-governmental organisation
NIMBY	not-in-my-back-yard
NLD	National League for Democracy
NSM	New Social Movement
ONGC	Oil and Natural Gas Corporation Ltd (India)
PAD	People's Alliance for Democracy
PPP	People's Power Party (Thailand – formerly TRT)
PTP	Pheu Thai Party (Thailand – formerly PPP)
PTTEP	Petroleum Authority of Thailand – Exploration and Production
PWYP	Publish What You Pay
RNDP	Rakhine Nationalities Development Party
RNP	Rakhine National Party
SAIN	Southeast Asian Information Network
SGM	Shwe Gas Movement
SLORC	State Law and Order Restoration Council
SPDC	State Peace and Development Council
TERRA	Towards Ecological Recovery and Regional Alliance
TNC	transnational corporation
TRT	Thai Rak Thai (Thailand)
TTM	Trans Thai-Malaysian (Pipeline)
UDD	United Front for Democracy Against Dictatorship
UN	United Nations
USDA	Union Solidarity and Development Association
USDP	Union Solidarity and Development Party
WCS	Wildlife Conservation Society (US)
WWF	World Wide Fund For Nature

Notes on Language and Terminology

Throughout the book I have used the Myanmar government's new terminology for most names and places. This is not intended as a political statement on the validity of these changes, which were made under the previous military government, but simply a recognition that many names are unlikely ever to revert back to the original, although some old terms are still in common usage and employed here. In particular, I use the term 'Salween River' throughout the book instead of the official 'Thanlwin River' as it is the subject of one of the campaigns and the new term is rarely used. Below are some of the new terms used throughout this book:

Myanmar – formerly Burma
Yangon – formerly Rangoon
Kayin – formerly Karen
Kayah – formerly Karenni
Rakhine – formerly Arakan
Bamar – formerly Burman
Tanintharyi Region – formerly Tenasserim Division
Ayeyarwady River – formerly Irrawaddy River

Thai family names are often long and unfamiliar, even to other Thais, and both ethnic Bamar (Burmans) and most ethnic minorities in Myanmar (except the Chins, Kachins and Nagas) have no family name. I have therefore followed the custom adopted by other academics specialising in this region by citing Thai and Myanmar authors in the text and reference list by their first, rather than last, names (Brown 2004; Fink 2009; Hewison 2005; Lintner 1999: 496; McCargo and Ukrist Pathmanand 2005; Warr 2005: xv). Romanisation of Thai and Myanmar names and words can result in several different spellings. I have endeavoured to maintain consistency throughout the book but in cases with various spellings in common usage I have noted the alternative spellings.

Series Editors' Preface

The beginnings of this series emerged at Keele University in a collaboration between Tim Doyle and Phil Catney. Since the late 1970s, Keele has been renowned across the globe as one of the leading universities engaged in teaching and research into the politics and international relations of the environment.

Our initial conversations with Ashgate were around two objectives. First, we wanted to transform the rather narrow, dominant conceptions of environmental politics and policy – particularly in the global North – by opening it up to include issues more central to traditional human politics and policy making. Gone are the days when the 'environment' is something that people engage in (and with) as some kind of 'luxury' pursuit, when and if they have the time and the resources to do it. Nowadays, environmental issues – both in the North and the South – are front and centre. Secondly, we strongly felt that much needed in environmental debates was for theoretical discussions to be rooted in policy outcomes and service delivery. In short, the series would look at the exciting ways in which environmental politics and policy can transform governance, in all its forms.

Timothy Doyle, Keele University, UK and University of Adelaide, Australia
Philip Catney, Keele University, UK

Acknowledgements

Many people have contributed to the development of this book in a variety of ways and over many years. Tim Doyle at the University of Adelaide, in particular, has been my central academic mentor, good friend and collaborator since the mid-1990s when he guided me through an entertaining Masters degree at the Mawson Graduate Centre for Environmental Studies. Elaine Stratford and Doug McEachern also provided mentoring during those years. Following an academic interregnum in London I returned to the School of History and Politics at Adelaide where Tim and Juanita Elias provided indispensable supervision and assistance during my time as a doctoral student and Associate Lecturer. Clem Macintyre, as Head of Politics, was always tremendously supportive.

Thanks are also due to my colleagues at the University of South Australia for the mid-corridor debates, particularly Lis Porter who as my original Head of School showed faith in me and an interest in my research. Unstinting support from subsequent School Heads Kerry Green and Clayton MacKenzie was also much appreciated. Kate Leeson from the Hawke Research Institute provided invaluable editorial assistance. There are many other academic colleagues around the world who have provided feedback and guidance over the years. If I start listing some others will be omitted so let me just send out a general thankyou to all the academics who have helped me along the way.

The editorial team at Ashgate, led by Rob Sorsby, were always helpful and thanks are also due to Rosaleen Duffy and Giorel Curran who provided constructive comments on earlier work related to the manuscript.

My gratitude must also go to the environmental activists across the South, particularly in Thailand and Myanmar, who helped me in my research. In particular, the activists in the NGO EarthRights International, including the co-founders Ka Hsaw Wa, Katie Redford and Tyler Giannini, found time for me in their busy schedules for numerous interviews and communications over many years, for which I am enormously indebted. Many other activists cannot be identified for their own security, but those who are listed in the interviews, whether under pseudonyms or not, retain my utmost appreciation. I hope this book ends up holding something of interest for them after all those hours of gruelling questions.

Many good friends have provided sounding boards for late night political discussions over many years during this research and I thank all these collaborators for their friendship and sometimes feisty contributions. Lastly, Lisa and Kyela have provided me with love and support throughout the somewhat arduous and convoluted journey on which this book has taken us, and for that I am eternally grateful.

Adam Simpson
September 2013

Chapter 1
Introduction

Introduction

Access to cheap and plentiful energy is the foundation of modern economies and the search for energy security is one of the key dynamics that is re-shaping global politics, governance and security in the twenty-first century. With the energy needs of Asia rapidly increasing centrally placed Thailand and Myanmar (Burma) are taking on greater significance in the emerging energy supply and demand chains that criss-cross the region. Contracts for hydropower dams, gas pipelines and other large energy projects have been signed at a furious rate in Beijing, Delhi, Bangkok and Naypyidaw, the new Myanmar capital.

While helping alleviate energy security concerns in relatively affluent states the growing global reliance on energy sources from energy-rich states in the less affluent South has also resulted in detrimental effects on the environmental security of marginalised communities across the South. Effective environmental governance of energy projects, particularly those that cross national borders, is therefore necessary to ensure that the pursuit of energy security does not exacerbate local injustices or fuel localised environmental insecurity. Globally, an increasing understanding of environmental concerns has led to improved environmental governance at many levels but often the most important issues remain the least well governed; energy – and the impacts of its production, trade and consumption – provides a key example. The centrality of energy security to modern states and economies ensures that it is often a key focus of foreign policy activities but there have been limited attempts to construct an effective global energy governance system and those that do exist have often bypassed the United Nations, the central global governance institution (Florini and Sovacool 2011; Karlsson-Vinkhuyzen 2010; Lesage, Van de Graaf and Westphal 2010). In the absence of a coherent global system, formal governance is largely undertaken in an ad-hoc manner at regional or national levels. Although this arrangement is far more subject to the vagaries of national political regimes it can also allow for less powerful non-state actors to influence local or regional outcomes.

The formal governance of transnational energy projects is usually undertaken by an array of administering bodies attached to the governments of the states involved in the projects. Environmental activists can play an important informal role in communicating community concerns to these bodies, as well as to transnational corporations (TNCs) and their governments, but the extent of their influence depends on the nature of the political regimes under which they operate. This activism is most efficacious under democratic systems where domestic popular opinion is regularly tested in free and fair elections, although it can also sway more authoritarian regimes (Mertha 2009: 1002–6). Despite regular opposition by powerful business and political interests, this activism, particularly in its emancipatory form, is a potentially significant tool in contributing to the environmental governance of transnational energy projects.

The history of the two core states in this book, Thailand and Myanmar, contrasts the opportunities and openings available for engaging in this activism, which can be defined as 'activist environmental governance', under two very different political regimes. Despite the democratic limitations in Thailand's political landscape, including a recent intervention by the military, in general there have been significant opportunities for political dissent and debate. As a result Thailand developed a dynamic, if fragmentary, domestic environment movement that played a key role in the environmental governance of its transnational energy projects. In contrast, enduring military rule in neighbouring Myanmar provided few opportunities for domestic activism. Until the new quasi-civilian government was formed under President Thein Sein in 2011 activists held no hope of directly influencing their own government. As a result they focused almost entirely on transnational modes of environmental governance, particularly those exiled activists who removed themselves from the military's sphere of influence to the contested border regions.

This activist environmental governance is particularly important in the South, where environmental security is most precarious and energy-rich states are often ruled by authoritarian or illiberal regimes. Due to either limited will or governance capabilities, or both, the effectiveness of formal environmental governance institutions and regimes in these regions is particularly lacking. For states in the South with plentiful energy resources the export of energy via transnational energy projects takes on a high priority, either as a source of government revenue for development or a stream of rent that facilitates corruption. Under military rule in Myanmar – a state with few established democratic institutions – five decades of authoritarianism and relative international isolation ensured that rent seeking was the norm, leaving much of the country in poverty. In Thailand, characterised by a more dynamic economy and civil society, corruption and rent seeking still influenced decision making, although the benefits of development were more widely distributed.

As the analysis in this book suggests, the pervasiveness of environmental insecurity within a country often mirrors the degree of authoritarianism that characterises its domestic political regime. In situations where states are either unable or unwilling to provide environmental security for their citizens, environmental activists often provide the most effective environmental governance of cross-border energy projects. The conditions that face environmental activists in the South are, however, fraught with risks and hazards that are entirely foreign to most activists in the North and which provide significant impediments to engaging in activism. In Thailand activists faced harassment, and occasionally assassination, by developers and the state but in Myanmar civil conflict between the Myanmar military and ethnic minorities, widespread poverty and, until recently, a repressive authoritarian state, combined to stifle domestic dissent and significantly magnify the hurdles to undertaking activism.

These circumstances can be illustrated by the situation on a remote conflict-ridden stretch of the Salween River where it forms the border between Thailand and Myanmar.[1] The Ei Tu Hta camp for ethnic Kayin (Karen) internally displaced peoples (IDPs) was established in 2006 on the river between the proposed Dar Gwin and Wei Gyi Dam sites in Karen National Union (KNU)–controlled Myanmar. In 2009 Hsiplopo, the camp leader, was unable to visit his family. Although they only lived three hours walk away, the camps of the Tatmadaw, the Myanmar military with which the KNU was engaged in the world's longest-running civil war, lay in between.[2]

Boxed in against the western shore of the Salween River, the camp was also built on the steep hillsides of a valley, denuding the limited forest cover to provide accommodation in the narrow area available. Due to poor soils and limited space the residents were unable to grow their own rice, relying instead on regular donations from the UN and other non-governmental organisations (NGOs) shipped upriver by longtail boat.[3] This type of insecurity coloured the daily existence of both the Kayin people in this camp and many other ethnic minorities in Myanmar. Nevertheless, despite these conditions, Hsiplopo's commitment to the campaign against the nearby dams was resolute: 'we don't want dams ... the military cannot build the dams because the KNU will not let them while the people do not want them'.[4]

Hsiplopo's stance reflected that of many environmental activists and groups who inhabited the nebulous and dangerous borderlands of eastern Myanmar. The dams were opposed for many reasons: they were likely to require forced labour from local ethnic minority communities; they would submerge villages and large areas of pristine forest and arable land; they would adversely impact food security and fisheries; they would cut off a major route for refugees fleeing repression into Thailand; and they were unlikely to alleviate energy insecurity for the local ethnic communities. While the campaign against the dams emphasised the universal human rights of the affected ethnic minority communities in Myanmar, it also promoted their culturally specific identities and was emancipatory in its outlook. This cultural particularism extended into the ecological realm where the activists highlighted the importance of indigenous knowledge of biodiversity, making a direct connection between environmental and political concerns (KESAN 2008: 5). Despite the civil conflict, exiled Myanmar environmental groups undertook

---

1 The Salween River is now officially known as the Thanlwin River but this term is still rarely used so throughout this book I have continued to use the term 'Salween'.

2 Hsiplopo, interview with author, Ei Tu Hta Camp, KNU-controlled Myanmar on the Salween River, 6 January 2009.

3 Nay Tha Blay, interview with author, Mae Sariang, Thailand, 7 January 2009.

4 Hsiplopo, interview with author, Ei Tu Hta Camp, KNU-controlled Myanmar on the Salween River, 6 January 2009.

perilous work with the KNU in this region to promote human and environmental security for the Kayin people. As an exiled activist from the Karen Environmental and Social Action Network (KESAN) explained: 'KESAN's programs are in the KNU area [in Myanmar] so we have a close relationship with the KNU leaders'.[5]

It can be difficult for environmental activists from the affluent North, unfamiliar with this precarious existence, to fully comprehend the existential struggle that dictates much environmental activism in the South. As a result, many Northern environment movements, and the American environment movement in particular, have been largely apolitical, with the issues of 'human health, shelter, and food security' traditionally absent from their agendas (Doyle 2005: 26). Despite increased attention from the North much more research is required to provide a more robust and nuanced picture of environmental activism in the South.

Rationale for this Book

This book developed during a decade and a half of research on environmental activism in Thailand and Myanmar.[6] Its origin can be linked to a residential course I was attending on Buddhist economics in 1998 at Schumacher College in the UK where one of the course teachers, Sulak Sivaraksa, a renowned Thai social activist and advocate of Engaged Buddhism, told me about forest protests that he was participating in over the Yadana Gas Pipeline Project that was to carry natural gas from Myanmar to Thailand.[7] Later that year I travelled to Thailand to make contact with the major environmental actors involved with the protests including the transnational NGO EarthRights International (ERI) and the local Kanchanaburi Conservation Group (KCG). Many of the issues that activists were addressing In this campaign were quite different from the ones often examined by scholars from the North, including my own previous research (Simpson 1998). The Yadana Pipeline was to transport gas through the Thai–Myanmar borderlands populated by the ethnic Kayin people. Decha Tangseefa described the experience of the people living in this region, many of whom, such as the IDPs at Ei Tu Hta camp, had been displaced from their homes in attacks by the Tatmadaw:

> Although these people are living in danger zones, the territorial sovereignty
> of the despotic state renders them imperceptible to the 'outside' world. Their

---

5   Alex Shwe, interview with author, Chiang Mai, Thailand, 8 January 2009.

6   Some of the research in this book has already appeared in various forms including in the Taylor & Francis journals *The Pacific Review* (Simpson 2013b), *Third World Quarterly* (Simpson 2007) and *Environmental Politics* (Doyle and Simpson 2006). Taylor & Francis was also generous enough to allow publication of research that appeared in a chapter in a book edited by Francesco Cavatorta on *Civil Society Activism under Authoritarian Rule: A Comparative Perspective* in the Routledge/ECPR Studies in European Political Science Series (2013a).

7   Sulak Sivaraksa, interview with author, Devon, UK, 25 January 1998.

sufferings have rarely been accounted for by the international community. Most of their stories have never been disclosed, and even when they have, they have often been ignored. No matter how loud they have screamed, a large number of forcibly displaced peoples 'inside' the Burmese nation-state have been tortured and killed without being heard as they dissolve back to the soil they hoped would be their homelands. (Tangseefa 2006: 405)

As my research project developed it became apparent that insecurity in these communities was exacerbated by the civil conflict and environmental degradation that accompanied large-scale energy projects. The research for this book therefore coalesced around the attempts by environmentalists to improve human and environmental security for local communities by contributing to the environmental governance of four transnational energy projects based in Thailand and Myanmar. It became clear that the extent and nature of the environmental campaigns against these projects was highly dependent on the level of authoritarianism of the political regimes under which the activists operated, and that this affected local and transnational activism differently. Local and transnational business interests that supported the energy projects also collaborated with illiberal political regimes in the pursuit of profits and rents. It became apparent that, while the proponents often cited improved energy security as a rationale for pursuing the projects, the actual impacts on the environmental security of local communities were often detrimental. This paradox drove the research project from its inception.

By examining the campaigns against these energy projects in Thailand and Myanmar I focus on how environmental politics is played out in both the states and transnational spaces of the less affluent South. Throughout the book I use the terms 'North' and 'South' as useful shorthand to distinguish between states, regions or communities that differ markedly in affluence. Interests in particular countries are far from homogenous, however, and throughout the countries of the South 'one can find dominant "local" elites supporting and sustaining global capitalism' (Chaturvedi 1998: 704), while it is also challenged by counter-hegemonic forces allied to the marginalised and exploited (Gramsci 1971; Harvey 2005). As a result there is a North (affluent class) in what is generally termed the South (poor states) and vice versa. While using these dualisms indiscriminately can be problematic (Eckl and Weber 2007), they can be usefully employed if their shortcomings are acknowledged and understood.

The North and South differ not only in levels of affluence but also, as a result, in the issues on which their environment movements tend to focus. Southern movements are often more concerned about immediate existential 'environmental security' priorities, such as access to food and water, while Northern movements are often motivated by post-materialist or longer term issues such as wildlife conservation and climate change. These differences can also be discerned between countries within the South that exhibit relative disparities in wealth (Doyle and Simpson 2006). Although some environmental movements in the North have shifted their focus over the last two decades to include social justice issues, differences in foci between activists based in the South and those in the North remain. These differences are also reflected in academia, which is dominated by scholars in the North.

Despite an increased focus on environmental issues over the last two decades, most book-length approaches to environmental politics still examine predominantly ecological issues or regulatory regimes and focus particularly on the affluent states of the North (Howes 2005; Kutting 2000; Paehlke and Torgerson 2005). Although there has been increased attention on environmental movements in recent years, much of the material still focuses primarily on movements within the North (Bomberg and Schlosberg 2008; Carter 2007; Connelly et al. 2012; Doherty 2002; Doyle 2000; Dryzek et al. 2003; Gottlieb 2005; Hutton and Connors 1999; Paterson 2000; Rootes 2007; Sandler and Pezzullo 2007; Shabecoff 1993; Wapner 2010). Large business interests play a significant role in pursuing inappropriate development in the South, yet studies that examine the role of business in environmental politics also tend to focus on the business interests of the North (Blair and Hitchcock 2000; Doyle and McEachern 2008). There has been some analysis of environment movements in the South (Doherty 2006; Doherty and Doyle 2006; Doyle 2005; Duffy 2006; Dwivedi 1997; 2001), and various studies of transnational activism more generally (Atkinson and Scurrah 2009; Bandy and Smith 2005; Cohen and Rai 2000; della Porta et al. 2006; Edwards and Gaventa 2001; Eschle and Maiguashca 2005; Keck and Sikkink 1998; Khagram, Riker and Sikkink 2002; Reitan 2007; Routledge, Nativel and Cumbers 2006; Rupert 2000; Tarrow 2005), but few comparative studies examining how authoritarian regimes in the South impact on environmental activism or policy (Doyle and Simpson 2006; Fredriksson and Wollscheid 2007). There are numerous studies that examine civil society under authoritarianism more broadly but these tend to focus on more traditional and formalised civil society organisations (Jamal 2007; Liverani 2008; Sater 2007). Some studies have demonstrated the importance of domestic environmental movements in undermining authoritarian regimes, particularly in the former communist countries in the Soviet bloc (Galbreath 2010; Kerényi and Szabó 2006: 805), but the role of exiled environmental movements in particular remains understudied.

It is also rare to see book-length analyses of environment movements or campaigns using a multilevel (Dwivedi 2001) or multiscalar (Kaiser and Nikiforova 2006) approach. Most studies of activism tend to focus on the local (Ford 2013; Rootes 2008), or the transnational level (Reitan 2007), although a recent edited collection takes an innovative look at local activism in the South against transnational environmental injustices (Carmin and Agyeman 2011). None, however, undertake comparative analyses of activism within the same campaigns at both local and transnational scales. An edited collection by Piper and Uhlin (2004) considers transnational activism in Asia, with each case study providing some linkages to national activism in a different country, but, as with other edited books, it is a disparate collection of case studies by a variety of authors rather than an integrated book-length analysis.

A book by Forsyth and Walker (2008) provides useful analysis that complements the research I undertake here. It examines the construction of environmental knowledge in northern Thailand, a cultural and geographical territory that overlaps with the case material of this book, and is focused on the environmental narratives deployed by various actors to underpin their arguments over contested land use. It provides a juxtaposition of the mythical social construction of different upland ethnic groups as either 'forest guardians' or 'forest destroyers' and provides compelling arguments regarding the implications for conservation or development policies but, as with many other environmental works, it avoids mention of energy issues, focusing instead on forests, water and agriculture, and is a single country study.

In this book I contribute towards filling these gaps in the literature by adopting a comparative approach in the analysis of the strategies, tactics and organisation of local and transnational environment movements under two illiberal, yet distinct, political regimes in the South to develop a model of 'activist environmental governance'. I undertake a multilevel, multiscalar analysis that examines both the various levels of environment movements – individuals, groups, NGOs, coalitions and networks – and also the various scales at which activism is undertaken, particularly the local and transnational dimensions.

In addition to the focus on governance these environmental campaigns provide an opportunity for the theoretical development of critical approaches to energy and environmental security. The concept of energy security has an uneasy place within the environmental politics literature. While it is driving the transformation of relationships within global politics, particularly throughout Asia and the emerging economies of the global South, its customary connection to the more traditional field of security studies has left it under-analysed by more self-consciously critical approaches. With increasing rates of energy consumption predicted for many parts of the world there is an urgent need to critically re-assess the more traditional approaches to energy security and the assumptions on which they are based. Although there exist various studies of human and environmental security whose foci have shifted away from the state (McDonald 2012; Thomas 2000), the importance of energy to the military and economic power of modern industrialised societies has resulted in the concept of energy security remaining one of the last bastions of predominantly state-centric analysis. As with other aspects of security, however, the state is often not the best means of pursuing energy security for marginalised individuals or communities, particularly in non-democratic states (Bellamy and McDonald 2002). Furthermore, state-centric approaches tend to preclude both an emphasis on more localised communities and a normative emphasis on justice.

The concept of energy security plays a dual role throughout this book as proponents in receiver countries often cite it as the rationale for pursuing the energy projects while marginalised communities in the vicinity of the projects often remain energy insecure, even following the project's completion. The impacts of this energy exploitation are felt most acutely in the environmental capital and processes that these communities rely on, such as food, water and more sustainable localised energy sources. The research in this book demonstrates that this relationship was particularly relevant for ethnic minority communities in Myanmar, on whose land energy projects were often sited. Most communities in Thailand were more energy secure, due to more advanced energy and economic infrastructure, but energy projects still had the potential to adversely affect other aspects of their environmental security.

The social and political context within which these communities existed played a significant role in determining the specific outcomes of the projects, and whether they proceeded at all. A critical approach to environmental security, such as the one adopted by Barnett in his seminal book *The Meaning of Environmental Security* (2001), which acknowledges the relationship between these communities and both their environment and the socio-political structures they inhabit, is therefore best placed to capture the significance of these impacts. Barnett argued that environmental security should be defined as the way in which 'environmental degradation threatens the security of people' (2001: 12). This approach adopts a human security standpoint and focuses on the inequitable distribution of degradation resulting from unequal social structures; '[a] human-centred environmental security concept places the welfare of the disadvantaged above all else' (2001: 127). Although the concept of environmental security is also relevant for people in the North, in the South it tends to embody more immediately existential threats, with precarious living conditions due to poverty and authoritarian governance. In addition, the North can usually afford to mitigate detrimental environmental impacts but in the South perilous living conditions render environmental security and the struggle for justice inseparable.

Within the environmental politics literature there are, however, few studies that link environmental activism to critical approaches to environmental security. Most studies tend to focus primarily on the actual threats posed by environmental change rather than the response of activists and communities (Dodds and Pippard 2005; Doyle and Risely 2008; Floyd and Matthew 2013; Liotta et al. 2008). Although this book does not provide a detailed analysis of the environmental security implications for each energy project examined, the threat of environmental insecurity was a key rationale for the activism generated.

Due to its centrality to all life, energy security should be a central fixture within the critical environmental security literature but it has been largely overlooked, with the energy security debate dominated by the realist (Klare 2012) or liberal (Yergin 2011) state-centric streams of security studies. The concept can, however, be critically re-imagined by adopting a justice focus with a particular emphasis on marginalised individuals and communities. In recent articles Mulligan (2010; 2011) has defined his energy security approach as 'critical' but, although these are valuable additions to broader energy security debates, there remain aspects of his work that could benefit from more overtly 'critical' analysis. As Nunes (2012) has noted, the proliferation of the 'critical' label in security studies has led to critique being sometimes 'blunted'. Although in part this is an inevitable result of the popularity of alternative approaches, as there is less traditional analysis to critique, in the energy security literature there is no shortage of the more traditional analysis. Although Mulligan employs critical tools to conclude that security should shift from a state-centred military focus to one 'grounded in discourses of global and human security' (2010: 85), he tends not to address issues specific to the global South, where existential energy shortages are so prevalent. He also focuses primarily on fossil fuels, and peak oil in particular, which is not, in general, a critical concern; as Dalby notes 'oil is not a resource that the marginalised peasantry of the Third World are directly fighting over; it's a matter of superpower competition' (2009: 75).

In Mulligan's analysis of the security literature he focuses on the Copenhagen School's concept of 'securitisation', which can result in authoritarian responses by the state: 'securitisation is thus a tool that enables states to take exceptional measures, including repression or the suspension of the public freedoms considered normal in the West' (2011: 639). While the Copenhagen School is in some ways clearly constructivist, and it helped broaden the security agenda away from purely militaristic national security approaches, its deployment within 'critical' literature can have a limiting effect. As Browning and McDonald (2011) note, the logic of security in the Copenhagen School is inherently pernicious while within the Welsh School of Critical Security Studies, for example, it is inherently progressive. In comparison to the 'panic politics' that accompanies securitisation under the Copenhagen School, in the Welsh School approach 'true security refers to the emancipation of the poor and disadvantaged' (Floyd 2010: 48), which is a far more critical conception of security. As Nunes argues, the Copenhagen School and other approaches that conceive security as having an 'undesirable logic' have been detrimental to the 'commitment to politicisation that constitutes the cornerstone of critical security studies' (2012: 357).

As it was the more critically normative arguments in favour of equity and justice that drove the politicised activism in this book, in the last chapter I use the campaigns and the projects they oppose to develop the criteria for a model of both critical energy *and* environmental security that challenges the more traditional security studies approaches. These models develop and systematise some of my earlier thoughts on critical approaches to energy security (Simpson 2007; 2013c) and aim to clarify the key contributing elements of a genuinely critical analysis. This analysis can then be deployed in the evaluation of energy or other development policies to promote outcomes focused on sustainability and justice and can therefore provide guidance on manifesting 'emancipatory activist environmental governance'.

Research Methodologies

This book is fundamentally about environmental politics but it draws on the fields and subfields of political science, international relations, international political economy, sociology and environmental studies. It is also written in the same tradition as other practitioners and theoreticians within the academy who consider themselves to be scholar-activists engaging in emancipatory research (Humphries, Mertens and Truman 2000), who contribute to academia while also providing analysis that is useful for activists (Reitan 2007: 26–32). There is a parallel here between environmental politics and feminist writings, with both fields seeking to bridge the gap between an academic discipline and 'street activism'. Eschle and Maiguashca seek to find ways of converging towards this goal through

> 'politicised' or 'critical scholarship', that is, research that explicitly recognises and takes responsibility for its normative orientation; that aims to empower a marginalised and oppressed constituency by making them visible and audible; and that attempts to challenge the prevailing power hierarchies, including in terms of the construction of knowledge. (Eschle and Maiguashca 2006: 120)

This approach can also be found in critical international political economy where authors are committed to seeking 'emancipatory forms of knowledge' (Gill 2008: 6), and to promote change through 'more progressive, emancipatory values' (Bruff and Tepe 2011: 355). A critical approach to international political economy in Myanmar therefore prioritises the welfare of marginalised local communities (Simpson and Park 2013). Within environmental studies in general there exists a strong notion of 'advocacy, ... of critical and at times radical thought and propositions. Environmental studies has been for the environment in all its diverse and ambiguous orientations. In environmental studies there is a strategic goal: to bridge the gap between theory and practice' (Doyle 2000: xxx).

The research in this book focuses predominantly on radical groups and actors, although this radicalism is highly dependent on the political milieu in which they operate. It aims 'to empower a marginalised and oppressed constituency' (Eschle and Maiguashca 2006: 120), and call into question existing 'institutions and social and power relations' (Cox 1981: 129). As with other critical approaches this book is written with a 'shared commitment to human emancipation and a common concern to analyse the causes of, and prescribe solutions to, domination, exploitation, and injustice' (el-Ojeili and Hayden 2006: 10).

These disciplinary considerations have also had impacts on the methodological approach adopted. As Reitan argues, '[i]f the personal is political, I would also say it is methodological' (2007: 26). As a result I adopted an emic approach that values the insider's perspective and which Harris argues makes the research subject the 'ultimate judge' of the adequacy of an observer's description and analysis (1979: 32). I treated the accounts by interviewees as 'part of the world they describe' (Silverman 2001: 95). My main form of field research was informal semi-structured interviews conducted primarily in person, but also by phone or email. The main participants were environmental activists as the core focus of this book is the strategies and tactics of activists rather than the management of environmental issues by governments or the for-profit sector.

This book is therefore somewhat based on how the activist community sees themselves and their role in society. As Neuman notes, this focus on one perspective may draw misplaced accusations of bias, related to the 'hierarchy of credibility', which are often levelled for giving 'voice to parts of society that are not otherwise heard' (Neuman 2000: 377). Compounding this problem is that most of the case study organisations examined in this book are relatively small; EarthRights International (ERI), the largest organisation studied, had only approximately fifteen employees in its main Southeast Asia office. As a result, many of the interviewees were either the founders of their organisations or employees with significant responsibility. These interviewees may have had vested interests in characterising their organisations in a particular way, but they were also in the best position to provide the rationale for the organisation and its founding and operational values and philosophies. Indeed, as Yin notes, insights gained from interviews 'gain even further value if the interviewees are key persons in the organisations ... and not just the average member of such groups' (2009: 264).

During the research I undertook fieldwork in both Thailand and Myanmar as well as neighbouring countries in the region. There were particular security difficulties and ethical dangers in undertaking this type of research in and around Myanmar, where 'the experience of the researcher and the researched' (Lee-Treweek and Linkogle 2000: 5) were tied together; although my Australian and EU citizenship provided me with a level of security not enjoyed by the research subjects. Many of the interviewees had experienced trauma in various guises, such as being tortured by the military or participating in insurgent conflicts. Interviewing these subjects therefore raised ethical issues relating to objectivity and the relationship with subjects that researchers in other conflict-ridden societies have faced (Clark 2012).

I conducted most of the fieldwork during the period of military rule, although follow-up fieldwork was also undertaken in the more permissive atmosphere that emerged after the new 'civilian' government came to power in 2011. While some interviews were conducted in Myanmar proper and the insurgent-controlled 'liberated area', most of the actual sites of the energy projects were in inaccessible areas characterised by civil conflict or high security that were strictly off limits to foreigners. As an example of the difficulties undertaking research in this area, I had arranged, through various contacts, to stay two nights in a hotel in Kyauk Phyu, the main site for the Shwe Gas Project, in the last days of direct military rule soon after the 2010 election. As I clambered aboard the ferry from Sittwe at dawn the immigration officer announced that I was refused entry to Kyauk Phyu and was instead required to continue on to Taunggok. The local travel agent, who told me that 'they have been asking about you since you got here', pleaded with me to follow these instructions as otherwise 'it would cause trouble' for him. My fieldwork in Kyauk Phyu was therefore restricted to scouring the port area for the half hour that the ferry stopped for lunch, with various security personnel in hot pursuit.

Fortunately, for the research purposes of this book anyway, Myanmar's authoritarian rule and civil conflict resulted in what I argue was an 'activist diaspora' so that interviews could be more freely undertaken in neighbouring countries. As a result, while my fieldwork in Myanmar under military rule was extremely limited, I could readily undertake research in India for the Shwe Gas Project and in the border regions of Thailand for the Salween Dams and Yadana and Thai–Malaysian Gas Projects. My research in the Thai–Myanmar borderlands therefore answered the appeal by Tangseefa who called attention 'to the necessity and urgency of conducting academic field research in the dangerous areas in the Thai–Burmese border zones, in "the condemned grounds"' (2006: 406). The limitations on the ability to conduct research inside Myanmar itself also provided insights into the restrictions that face environmental activists, not only in that country but also under other similar authoritarian regimes such as Iran (Doyle and Simpson 2006).

Despite being interviewed in the relative safety of Thailand and India, exiles from Myanmar often preferred to remain anonymous, either due to security concerns for family members still in Myanmar or sometimes their own precarious residency status. In these relatively democratic countries, however, even Northern activists had security concerns and some requested anonymity. Many exiled Myanmar activists already operated under pseudonyms even within formalised NGOs. Even Ka Hsaw Wa, the well-known name of the co-founder and executive director of ERI, was a pseudonym, which, for several years, allowed him to travel freely under other names in the Thai–Myanmar borderlands.[8]

The Energy Projects

In this book I examine the campaigns against four transnational energy projects based in Thailand and Myanmar as case studies of activist environmental governance seeking to improve environmental security for marginalised communities. Three of the projects were transnational gas pipelines that were originally planned to link together the proposed South Asian regional energy grid with the proposed trans-ASEAN gas pipeline grid (ACE 2003; Chaturvedi 2005: 125; Simpson 2008; Sovacool 2009). The other project was actually a series of proposed large hydropower dams on the Salween River where the electricity was to be exported across the Thai and Chinese borders for foreign exchange. These dams were part of either the proposed Mekong Power Grid, a project promoted within the Greater Mekong Subregion forum with the backing of the Asian Development Bank (ADB), or the ASEAN Power Grid, which had the ambitious and ultimately unsuccessful goal of supplying electric power from a grid serving all ten member countries by 2011 (Osborne 2007: 5). All the projects were opposed by environmental activists for a variety of reasons related to adverse impacts on environmental security for local populations, in large part due to the lack of democratic participatory governance processes.

---

8   Ka Hsaw Wa, interview with author, Chiang Mai, Thailand, 14 January 2004.

The first project was the Yadana ('Jewel') Gas Pipeline, the first major cross-border gas pipeline in Southeast Asia and, in many ways, prototypical for this sort of project. It transported gas from the Gulf of Mottama (Martaban) across Tanintharyi Region (formerly Tenasserim Division) in southern Myanmar and across Kanchanaburi Province in Thailand. The Yetagun Gas Pipeline was constructed soon after the Yadana and followed a similar route across Myanmar with the two pipelines joining together at the Thai border with a 1.25 billion cubic feet per day (Bcf/d) capacity pipeline carrying the gas on to Ratchaburi (Rodger 2013). Although the issues arising from the Yetagun Pipeline were similar to the Yadana, the latter, as the original and most prominent project, was focused upon in the campaigns and is therefore also focused upon in this book. It was planned and built throughout the 1990s, with gas deliveries commencing eighteen months late in December 1999. It was a contentious project in both Myanmar and Thailand, being the subject of long-running human rights court cases against the oil and gas TNCs Unocal and Total, in the US and France respectively. The Petroleum Authority of Thailand Exploration and Production (PTTEP)[9] and Myanmar's state oil company, Myanmar Oil and Gas Enterprise (MOGE), were also partners in the project. The Electricity Generating Authority of Thailand (EGAT) purchased the gas and converted it into electricity at its Ratchaburi power plant (see Table 1.1). The authoritarian rule of the Myanmar military regimes at the time, the State Law and Order Restoration Council (SLORC) and, from 1997, the State Peace and Development Council (SPDC), ensured a lack of community involvement in this project although the TNCs and the Thai governments of the 1990s were also complicit.

Throughout the project activists focusing on the Thai side of the border argued that the project was unnecessary for Thailand's energy needs, that the Thai section of the pipeline route caused environmental damage and that Thailand's environmental laws had been broken. Activists and local ethnic minorities from Myanmar argued that the project exacerbated environmental degradation and military repression along the Myanmar section of the route, including increased forced labour and systematic rape in ethnic minority communities. After several years of litigation in the US by ERI and others on the basis of these claims, Unocal announced an out-of-court settlement with the Myanmar plaintiffs in early 2005. In June 2005 CNOOC, one of China's national oil companies, announced a takeover bid for Unocal that was blocked in the US, largely for national energy security reasons (ICG 2008b: 10; Miller 2010: 103). Chevron then launched a successful counter-bid for Unocal, taking up Unocal's 'minority non-operating interest' in the Yadana project (Chevron 2007). From that moment on Chevron became the focus of the US arm of the campaign.

The Thai government's support for the project on the basis of national energy security, along with the acquiescence of other states, resulted in increased

---

9 PTTEP is the exploration arm of PTT but PTT is sometimes used throughout this book to represent the activities of both PTTEP and PTT.

Table 1.1   The primary TNCs involved with the energy projects and their home states

Project	Primary TNCs (with home country governments also as a potential target for activism)
Yadana Gas Pipeline	PTT (Thailand), EGAT (Thailand), Unocal [now Chevron] (US), Total (France), MOGE (Myanmar)
Thai–Malaysian Gas Pipeline	PTT (Thailand), Petronas (Malaysia)
Shwe Gas Pipeline	Daewoo International (South Korea), PetroChina (China), Kogas (Korea), ONGC Videsh (India), GAIL (India), MOGE (Myanmar)*
Salween Dams	EGAT (Thailand), MDX (Thailand), Sinohydro Corporation (China)

* PetroChina was buying the gas from the other TNCs who owned the A1 and A3 offshore blocks. Although not initially a partner, MOGE used its 'step-in' rights in the Production Sharing Contract to take a 15 per cent stake in the blocks in 2008 after the discoveries were made (*The Economic Times* 2008).

environmental insecurity for Myanmar's ethnic minorities along the pipeline route under the ruling military regime. Promises of improved access to electricity for local communities remained unfulfilled, however, and figures from the state-owned newspaper, the *New Light of Myanmar*, showed that the per capita usage of electricity in Tanintharyi Division was still the second lowest in Myanmar six years after the pipeline's completion (Simpson 2007: 545; Thiha Aung 2005b).

The second energy project was the Trans Thai–Malaysia (TTM) Gas Pipeline Project (the 'Thai–Malaysian Pipeline'). In 1994 Thailand's PTTEP and Petronas of Malaysia signed Production Sharing Agreements (PSA) to exploit natural gas in the Malaysia–Thailand Joint Development Area in the Gulf of Thailand; another was signed a decade later (Petronas 2004; PTTEP 2004; TTM 2012). The TTM project called for offshore drilling, the construction of two gas separation plants (GSPs) in the predominantly Muslim Chana District of Songkhla Province and the laying of a gas pipeline from the GSPs on the coast to the border with Malaysia where it joined with Malaysia's pre-existing network. The concerns held by environmentalists, academics and local communities included an increase in air pollution, changes to the rural lifestyle of the local inhabitants and increased coastal pollution, including mercury from the drilling in an area predominantly populated by small-scale fishing families (Penchom Tang and Pipob Udomittipong 2003). There was ample evidence that the pipeline would not benefit the local communities, but peaceful attempts to influence decision makers were met with either indifference or violent repression. The government and developers initially argued that the project would only include a pipeline and GSP, but as these projects neared completion in 2007 a power plant and other industries were proposed to create demand for the gas in Thailand (Allison 2000; Kamol Sukin 2007; Lohmann 2007: 15). This was precisely the toxic industrialisation that occurred in fishing villages in Rayong and Chonburi on Thailand's eastern seaboard, which activists argued would occur again in Chana, despite protestations to the contrary by state authorities (Supara Janchitfah 2004: 27–33).[10] The historic marginalisation felt by Muslims in southern Thailand was exacerbated during the government of Thaksin Shinawatra by a crackdown on Muslim separatists in Songkhla's three neighbouring provinces, including two events in 2004 in which Thailand's security forces killed over 130 Muslims (Baker and Pasuk Phongpaichit 2005: 228; Gilquin 2005: 129–36; McCargo 2009; Simpson 2008: 218–20; Stein 2004; Supalak Ganjanakhundee 2005).

The third 'project' was a series of proposed hydroelectric dams on the Salween River in Myanmar, which included the Tasang, Kunlong and Nawngpha (or Nao Pha) Dams in Shan State, the Yawthit Dam in Kayah (Karenni) State, the Hat Gyi (or Hatgyi) Dam within Kayin (Karen) State and the Wei Gyi and Dagwin Dams on the border between Thailand and Kayin State. The dams were at various stages of development, with the Hat Gyi and Tasang the most advanced, but all faced campaigns over similar issues relating to environmental degradation and the military repression of ethnic minorities that accompanied the Yadana Project. The projects were scheduled to export most of the resultant electricity to Thailand and to divert water into Thailand's Bhumiphol and Sirikit reservoirs (Piya Pangsapa and Smith 2008: 493). In June 2006 China's largest hydropower company, Sinohydro Corporation, agreed to partner EGAT and build the $1 billion 1,200 MW Hat Gyi Dam Project (Corben 2006; Osborne 2007: 11). The previous December Myanmar's Department of Hydroelectric Power had signed a Memorandum of Understanding (MoU) with EGAT for the 'development, ownership and operation' of the Hat Gyi Hydropower Project (EGAT and DHP 2005). In April 2006 the Thai construction company MDX Group formed a $6 billion joint venture with the department to build the 7,110 MW Tasang power plant and in March 2007 China's state-owned Gezhouba Group announced that it had won a contract for the diversion tunnel as part of the dam construction (AP 2006; Sapawa 2007).

The campaigns against these dams were representative of broader campaigns by activists and ethnic minorities against environmental devastation and political repression in Myanmar. As Vandana Shiva notes of the resistance against the Narmada and Tehri Dams in India, the struggle of local communities was not just to preserve their homeland; it was against the destruction of entire civilisations and ways of life (Shiva 1989: 189). The Hat Gyi Dam site, in an ethnic Kayin region that was prone to civil conflict between the KNU, the Myanmar military

---

10   Prasart Meetam, interview with author, Hat Yai, Thailand, 11 February 2005.

and the Democratic Karen Buddhist Army (DKBA), reflected these issues.[11] Despite attempts to work on the Hat Gyi Dam, security was still tenuous, and EGAT suspended the project in September 2007 after two employees died from wounds associated with civil conflict (TNA 2007). Ethnic conflict in the area had already displaced 500,000 ethnic Kayin, and 140,000 refugees were registered in Thai refugee camps along the border. Various estimates suggested that between 75,000 and 100,000 further ethnic minority people would be displaced by the Hat Gyi and other Salween dams (DPA 2006; McLeod 2007; Pianporn Deetes 2007).

As with all the projects, the impacts on local communities were often detrimental while the benefits were accrued further afield. Large dams in the South tend to adversely impact local and marginalised communities as their livelihoods are disrupted with a World Bank–funded ecologist arguing that many large dams 'exacerbate poverty by damaging the fisheries and wetlands on which the poorest people depend most' (Pearce 2006: 10). The case of the Lawpita Hydropower Project in Kayah State is also instructive here. It was the first large hydroelectric project in Myanmar but three decades after its construction the power lines transported the electricity to distant sites while the local villages still lacked electricity (KDRG 2006: 33; KRW 2004: 68; Salween Watch and SEARIN 2004: 43).

The fourth project was the Shwe ('Gold') Gas Pipeline Project, which emerged in 2004 as a tri-nation project to pipe natural gas from the Bay of Bengal off Myanmar's Rakhine (Arakan) State to India via Bangladesh. The main partner corporations in the venture were South Korea's Daewoo International with a majority interest, Korean Gas Corporation (Kogas) and the Indian corporations Oil and Natural Gas Corporation (ONGC) Videsh Ltd and Gas Authority of India Ltd (GAIL). In January 2006, however, media reports emerged that the vice chairman of PetroChina and the Myanmar Ministry of Energy had signed an MoU in which the ministry agreed to sell gas from the offshore A1 Block through an overland pipeline to Kunming, and then Nanning, in Yunnan Province, China, for 30 years (*Financial Express* 2006; *India Daily* 2006; Tin Maung Maung Than 2005b: 265; Turnell 2007: 123; 2008: 962). A year later China National Petroleum Corp (CNPC), the state-owned parent company of the listed PetroChina (Newmyer 2008: 191), announced it was launching a feasibility study on a gas pipeline that would follow a proposed 1,250 km oil pipeline between Sittwe and Kunming. The gas pipeline would travel across the Rakhine Roma Range, central Myanmar and northern Shan State (AFP 2007a; *Pipeline and Gas Journal* 2007; Xinhua 2007). In August 2007 the Indian government and a senior energy ministry official from Myanmar finally announced that the gas from Myanmar's A1 and A3 blocks would be sold to China through PetroChina but it was not until later that year that Daewoo International, as the majority operator of the gas fields, confirmed the announcement (Mukul 2007; Reuters 2007a; 2007c; Simpson 2008: 221; Verma

---

11   The Karen National Liberation Army (KNLA) is the armed wing of the KNU but, as the actions of the KNLA largely represent KNU policy, references to the KNU throughout this book include the KNLA.

2007). CNPC and the Myanmar government signed further MoUs in 2009 that set the terms for the gas pipeline and the parallel overland oil pipeline that was to transport oil from the Middle East to Kunming, thus avoiding the congested and pirate-infested Straits of Malacca. The South East Asia Pipeline Company Ltd was then founded in June 2009 as a joint venture between CNPC and the state-owned Myanmar Oil and Gas Enterprise (MOGE) to build both pipelines with construction commencing in mid-2010 (Bo Kong 2010: 57). The gas pipeline was finally finished in May 2013 and in July the first deliveries of the contracted 400 million cubic feet per day (MMcf/d) were sent to China with another 100 MMcf/d of the gas reserved for domestic use (*The New Light of Myanmar* 30 July 2013).[12]

With the project lacking transparency and democratic oversight the campaign centred on concerns that the same environmental destruction and contempt for human rights that occurred in the Thai–Myanmar border region during construction of the Yadana Pipeline would be meted out to local peoples and ethnic minorities during this project (ERI 2004a). Rakhine State was one of Myanmar's poorest, with, according to the Myanmar regime, the lowest per capita electricity usage in the country, but there were no plans to use the gas to provide local electricity (Thiha Aung 2005a). Rather, 80 per cent of the gas was to be exported while the remainder was to be siphoned off by the military and its business associates.

The campaigns against these four energy projects were chosen for this book as they provided exemplars of transnational and, where political conditions allowed, local activist environmental governance in the South. The projects were focused on two states, Thailand and Myanmar, which shared a border and could therefore easily cooperate on transnational energy projects. Despite their proximity, however, they had very different political regimes and levels of environmental security, which allowed a direct comparison of the impacts of these variables on environmental activism at various scales. The linkages between the campaigns in the two countries were enhanced as much of the transnational activism over the Myanmar projects was undertaken in Thailand by Myanmar's activist diaspora. Despite differences in the structure of the two economies and the role that business played, transnational capital was central to the projects in each country, which provided an opportunity to examine the repertoires of action adopted by activists to target business actors under both authoritarian and more competitive regimes. Both the Yadana Pipeline and most of the Salween Dams were designed to supply Thailand with energy so an important distinction between the two countries in this research was that Thailand was largely an energy consumer, although it also exported gas to Malaysia through the Trans Thai–Malaysian (TTM) Pipeline, while Myanmar was solely an energy supplier and exporter. Elsewhere I have argued that through the Yadana Pipeline and Salween Dam Projects, along with the Nam Theun 2 Dam in Laos, 'the dominant

---

12   Analysts Wood Mackenzie estimated that the border price for gas in 2013, including tariffs, was approximately US$12 per thousand cubic feet (/Mcf) at the Thai border for the Yadana and Yetagun gas and $10.50/Mcf at the Chinese border for Shwe gas (Rodger 2013).

Table 1.2   Energy suppliers and receivers

Project	Supplier	Receiver
Yadana Gas Pipeline	Myanmar	Thailand
Thai–Malaysian Gas Pipeline	Thailand	Malaysia
Shwe Gas Pipeline	Myanmar	China*
Salween Dams	Myanmar	Thailand, China

* Initially this gas was destined for India and Bangladesh but during 2006–7 it emerged that the pipeline for this project would send gas to China.

classes have created an energy "love triangle" whereby Thailand exports the many problems associated with cross-border energy projects to its more authoritarian neighbours while importing the resultant energy' (Simpson 2007: 539). The problems were compounded in the projects examined here as the contracts specified that the supplier country was allocated only a small fraction of the energy produced, with the vast majority to be exported (see Table 1.2). Competition in the region for energy resources is only likely to increase with both India and China, in addition to Thailand, looking to Myanmar as the new 'crossroads of Asia' for future energy supplies, so the issues raised during these projects are likely to recur (Steinberg and Fan 2012; Thant Myint-U 2011).[13]

Despite Thailand's future role as the customer for the Salween Dams electricity, most of the Thai campaigns had wound down by the end of the Thaksin Shinawatra government in 2006. The Thai activism in the research was dominated by the Yadana and TTM campaigns and therefore most of the activism over these energy projects traversed the political epochs of the second Chuan government (1997–2001) and the Thaksin ascendancy (2001–6). As would be expected with development projects of this size and complexity, many were actually proposed much earlier and may continue operating for much longer than these specific periods but these eras were the most significant in terms of environmental activism. Over this decade activism against the Yadana and TTM pipelines was energetic and successful in drawing attention to the issues, with regular coverage in the Thai English-language newspapers, the *Bangkok Post* and *The Nation*. The Thai component of the Yadana campaign was largely finished by the time gas started flowing in 1999 and between this point and the end of the Thaksin years

---

13   In 1995 India's consumption of natural gas was 0.6 trillion cubic feet (Tcf). By 2010 it was 2.3 Tcf and likely to continue increasing each year. Despite a major domestic find of an estimated 7 Tcf offshore from Andhra Pradesh in 2002 India will continue to look for natural gas supplies abroad in the foreseeable future (EIA 2011; Misra 2007: 70–71). Despite China's search for greater energy security there is large potential for energy efficiencies. Since 2001 energy demand has grown by 1.5 per cent for every 1 per cent increase in GDP and it requires up to 5 times as much energy per unit of GDP than the US and 12 times that of Japan (Yi-chong 2007: 47).

the TTM Pipeline campaign was the most prominent environmental campaign in the country. As the opposition to Thaksin grew in 2006 and Thai politics became more polarised between 'yellow' and 'red' activists (Forsyth 2010; Hewison 2010; Montesano et al. 2012), emancipatory environmental activists became less visible. Although environmentalists in Thailand continued to do useful work, their ability to mobilise popular support decreased. As some authors have argued, the lack of opposition to the 2006 coup by activists and organisations other than the 'red shirts' was a 'rather tragic testimony to the degeneration of Thai social movements' (Glassman, Park and Choi 2008: 341). The previous decade can therefore be seen as the apogee in Thai environmental activism and provides the best case studies for assessing activist environmental governance.

Significant campaigning against energy projects in Myanmar also began with the Yadana Pipeline, Myanmar's first transnational energy project, although the nature of this activism was quite different. There was virtually no domestic activism at this time due to the oppressive military rule of SLORC/SPDC. While the seeds of the activist diaspora were sown by the exodus of students to Thailand and the border regions following the 1988 crackdown, knowledge of environmental issues in the exiled community was limited and grew slowly throughout the 1990s. It was not until the late 1990s that environmentally focused exiled groups emerged, although ERI had been formed by its US and Kayin founders in 1995 to launch the transnational campaign against the Yadana Pipeline. The campaign against the Yadana Project continued at a high level with international prominence for the next decade and only slowed following the out-of-court settlements by Total and Unocal in 2005. In the meantime the Salween Dams and the Shwe Gas Pipeline had emerged as the central campaign foci of the activist diaspora and the sophistication of the campaigning gradually increased. The epoch I examine here to assess activist environmental governance in Myanmar therefore covers the entire period of environmental activism under traditional authoritarianism from its emergence in the mid-1990s until the end of direct military rule in 2011.[14]

Although this study is therefore primarily concerned with the impacts that traditional authoritarianism had on local and transnational activism, I also analyse the implications of the new government's political and economic reforms for activism in Myanmar. Although the military still retained its position of unchallenged primacy following the elections and a genuine process of 'democratisation' would necessarily include significant constitutional amendments, the end of direct military rule in 2011 was, nonetheless, a significant historic moment, accompanied as it was by increased political freedoms and the eventual entry of the National League for Democracy (NLD) and its leader, Aung San Suu Kyi, into parliament. The activism in this book undoubtedly contributed in some way to the reform agenda, even if only indirectly through pressure on Western governments and TNCs, but there was no single cause for the reform process as a

---

14   The typology of political regimes used in this book is discussed further in Chapter 3 and comprises liberal democracies, traditional authoritarian regimes and hybrid regimes.

whole. While the elected government was the final phase in the military's seven point 'Roadmap to Discipline-Flourishing Democracy' of 2003 (Holliday 2011: 81–86), most Myanmar watchers, activists and researchers were surprised by the pace of political change that occurred under the new president, Thein Sein.

In the interviews I conducted in Myanmar and Thailand during the new government's first year there were seven popular viewpoints on why the military had allowed the political changes to occur so fast: first, the Arab Spring and the fall of President Mubarak in Egypt in early 2011 shocked the military into action (this was the favourite of activists); second, Senior General Than Shwe, the former leader of the SPDC, wanted to protect himself and his descendants in his old age by diffusing power amongst new institutions and players so that the treatment he had meted out to former dictator Ne Win and his family could not be repeated; third, the influence of engagement by ASEAN, but primarily through exposure to Singapore, Bangkok and other wealthy cities, demonstrated to the military leadership just how far behind Myanmar had become; fourth, the military regime wanted to end the international isolation and pressure that was brought to bear by the Western sanctions regime, which had been supported by the work of activists campaigning against energy projects; fifth, the regime was attempting to reduce the country's reliance on China by re-engaging with the West; sixth, the reform process was simply part of the plan from the announcement of the Roadmap in 2003; and seventh, the military was not monolithic and the reforms were the result of competition in the military between 'reformers' and 'hardliners'.[15]

It is likely that a combination of all seven factors contributed to the reform process so, while activism in and around Myanmar certainly played a role, it was undoubtedly only one of many. A full analysis of the emerging civil society activism during the transition process must be left to another book but the voluntary end of direct military rule provided a fascinating coda to a repressive era dominated by a largely incompetent regime. The overlapping epochs of the different regimes in Thailand and Myanmar therefore provided a useful contrast, reflected also in the different stages of the associated environment movement life cycles, with which to compare the nature and impact of activist environmental governance.

Book Structure

The body of this book can be thought of as comprising two main sections with Chapter 2 developing the theoretical model and the succeeding chapters applying the model to the environmental activism undertaken in the campaigns. Although

---

15   While most interviewees thought that Thein Sein was genuine in his reform agenda, there was near universal acceptance that the process could not have been initiated without the imprimatur of Than Shwe. Nevertheless, it was stunning how quickly political conversations in Myanmar changed from a focus on Than Shwe as the undisputed dictator of the country in January 2011 to his almost complete absence from most discussions a year later.

the activists quoted in this book provided a wide variety of justifications for their activities there were common themes that emerged, such as the linkages between environmental protection and human rights and the injustices felt by marginalised communities that were exacerbated by the transnational energy projects. The emancipatory outcomes that these activists pursued can be encapsulated as the pursuit of energy and environmental security for these communities. The pursuit of security therefore provided a rationale for the environmental activism against the energy projects that formed the basis for the model of 'activist environmental governance' developed in Chapter 2.

This model, building on work by Doyle and Doherty (2006), suggests that emancipatory environmental groups can play a crucial role in the emancipatory environmental governance of transnational energy projects provided they adhere, in both their organisational structure and their aims and activities, to the four core pillars of green politics: participatory democracy, ecological sustainability, social justice and nonviolence (Carter 2007: 47–48). These emancipatory governance groups (EGGs) are emancipatory social movement organisations and, although all EGGs adhere to the green pillars, EGGs in the South may interpret these concepts differently to those in the North due to differing environmental security concerns. This model also includes compromise governance groups (CGGs), which have emancipatory aims but conservative organisational structures and are, therefore, less effective in achieving emancipatory outcomes. The last category encompasses the organisations of the environmental governance state (EGS), predominantly Northern conservation groups that have both conservative aims and structures resulting in conservative rather than emancipatory outcomes. Due to the broadly emancipatory nature of the campaigns against the transnational energy projects most of the organisations in this book qualified as EGGs, although I have also included brief examples of CGGs and organisations of the EGS. It must be understood, however, that the emancipatory inclinations of movements or groups may be qualified by the cultural and political milieu in which they arise. External analysis of these movements should therefore both understand this location but also highlight inconsistencies in the exercise of emancipatory values.

Chapters 3 to 6 contain the case study analysis. In Chapter 3 I introduce environmental politics in the two core countries, Thailand and Myanmar, with a particular focus on activist environmental governance and the impacts of political regimes and business interests on environmental security. The possibilities for environmental activism in these countries are largely determined by their domestic political regimes, and I use Levitsky and Way's (2010) model of competitive authoritarianism to assess the competitive and authoritarian aspects of the regimes in these countries. As well as framing the domestic environment, the nature of these regimes also influences the level of transnational activism. In particular, traditional authoritarian military rule in Myanmar resulted in an activist diaspora that assisted in transnationalising the campaigns.

The relationship between business and governments in power is often crucial to understanding how a political regime functions. Thailand and Myanmar had very different political regimes and this diversity was also reflected in the structures through which business power was represented. Nevertheless, the deployment of power by domestic and transnational business in both countries played a significant role in the attempted de-legitimising of environmental activism. The extent of environmental insecurity in both countries was reflected in the level and type of environmental activism, with heightened insecurity resulting in increased transnational activism. This chapter, therefore, provides an important context for the analysis of the specific campaigns against the transnational energy projects in the rest of the empirical analysis.

In Chapters 4 to 6 I provide a multilevel analysis of these campaigns addressing the particular geographic scales of the campaigns. Chapter 4 examines the local, or domestic, aspects of the activism while Chapters 5 and 6 relate to transnational components. The diversity of campaign strategies and tactics reviewed in these chapters reflects the differing stages of the projects and the variety of states, TNCs, communities and environmental actors involved. Campaigns are useful ways to analyse activism because, as Keck and Sikkink (1998: 7) demonstrate, they highlight the relationships, resources and institutional structures that facilitate or impede this activism. Chapter 5 is a case study of the central EGG of this book, EarthRights International (ERI), a transnational NGO that straddles the North and South with a particular interest in the nexus between human rights and environmental protection in Myanmar. As I suggest in this chapter, the implementation of the four green pillars both in ERI's organisational structure and its aims and activities may well be linked to its broader success in achieving its goals. While I examine various levels of the campaigns in Chapters 4 and 6 – including informal groups, NGOs and coalitions – Chapter 5 therefore focuses particularly on a transnational EGG.

I have separated my discussion into local and transnational activism to isolate activities that were culturally significant to particular localities and to identify differences between approaches at local and transnational levels that were often aimed at different audiences. Difficulties arose, however, in categorising some aspects of campaigns as either local or transnational when the division was not clearly defined in practice. Indeed Keck and Sikkink argue that the divide between the international and national realms is becoming 'increasingly artificial' (1998: 4). This was particularly so in Myanmar where local or domestic protests were rare due to repressive authoritarian governance. While rare dissenting activities undertaken in Myanmar by domestic activists were clearly local, defining other aspects of activism as local or transnational was more complex.

Throughout the eastern border regions of Myanmar adjoining Thailand activists, like other refugees, lived or operated in a kind of citizenry limbo, neither clearly in Thailand nor Myanmar. Some activists crossed these borders at will and lived and operated under pseudonyms. Some ethnic minority communities lived on the Myanmar side of the border in areas under the control of insurgent ethnic groups, and therefore beyond the reach of the Myanmar military, or in refugee camps nominally on the Thai side of the border that faced military incursions from

Myanmar.[16] The borderlands were especially useful for activists and insurgent ethnic groups, as insurgent groups often 'specifically target border regions for their intrinsic, tactical and material importance' (Acuto 2008: 33). As Giddens has noted (1987: 18–19), in Weber's definition of a state the territorial element of a claim to a monopoly of violence over a given territory may be 'quite ill-defined' and this 'claim' may well be contested. The Thai-Myanmar borderlands epitomised this zone of contestation. The region was particularly important to this book as it played host to two of the transnational energy projects considered here, with the impact of the Salween Dams expected to be even greater than that of the Yadana Project (Piya Pangsapa and Smith 2008: 493). In her study of women activists in this region O'Kane describes the state of flux in these borderlands:

> Numbers of border arrivals and crossings fluctuate in relation to military operations, economic deterioration inside Burma and the continued possibility of sanctuary in Thailand. Each location has its own historical, cultural and geographical characteristics and people's semi-permanence there has complicated and re-constituted the borderlands in various political, social, economic, cultural and environmental ways. In this way Burmese political opposition groups have also become established components in this complex human milieu. (O'Kane 2005: 14–15)

Similarly, in Myanmar's western border regions adjoining Bangladesh and India, refugees and insurgents populated both sides of the mountainous borders, although expatriate activists tended to congregate in the major cities (Egreteau 2008). These borders were, therefore, relatively fuzzy (see Chaturvedi 2003; Christiansen, Petito and Tonra 2000; Gleditsch et al. 2006), rather than hard and well defined. Borderlands are grey zones that, particularly in times of conflict, acquire numerous meanings beyond mere legal boundaries (Acuto 2008: 32; Gainsborough 2009). Kaiser and Nikiforova argue that borderlands are in themselves central to the forming of identity, being

> central nodes where the intersections of power, place and identity are made visible. As both zones of contestation and spaces of becoming, borderlands are fundamental sites in the multiscalar reconfiguration of the sociospatial imaginary. (Kaiser and Nikiforova 2006: 952)

Expatriates of the activist diaspora in contested border zones or neighbouring countries often undertook fieldwork inside Myanmar incognito while writing and publishing reports from their offices in exile. In these cases the activism was transnational, as was the intended audience.

Local activism within the vicinity of a project itself or in a town or city nearby was usually, but not always, focused on the energy supplier country. In the case of

---

16   K. Redford, interview with author, Chiang Mai, Thailand, 15 January 2004.

the TTM Project, this approach was self-evident as most of the issues relating to the project affected Songkhla Province in southern Thailand, which supplied the gas, while the pipeline linked up with the existing Malaysian pipeline grid. In the case of the Yadana Project, however, Thailand was the receiving country but most of the local activism – as defined here – during the pipeline's construction occurred in Thailand since overt activism at this time within Myanmar was extremely rare. As the Salween Dams and the Shwe Gas Pipeline were still at the early stages of their development during much of this research – with the transnational infrastructure of the project not yet completed – activism in the receiver countries was still considered transnational. As a result of this approach, most of the local activism I examine in Chapter 4 occurred in Thailand in response to either the Yadana or TTM projects. As a general rule, therefore, I considered activists to be involved in local activism when they operated primarily in their home country for a domestic audience with that country also the physical location of the project (usually the supplier country). Otherwise I considered it transnational, examining it in Chapters 5 and 6.

The case study material provides detailed insights into the operations of both local and transnational campaigns against transnational energy projects in the South and these insights inform the conclusions in Chapter 7. This final chapter suggests four criteria for a new model of critical energy security, based on a critical analysis of the empirical data, and by generalising this model I also propose a critical environmental security framework. The conclusions in this chapter also suggest that, despite opposing powerful political and business interests, local and transnational activist environmental governance experienced some success in promoting environmental security for marginalised communities in Thailand and Myanmar. It should be noted that although the overt outcomes for the campaigns – usually for the project to be halted or for much greater public participation and involvement – were often not achieved, I also measured success in terms of the development of emancipatory processes within social movements through an emphasis on participatory democracy, social justice, ecological sustainability and nonviolence. These processes provide the foundations for new and complex activist networks that may well improve future governance and security for communities across the South.

Chapter 2
Activist Environmental Governance

Introduction

With the demand for energy in the region surrounding Thailand and Myanmar increasing at a rapid rate there is a pressing need for meaningful environmental governance of transnational energy projects that have the potential to inflict serious and widespread social and environmental impacts. While under a democratic regime of sensitive and comprehensive environmental governance some of these energy projects could potentially benefit the whole population, a lack of democratic participation too often results in severe adverse impacts on marginalised groups whose voices are silenced. In general, environmental governance of large projects by governments and their agencies in the South is poorly executed but projects involving Myanmar have been particularly notable for both environmental destruction and human rights abuses. Thailand has a more effective government and bureaucracy with a more substantial democratic tradition but its political and legal structures are similarly afflicted by persistent authoritarianism. In the absence of an effective democratic state with rigorous formalised processes it often falls to non-state actors to provide substantive environmental governance of development projects in the South. Critical models of energy and environmental security can also be deployed to analyse these projects, resulting in a process of environmental governance that has sustainability and social justice for marginalised populations at its core.

In this chapter I provide a theoretical overview of the nature of environmental movements and the challenges that they face, particularly those based in, or focused on, the South. I critique a classification model of environmental organisations and governance proposed by Doyle and Doherty (2006), and suggest refinements that result in a more nuanced interpretation of the nature of environmental governance and the prospects for effective contributions by environmental movements. I argue that emancipatory actors have a crucial role to play in processes of environmental governance, and that this role is particularly important under illiberal regimes. Emancipatory environmental movements in the South are driven by similar values and world views to emancipatory actors in the North, although as a consequence of the often extreme life experiences of Southern activists these values may be interpreted differently. It should be acknowledged, however, that environment movements, as social movements, are not simply comprised of NGOs, their most visible component; they also encompass individuals, informal groups, coalitions and the networks that connect these actors. This chapter therefore explores these components and their potential contributions to both environment movements and environmental governance in the South.

Environmental Governance

The nature of environmental governance, like most concepts in the social sciences, is ultimately contested and there are a variety of interpretations of what can or cannot be included or achieved under its banner. While some authors have argued that environmental movements can engage in forms of environmental governance (Bretherton 2003; Elliott 2004; Kutting 2011; Lipschutz and Mayer 1996; Paterson, Humphreys and Pettiford 2003: 2), the focus is often on formalised activities that do little to challenge existing social relations; the Forest Stewardship Council is an oft-cited example (Garner 2011: 171). Such schemes are often supported by TNCs and 'negotiated, implemented and enforced across national borders by non-state actors' without the involvement of states (O'Neill 2009: 188). Due to the lack of critical analysis in such schemes, however, this activism can be seen as a counterproductive, neoliberal form of governance, which dominates non-state environmental governance. This tendency is evident in the concluding article of a special issue of *Environmental Politics* edited by Brian Doherty and Tim Doyle, which divides environmental organisations – informal groups and NGOs – into either emancipatory groups (EGs) or part of an environmental governance state (EGS). The authors imply that environmental governance is now largely limited to actions within the EGS as part of a global neoliberal project that offers 'no challenge to environmental injustice' (Doherty and Doyle 2006: 705). This neoliberal form of environmental governance can also be considered a form of eco-imperialism (Dyer 2011).

Assessing Doyle and Doherty's model is particularly important due to their prominence within the literature on environmental politics, and environmental movements in particular, from which this book draws. Their writings on transnational environmentalism (Doherty and Doyle 2008), environmental activism in the global South (Catney and Doyle 2011; Doherty 2006; Doyle 2005; Doyle and Risely 2008), and environmental activism in general (Doherty 1999; 2002; Doyle 2000; Doyle and McEachern 2008) provide essential analysis of the nature of modern environmental activism. Despite their considerable contributions to the field, however, their model of environmental governance severely limits the constructive role that emancipatory environment groups and movements can play.

Doherty and Doyle's concept of the EGS draws heavily on Duffy's (2006) contribution to the same *Environmental Politics* issue examining Madagascar, which itself draws on Harrison's (2004) concept of the governance state. Duffy argues in her Madagascar case study that the power of the country's Donor Consortium – which included the environmental NGOs Wildlife Conservation Society (WCS), WWF and Conservation International as well as governments from various countries and the World Bank – compromised Madagascar's sovereignty, resulting in a 'conditioned form of autonomy [which indicates it was] a good example of a governance state' (Duffy 2006: 746). She also makes the point, however, that it was important not to overstate the consortium's power, as it did not have a unified view on environmental management. Local organisations adopted donor language to attract external funds but then pursued local agendas; it was clear that they used the existing 'neoliberal institutionalist framework' (Doyle and Doherty 2006: 883) for their own ends. Although Okereke (2008: 26), too, argues that the 'most important determinant' of the success of social equity norms in environmental governance is their 'fit' within this framework, it is evident that a more nuanced approach to defining environmental governance is required.

As well as the formal institutionalised processes of environmental governance, such as the aforementioned Forest Stewardship Council, the World Commission of Dams (Khagram and Ali 2008) and the Extractive Industries Transparency Initiative (EITI) (Haufler 2010), the broad possibilities emerging for more informal manifestations of environmental governance are illustrated in an edited collection by Kutting and Lipschutz (2009). Although the editors acknowledge that 'environmental governance in its current discourse is about environmental management and not about attaining local ecological democracy globally' (2009: 6), the volume also provides innovative and alternative conceptions of environmental governance. An example is a chapter on the global ecovillage movement (Litfin 2009), which shifts environmental governance away from state-based and neoliberal forms towards interactive and localised, but simultaneously globalised, interpretations. As the editors acknowledge, their goal is 'not to offer definitive "solutions" ... but, rather, to suggest "processes" that might point agents toward knowledge-based strategies that foster effective forms of social power' (2009: 9).

Evans (2012), in the first textbook on environmental governance that analyses the concept of governance from first principles, supports this approach. He includes 'monkey wrenching' by radical activist group EarthFirst! (Foreman and Haywood 1993) as a form of environmental governance (2012: 200). The importance of maintaining this radical edge in creating transnational or global forms of environmental governance is now well acknowledged:

> Some radical movements are pointing out the dangers of cooptation through global civil society, as well as the dangers of adopting orthodox discourse. They are recognising the need for the engagers to retain links with the grassroots in the battle over the agency of global civil society and attempts to radicalise and expand it. (Ford 2003: 132)

Various forms of environmental governance therefore challenge the neoliberal or market-oriented forms of environmental governance prevalent in, for example, the global climate change discourse. There are many non-market-oriented solutions to climate change but the widespread adoption of market-based tools by many of the governments that actually attempt to deal with climate change has ensured that the debate in the global media largely revolves around this strategy. This often limits the solutions that communities can imagine to issues such as climate change and can be thought of as an example of 'problem closure' (Forsyth and Walker 2008: 12). As Evans (2012: 202) acknowledges, non-traditional ways of thinking must be disseminated to, and pursued by, communities to provide more progressive forms of environmental governance.

Environmental governance in a globalised world is clearly, therefore, not simply involvement in transnational funding bodies, or even transnational institutions. It involves a complex interplay amongst state and non-state actors in both the North and the South (Arora-Jonsson 2013), and includes local and transnational processes of societal transformation through activism that acknowledge the intimate connection between ecological and social concerns. This activist aspect of governance can therefore be termed 'activist environmental governance' to distinguish it from more formalised state-based modes. All emancipatory actors, whether informal local groups or transnational coalitions, have a significant role to play in this process. By effectively labelling systems of environmental governance as neoliberal the Doyle and Doherty model precludes this function. This model therefore needs revision with the first step being to rename EGs as 'emancipatory governance groups' (EGGs) to better reflect the diverse possibilities available within formal and informal contributions to environmental governance. These EGGs may be individuals, informal groups, NGOs or formalised coalitions within emancipatory environmental movements.

The Doyle and Doherty model also requires a more rigorous definition of what being 'emancipatory' actually entails. I therefore argue that these EGGs must adhere, in both their organisational structure and their aims and activities, to what can be considered the four pillars or principles of green politics, first elucidated by the German Greens in 1980; namely participatory democracy, ecological sustainability, social justice and nonviolence. In contrast, as Doyle and Doherty note (2006: 888), organisations within the EGS often have aims that focus on wilderness, or post-materialist, concerns that exclude humans, even if at the expense of pursuing justice or democracy. Organisations within the EGS can therefore be clearly set apart from EGGs by examining their aims and activities but they can also be distinguished by their organisational structure which, as other studies have demonstrated, can impact on the longevity and salience of formalised NGOs in particular (Wong 2012). Doyle and Doherty say little about this aspect other than that emancipatory groups 'celebrate more non-institutional forms of organisation' (Doyle and Doherty 2006: 883). Here again, however, the four green pillars can be more usefully employed as a distinguishing characteristic: EGGs self-consciously adopt horizontal management structures that display elements of anarchist philosophy with high degrees of participation in decision making and an emphasis on internal justice, while organisations within the EGS are characterised by the vertical, top-down management structures that are anathema to green ideals. These four pillars, if enshrined both within aims and activities and as universal organising principles, can therefore be considered defining features of EGGs.

This definition suggests another limitation to Doyle and Doherty's model. Categorising these environmental actors with a simple dualism is too blunt an instrument to capture the subtle nuances that exist within the great diversity of environmental movements. Although endless categories could be established to take account of this diversity the resulting complexity would remove any benefit of the categorisation itself. There is one further grouping, however, that would add substantially to the analytical capacity of this model and this includes organisations that pursue emancipatory aims and activities but still retain a more conservative vertical management structure with little participatory democracy. These organisations do not completely qualify as either EGGs or organisations within the EGS so a third category is therefore required which comprises organisations characterised by a compromise between emancipatory aims and conservative structure; they can therefore be best described as 'compromise governance groups' (CGGs). The inverse configuration, with a self-consciously emancipatory organisational structure and conservative aims rarely occurs, because the aims generally reflect the more progressive world view; it is much easier to talk about emancipatory outcomes than to put it into practice. The typology of what can be considered activist environmental governance therefore consists of three categories – emancipatory governance groups (EGGs), compromise governance groups (CGGs) and organisations of the environmental governance state (EGS) – each defined by the nature of their organisation and aims and activities (see Table 2.1).

Table 2.1   Taxonomy of activist environmental governance

	Emancipatory governance groups (EGGs)	Compromise governance groups (CGGs)	Environmental governance state (EGS)
Aims and activities	Emancipatory	Emancipatory	Conservative
Organisational structure	Emancipatory	Conservative	Conservative

It should be noted that organisations which qualify as EGGs adopt an emancipatory structure not only because they believe it will be more successful in achieving their short-term aims but because they are more self-conscious of their location within a long-term justice-focused environment movement. They see the pursuit of justice as wider than simply the overturning of illiberal governance regimes; that engagement in activism itself is transformative and should be undertaken within an emancipatory and just environment. In this sense EGGs are more 'process-oriented' than 'outcome-oriented' CGGs and the EGS. As a result of their conservative structures CGGs are less effective in achieving emancipatory outcomes than EGGs. Organisations within the EGS, however, with both conservative aims and structures, are entirely counterproductive to emancipatory objectives. Due to the justice-oriented nature of the campaigns in this book there are few organisations analysed that qualify as either CGGs or as part of the EGS, although for completeness I briefly analyse an example of each type in Chapter 6. The organisations selected are transnational NGOs that operate in or on Myanmar and can therefore be considered as 'like' organisations to ERI for comparison.

Despite similar core values, EGGs in the North and the South can have differences in approach or emphasis that relate to differing political, cultural and environmental settings. In addition, the potential for EGGs in the South to undertake a two-track strategy, similar to that outlined by Duffy above, suggests that EGGs in the South may adopt different approaches to their campaigns at local and transnational levels, depending on the context they are in and the role they are playing. Whatever their tactics, EGGs in both the North and the South self-consciously place themselves within the global governance system to provide a more sensitive and interactive contribution to environmental governance while forming transnational networks and coalitions to enhance campaigns and overcome limited resources.

Most of the diverse organisations in the case studies of this book operated predominantly in the South and appeared to have both aims and organisation based on emancipatory principles that promoted environmental security for marginalised populations in Myanmar and Thailand. The main organisational case study, which I examine in Chapter 5, is EarthRights International (ERI), a transnational North–South NGO with Special Consultative Status to the UN which engaged in both local and global activities, including fieldwork in the forested conflict zones of Myanmar and representing ethnic minority villagers in US courtrooms. Other organisations involved in the campaigns included: Arakan Oil Watch (AOW), a small organisation with only three staff that produced detailed regular bulletins and larger reports related to the Shwe Gas Pipeline; Kanchanaburi Conservation Group (KCG), a small local group engaged in forest protests in Thailand to stop construction of the Yadana Gas Pipeline; the Alternative Energy Project for Sustainability, a small Thai NGO that placed Bangkok-based activists with local communities for two years at a time in the campaign against the Thai–Malaysian Gas Pipeline; Salween Watch, a coalition of NGOs and groups exiled from Myanmar campaigning against large hydroelectric dams on the Salween River; and the Shwe Gas Movement, a transnational coalition of NGOs that gathered and disseminated information in Myanmar at the local village level while petitioning governments and transnational corporations in international fora.

Despite the diversity of these actors and the multiplicity of their activities their campaigns all contributed to enhanced environmental governance of the energy projects in this book by monitoring the activities of key actors, providing information, support and training to those adversely affected, and lobbying public and private bodies at local, national and international levels. They were broadly emancipatory social movement actors who challenged existing social structures including the establishment of regional energy markets that benefited entrenched political and economic elites. Rather than relying on liberal markets to determine the distribution of energy, they aimed to promote natural resource rights and politicise resources in favour of ethnic minority or indigenous communities. They therefore rejected the dominant market-based approach to environmental governance (Evans 2012) and qualified as EGGs that played a constructive role in promoting a localised and emancipatory approach to activist environmental governance.

It is clear that EGGs both emerge from, and contribute to, diverse environment movements in and across the South that are centred on the four green pillars. To understand more fully the nature of activist environmental governance, in the rest of this chapter I explore the characteristics of these environment movements and the way in which they interpret these green ideals.

Environmental Movements

This activist environmental governance model was developed using a very specific understanding of the nature of environmental politics. Some analyses of environmental politics focus particularly on state-based and intergovernmental agreements on the regulation of the environment in which environmental movements play little part (Howes 2005; Kutting 2000; Paehlke and Torgerson 2005). Dryzek and Schlosberg widen this approach but limit environmental politics to

> how humanity organises itself to relate to the nature that sustains it [and only impinges on] other areas of political concern such as those related to poverty, education, race, the economy, international relations, and human rights inasmuch as what happens in these areas affects our environment (and vice versa). (1998: 1)

Doyle and McEachern (2008) take a more expansive view, however, including political issues that go beyond overt human–environment interactions. They consider humanity as a part of nature, recognising that human relationships with the non-human world are socially as well as biophysically constructed, resulting in an integrative human-centred approach whereby environmental politics is dominated by 'issues of social democracy (participatory and representative), nonviolence, social equity and justice as well as ecology' (Doyle and McEachern 2008: 22). These values are clearly recognisable as the four green pillars outlined above and add the emancipatory element to environmental politics that underpins the environmental governance model. This approach is strongly focused on the social movement aspects of environmental politics, which is not inconsistent with the 'sociological turn' in Australian international political economy since the turn of the century (Seabrooke and Elias 2010).

Academic attention to this area of environmental politics is still relatively recent. Princen and Finger noted in the mid-1990s that, while significant attention had been given to 'documenting environmental conditions and prescribing remedies to save the planet' (1994: x), little had been written on the activities of environment movements. As I noted in the introduction, while much more has been written on environment movements since then, relatively few books examine environment movements in the South and there remains a lingering bias in most environmental politics texts towards the North (Carter 2007; Connelly et al. 2012; Rosenbaum 2010).

This book, however, is focused almost entirely on environmental politics of the South. In the South, which is often characterised by illiberal governance and highly unequal resource ownership, environmental activism often means opposing large-scale development projects. In essence, these movements are based on fundamental struggles for justice for marginalised communities rather than the pursuit of incremental policy changes within an existing order. Central to the analysis of these environment movements, therefore, is the distinction between emancipatory social movement actors, who advocate significant societal transformation, and less radical actors who do not 'engage in ideological conflict with their opponents' (Doherty 2002: 14). This distinction follows from an established division in the environmental literature of the North between 'radical' greens and 'reformist' environmentalists (Dobson 2007: 5; Doherty and de Geus 1996: 2). In other words it could be argued that reformist environmentalists are part of a Gramscian civil society, which supports or forms part of the state, while radical greens are independent counter-hegemonic non-state actors who operate outside, and often in opposition to, the state (Cox 1993; Gramsci 1971: 263; Harvey 2005: 78). Although this distinction between green and environment movements is well established in the literature of the North, in the South almost all environmental activists challenge existing power relations and are radical within their own political milieu. Even nominally environmental or ecological disputes usually broaden over time to include radical social and political concerns. Often, however, these concerns are central to campaigns from their inception (Doherty 2002: 216; Haynes 1999).

Any thorough analysis of environment movements must begin by determining their composition. Doyle has generally adopted an emic approach for environment movements where they consist of everyone who considers themselves to be an environmentalist (Doyle 2000: xix; 2005: 6; Harris 1979: 32). This perspective can be seen as borrowing from Melucci's (1989; 1996) constructivist approach to social movements, which Eschle has also used to argue that 'we know that movements exist when activists claim that they are part of one and participate in efforts to define "their" movement in particular ways' (2005: 20). The emic approach is democratic and inclusive but for analytical purposes the distinction must be made between radical social movement actors and more conservative or reformist parts of the movement that 'function in a formalised, structured fashion, openly endorsing the existing status quo' (Doyle 2000: 8). As discussed above, however, most environment movements in the South can be seen as radical social movements so in this book, unless specified, I tend to follow Torgerson (1999: 2) in using the terms 'green' and 'environmental' interchangeably.

To clarify what distinguishes actors who belong to a social movement from those who do not, della Porta and Diani (2006), of the European or New Social Movement (NSM) school, provide a well-established definition. This European approach minimises the emphasis on movements' goals and concentrates on identity-forming networks, while acknowledging that these identities are fluid and heterogeneous (Melucci 1989; 1996). The dominant concepts in NSM formation in the North have been the post-materialist and post-industrialist theses (Catney and Doyle 2011), although, as Martinez-Alier points out, the term 'post-materialist' 'is a terrible misnomer [when applied to countries of the North] whose economic prosperity depends on their use per capita of a very large amount of energy and materials, and on the availability of free sinks and reservoirs for their carbon dioxide' (2002: 4).

While these theories have been applied extensively to the North, more homogenous, class-based theories are sometimes more appropriate for the South. Despite these differences, however, the NSM-influenced perspective has been successfully applied to case studies in India and the Philippines and I therefore also apply it here (Doyle 2005: 3). In assessing whether activists and groups are part of a social movement, della Porta and Diani (2006: 20–23) identify three key characteristics. As this book is primarily focused on EGGs as emancipatory actors I have adapted their framework to define *emancipatory* social movements. The first characteristic of social movements is that they must be involved in conflictive relations with clearly identified opponents. Social movements may therefore be conservative rather than emancipatory; the pro–fox hunting movement in the UK, for example, promotes the exploitation of animals and reinforces oppressive class structures. Emancipatory social movement activists, however, must challenge some feature of the dominant social or political structures or values of their society in the pursuit of justice. From Max Horkheimer's (1972) distinction between critical and traditional theory, and Robert Cox's (1981: 128–30) adaptation of the distinction in international relations, we could distinguish between traditional problem-solving reformist actors, who sustain the existing order and are not part of an emancipatory movement, and radical actors who adopt a critical approach and challenge the prevailing order in society. Movements that 'fail to recognise their own location within global hegemony ... may end up reproducing [it] rather than challenging it' (Ford 2003: 124), and are therefore not emancipatory social movements. In the environmental governance model above, organisations that are part of the EGS do not conflict with dominant structures – they are embedded within them – and they are therefore not emancipatory.

The second characteristic of social movements is that activists must be linked by dense informal networks, with no single organisation defining the movement. These networks are the central component of social movements and, according to Keck and Sikkink (1998: 2), are characterised by several defining features: the centrality of values or principled ideas; the belief that individuals can make a difference; the creative use of information; and the employment by non-state actors of sophisticated political strategies in targeting their campaigns. These features contribute to the third characteristic of social movements, which is that they must share a distinctive collective identity. This identity is based on ideas and practices that may be developed through collective action. I argue that within emancipatory environmental movements this identity must be based on emancipatory principles that eschew hierarchy and promote equity and justice in all activities, including the formation of activist groups. Welzel and Deutsch (2012) argue that the existence of emancipatory values in collective action elevates an individual's emancipatory values and that each individual's emancipatory values amplify the group's. The existence of emancipatory values therefore provides a positive feedback loop that results in a higher overall emancipatory identity. As organisational structure is a key indicator of inherent values, neither CGGs or groups in the EGS are, therefore, likely to satisfy this third criterion as their organisational structures are hierarchical with limited emphasis on the green pillars.

A fourth key element of social movements, which is central to their activist repertoire (Tilly 2004: 3), is engaging in public protest. While della Porta and Diani do not list protest as a core characteristic, they do acknowledge that social movements tend to invent 'disruptive forms of action' (2006: 29). Della Porta gives more prominence to protest in her more radical writings, where her sentiments concur with Melucci on the importance of protest in challenging power, arguing that social movements are antagonistic and that their use of protest distinguishes them from other political actors (della Porta et al. 2006: 19–20; Melucci 1996: 35). In their analysis of environmental movements Doherty and Doyle take this emphasis further, arguing that some form of public protest activity in part of the network is a fourth key characteristic (Doherty 2002: 7; Doherty and Doyle 2006: 702–3). The ability of movements to engage in public protest is a key indicator of the nature of the political regime under which they operate and the potential for effective domestic activist environmental governance. In Thailand street protests are a common occurrence but in Myanmar these protests have been rare due to the prevalence of more traditional authoritarian governance. Although rare protests occasionally erupted on the streets, notably during the national protests of 1988 and 2007, protest under military rule was more apparent via 'underground' media. Whether through transnational publications, online activism, public parody or nonviolent support for insurgent groups facing repression, emancipatory movement activity by Myanmar's activists still took place. In contrast, most organisations of the EGS do not satisfy this criterion as they are immersed within the bureaucratised institution-based decision-making structures that form the EGS and therefore eschew protest activities. By employing the four elements above as key characteristics of emancipatory environmental movements and testing them against the three environmental governance categories it becomes clear that only EGGs satisfy all four criteria, therefore qualifying as emancipatory environment movement actors.

While the movements in this book confirm that social justice and radical political reform are of central significance in emancipatory environment movements, some authors argue that the environment itself should be more central than challenging political norms (Rootes 2006: 779). While in the global North a focus solely on ecological issues may result in social injustices, in the South this approach, in situations such as the removal of communities from 'wilderness' areas, may result in life-threatening situations. While most activists involved in the campaigns in this book link some form of political reform to resolution of the dispute it should also be acknowledged that in Myanmar under military rule campaigning for a

Western-style liberal democracy, or even telling jokes about the military regime, was considered a radical form of insurrection punishable by hard labour and long prison sentences.[1] Activists in the South may therefore be radical in the context of their own political milieu but considered mainstream within a broader Northern political context. This caveat may also apply to their designation as emancipatory actors; while seeking emancipation from the structures that constrain their existence their interpretation of liberation may in some cases conflict with the global or universal values espoused by emancipatory actors in the North. Greater North-South collaboration within activist networks provide opportunities for the cross-fertilisation of ideas relating to emancipation in the pursuit of more mutually informed perspectives.

The Structure and Composition of Environment Movements

An analysis of the structure of both the movements and their constituent parts is required to fully consider the potential governance role that environment movements in the South can play. Environmental activism takes place in the South under living conditions and governance structures that may shift campaign goals towards social justice and human rights and the structure of both the movement and its components may help or hinder the achievement of these goals. In the case study chapters of this book I undertake a *multilevel* (Dwivedi 2001) analysis of the environmental campaigns opposing the various energy projects; these include individuals, informal groups, formalised NGOs, coalitions and informal networks. These multilevel constituent parts are introduced here.

The most formalised and visible component of environment movements are NGOs, which are formed by a constitution and 'involved in many different spheres of politics, from the local community level, through the politics of the nation-state, to international politics' (Doyle and McEachern 2008: 123). While they are important actors within movements, NGOs are often mistaken for the sole actors in a movement. In Newell's (2001) contribution to Edwards and Gaventa's *Global Citizen Action* he begins by stating that the chapter will examine the 'environment movement' but for the rest of the chapter he focuses entirely on environmental NGOs. The central EGG examined here, EarthRights International (ERI), is indeed a transnational NGO but many other, more informal, movement components are also examined. Although NGOs such as ERI have formal structures, studies have found that there is actually no relationship between the formalisation or institutionalisation of movement actors into NGOs and the levels of protest (Doherty 2002: 148–49). If NGOs maintain a collective identity compatible with the four core green ideals of democracy, sustainability, justice and nonviolence they retain a strong claim to being emancipatory environmental movement organisations. While ERI is formalised, with a budget that increased

---

1   Par Par Lay and Lu Maw, interview with author, Mandalay, Myanmar, 15 December 2003.

to over a million dollars, it maintains a strong commitment to these core green commitments and is involved in transnational emancipatory protests that critically challenge established power structures.

Informal groups are less formalised than NGOs and can have a broad range of characteristics. Doyle (2000: 34) proposes two categories of informal groups: introspective and non-introspective. The former are generally conscious of their political form, their ideology and their structure while the latter are not. While Doyle uses this categorisation only for groups, I argue that it can be applied to other more formalised EGGs as well. An examination of ERI demonstrates that these three organisational aspects can be equally important in more formalised organisations, effectively creating introspective formalised EGGs. ERI appears conservative in some respects – they have confidence in the 'power of law' and do not try to 'break down the capitalist system'[2] – but it is introspective and emancipatory in much of its organisation and activities. Non-introspective groups are unlikely to adopt radical or critical perspectives that challenge dominant social or political structures of their society and are therefore unlikely to be classified as EGGs or part of an emancipatory environment movement.

Nevertheless, these groups may change or transform over time as members are radicalised by their activism and the intransigence of the system in which they operate. This radicalising and transforming effect means that even not-in-my-back-yard (NIMBY) local environment groups, often considered a conservative element in the movement, may become social movement actors (Doherty 2002: 185; Wall 1999: 25), nurturing 'valid and valuable forms of gender and political emancipation' (Ford 2003: 128). Activists in the Kanchanaburi Conservation Group (KCG), which was originally a NIMBY group opposing the Yadana Gas Pipeline, were radicalised in this way to later become social movement actors who addressed more structural political issues in Thailand.[3]

Once formed by a particular issue, groups and NGOs tend to survive while informal networks, another component of environment movements, tend to disintegrate once their issue-oriented goals are achieved (Doyle and Kellow 1995: 106–10). Despite the sometimes temporary nature of networks, however, they are considered the defining feature of NSMs (della Porta et al. 2006: 20). The transnational effects of these movements and mobilisations have been noted since the 1980s (Melucci 1989: 88), and Castells (2000: 22–23) contends that the 'alternative networks' of social movements provide one of the few genuine avenues for challenging dominant power structures. Part of their potency is the ability of networks to be simultaneously local and transnational. Routledge argues that

---

2   K. Redford, Co-Founder/Director, EarthRights International (ERI), interview with author, Chiang Mai, Thailand, 9 January 2009.

3   Phinan Chotirosseranee, interview with author (translator: Ellen Cowhey), Kanchanaburi Province, Thailand, 5 October 1998.

when local-based struggles develop, or become part of, geographically flexible networks, they become embedded in different places at a variety of spatial scales. These different geographic scales (global, regional, national, local) are mutually constitutive parts, becoming links of various lengths in the network. (2003: 336)

Some authors have also applied this multiscalar analysis to coalitions, although they have been relatively overlooked in the analysis of environmental politics (see, for example, Doherty 2002; Doyle and McEachern 2008). Multiscalar coalitions form because a campaign 'in various countries and at various levels of politics, is one that is difficult for most individual NGOs to accomplish, because of the time required to coordinate such a campaign' (Yanacopulos 2005b: 106). Coalitions are central to the campaigns in this book, particularly at the transnational level. Smith and Bandy (2005: 7) suggest that transnational coalitions are particularly difficult to achieve because of the diversity of languages, political experiences and national cultures but in the case studies in this book these difficulties have been largely overcome and this diversity is actually seen by activists as a strength.[4] Coalitions may have more permanent links than single-issue networks, although Levi and Murphy (2006) demonstrate that shorter term 'event coalitions' also form. The coalitions in this book are, however, all more long term and can be considered what Tarrow defines as 'campaign coalitions' due to the combination of 'high intensity of involvement with long term cooperation' (2005: 168). Coalitions can harness expertise by pooling resources but this is particularly difficult if using only the internet, so face-to-face meetings are highly valued (Gillan, Pickerill and Webster 2008: 101–2). Once the difficulties of cooperation are overcome, however, coalitions often result in 'permanent staff members, a more permanent membership base, a headquarters or secretariat, and are organisations in and of themselves' (Yanacopulos 2005b: 95).

Later in her definition, however, Yanacopulos argues that coalitions 'have broader strategic aims than single-issue thematically focused networks' (2005b: 95). While the coalitions in this book may have broader strategic interests the focus is, nonetheless, on each particular energy project. Salween Watch, Karen Rivers Watch, the Burma Rivers Network and the Shwe Gas Movement are all actually coalitions – despite their names – even though they also link into global networks of activists. The individual members of the coalitions often have broader political aims related to Myanmar – primarily democratisation – but this does not detract from the limited rationale or stated aims of each coalition. Indeed when broader political activities create common positions, such as on democratisation in Myanmar, coalitions from different campaigns with different short-term aims may form 'discourse coalitions' as their messages reinforce each other (Hajer 1995: 65). Regardless of the differing campaigns, as EGGs working on similar issues their shared emancipatory values suggest they are likely to interact regularly as a result of value homophily (Di Gregorio 2012).

---

4   K. Redford, interview with author, Chiang Mai, Thailand, 9 January 2009.

These organisations have excelled in employing the recent improvements to communication technologies, such as the internet, mobile telephones and social media, which have been essential in the information-sharing role of networks and coalitions and assisted in the development of both transnational environment movements and other social movements around the world (Diamond and Plattner 2012; Yanacopulos 2005b: 102). Castells (2003: 187) argues that environmental movements in general have been at the cutting edge of employing new communications technologies for use as mobilising and organising tools. These technologies have assisted transnational activists in creating what Torgerson (1999: 19–20) suggests is an emerging green public sphere, a space of dialogue and debate with multiple concurrent transnational spaces. More importantly, in relation to this book, one of Torgerson's key green public spheres deals with a postcolonial environmentalism that not only examines elite domination but also the way groups in the South interact with other groups and cultures in the local–global nexus (2006: 717). The green public sphere created by the campaigns in this book therefore promotes environmental protection in both Myanmar and Thailand, but there is a strong underlying focus on environmental justice and security. These shared values resonate strongly with the four pillars of green politics that created the networks within this sphere to begin with (Keck and Sikkink 1998: 30).

The Four Pillars of Green Politics

The four core key concepts or pillars of green politics – ecological sustainability, participatory democracy, environmental justice and nonviolence – appeared initially in the 1983 political program of Die Grunen, the German Greens (Carter 2007: 47–48). They neatly synthesised social and environmental concerns and, despite amendments and additions over the years by various actors, these pillars still represent the core values embedded within emancipatory environment movements. The pillars represent both methods and goals, as emancipatory actors organise based on these values with the aim of achieving them in society. As critical reflections on the relationship between nature and society they can be considered the key conduits for achieving critical environmental security for marginalised populations, as defined in Chapter 7. Although the pillars emerged in the North they also guide emancipatory actors in the South, if sometimes implicitly. There may, however, be differences in emphasis or interpretation between activists in the North and the South because of the stark differences in their daily existence. Due to precarious living conditions in the South environmental struggles are often 'struggles for democracy and against the unequal distribution of power' (Doherty 2007: 80). There are inequalities in all societies but injustices in the South have the capacity to significantly affect livelihoods and even survival in marginalised communities. The strong justice focus that permeates the pillars ensures that they are key concepts in the development of emancipatory activist environmental governance in the South. In the rest of this chapter I therefore examine the philosophical underpinnings of the predominantly Northern writings on these concepts as they relate to emancipatory movements and explore the sometimes differing perspectives that emerge from the South.

Ecological Sustainability

Although ecological considerations drive wilderness or conservation groups within the EGS they are not the sole overriding concern of emancipatory environment movements. There is little doubt that developing an ecologically sustainable society is important for EGGs but this ecological component is addressed primarily in the context of the more social and political attributes of the other pillars: 'sustainability, like democracy, is largely about social learning, involving decentralised, exploratory, and variable approaches to its pursuit' (Dryzek 2005b: 158). While sustainability retains a central place in the pantheon of green philosophies, the term sustainable development has, in many environmental circles, lost much of its credibility. It is now largely seen as a means by which business can neutralise environmental critiques and as a way for the North to exploit the South (Beder 1996; Doyle 1998: 782; Howes 2005: 109). With rising environmental awareness throughout the 1980s, business rapidly came to understand that adopting 'sustainable development' was an effective response to environmental criticism, 'not in the sense that it changed how business conducted itself so as to do less harm, but because it provided a rhetoric to protect the continuation of business as usual' (Doyle and McEachern 2008: 206). While sustainable development is currently accommodated within the capitalist economic system there are still those who wish to stretch it as far as possible within these confines (Dryzek 1999: 270).

Although in recent years much government funding and research activity in the North has focused on sustainability in general, and climate change in particular, environment movements in the South have always tended to focus much more on the pressing concerns of poverty and short-term water and food security. When issues of sustainability arise in the South they are far more likely to be linked to immediate livelihood issues, such as the sustainability of migratory fish catches following the building of dams, than predictions of distant climate change. There is little doubt that concerns over climate change are increasing in the South, particularly as this is where impacts are likely to be most severe and least likely to be mitigated (Harris 2010; Hurrell and Sengupta 2012), but in many parts of the South, such as Myanmar and Thailand, sustainability is seen as something that would naturally follow from a more urgent focus on justice and more democratic access to political power for marginalised people.

Participatory Democracy

While democracy is a key value for environmental activists and academics in both the North and the South, in the South there is often a different emphasis, as with sustainability, due to the different political settings. Northern environmentalists operate in largely liberal democratic societies where the discussion is on greening and extending participation in democracy. While these writings are an important contribution to academic debate, many communities in the South, and particularly in Myanmar, face a daily struggle for survival where even basic democratic principles and human rights are ignored. The focus in these situations may be on achieving democratic fundamentals such as an elected government or a free press rather than more deliberative democratic processes. Nevertheless, there is an element of cross-fertilisation in democratic theory and practice between the North and South, and participatory budgeting in Porto Alegre in Brazil provides a more democratic template than most places in the North (Gret and Sintomer 2005). Although some authors argue that in the South, and East Asia in particular, environmental politics will be carried out under authoritarian governance for some time to come (Beeson 2010; Gilley 2012), the goal of greater participatory democracy remains central to emancipatory environmental movements.

While ecological authoritarianism was a strong element in much of the early environmental writings in the 1960s–70s (Pepper 1996), in recent years this authoritarian stream has largely fallen by the wayside. In most green literature the debate now is about how best to approach and deepen democracy (Dryzek 2006; 2010; Hay 2002: 303; Paehlke 2005: 25). With green political thought and action now virtually inseparable from democracy this connection has become an important methodological and epistemological approach for emancipatory environment movements. As Hardt and Negri (2004: 87) note of globalisation protest movements, '[d]emocracy defines both the goal of the movements and [their] constant activity'. At its most elemental level, the pursuit of democracy involves attempts at maximising respect for human rights and public participation in the governance of the political sphere. Democracy is achieved by a

> level of civil and political liberties – freedom of thought and expression, freedom of the press, freedom of assembly and demonstration, freedom to form and join organisations, freedom from terror and unjustified imprisonment – secured through political equality under a rule of law, sufficient to ensure that citizens can develop and advocate their views and interests and contest policies and offices vigorously and autonomously. (Diamond, Linz and Lipset 1988: 6–7)

Democracy is therefore conceptualised as lying at one end of a spectrum with the other end being a system of rule where citizens are totally excluded from the decision-making process (Beetham 1992: 40). In this book, therefore, the political regimes discussed include the traditional authoritarianism of military-ruled Myanmar, where people were totally excluded from the decision-making process with a complete lack of democracy, and the competitive regime in Thailand, primarily under Prime Minister Thaksin Shinawatra, where a framework of democratic institutions existed, but was undermined in practice.

Regardless of specific democratic attributes, there is a broad consensus that the possibilities for environmental justice and an overarching ecological rationality governing societies are greatly improved by strengthening democratic forces and increasing community involvement in decision making (Beck 1999: 152; Dryzek 1999; Giddens 1998; Mason 1999; Mitchell 2006). Environment movements can contribute to a process of democratisation simply by their formation, as participation and networking in the creation of social movements generates crucial sites for democratisation (Tilly 2004: 141). While accepting that conventional liberal democratic channels of public participation are better than the opportunities under authoritarian regimes, many environmentalists are nevertheless critical of the 'reactive and piecemeal environmental measures emanating from the liberal democratic parliamentary process' (Eckersley 1996: 216). Part of this response relates to the central role of business in the political process. As a result, many greens argue for more participatory organisational structures such as 'meshworks', networks that are decentralised and heterogeneous with a commitment to the locale/local (Escobar 2004: 353–55). Decentralisation is also considered a key goal, from both ecological and democratic perspectives (Carter 2007: 58; Doherty and de Geus 1996: 3), as there is a 'close link between local economic self-reliance and sustainability' (Douthwaite 1996: 58). Echoes of this approach towards decentralisation can be found in the Thai movement towards localism, which combines traditional Thai approaches to self-sufficiency with global environment movements that promote decentralisation (Connors 2005: 267).

Environmental impact assessment (EIA), which sometimes formalises a measure of public participation as part of the planning process (Harvey 1998), is central to environmental politics as it is anchored within an ecological rationality (Bartlett 2005: 48; Dryzek 1987; Howes 2005: 96–100). Undertaking these assessment studies, ensuring the veracity of the results and basing decisions on these outcomes are important elements in determining whether justice and sustainability are improved as a result of a particular project. Nevertheless, a problem inherent in many of the world's EIA systems, and particularly in countries of the South such as Thailand, is that it is often the project's proponent, who has a vested interest in the outcome, who also undertakes the study with the result that impacts are often underestimated (Fahn 2003: 204). Another advantage for the proponents, whether they are governments or corporations, is not that EIA necessarily makes the decision more rational, but that it appears more rational, undermining the credibility of public opposition (Carter 2007: 303). While always imperfect in practice, the evidence from projects in Myanmar, where EIAs and public participation are largely absent, supports the conclusion that 'it is better to have some kind of EIA system than nothing' (Howes 2005: 108).

Environmental Justice

Central to both green philosophy and politics, according to Dryzek, is the view of a natural relationship of equality among individuals and, therefore, 'hierarchy that both pre-dates and is reinforced by modernity is recognised and condemned' (2005b: 216). As a result, ideals of justice and equity within environmental discourse, or what Torgerson defines as the green public sphere, now include 'green critiques of racism, patriarchy, sexism, heterosexism, homophobia and class exploitation. These critiques have drawn detailed attention to the various and complex ways in which the project of dominating nature is also a project of dominating human beings' (2006: 716).

In this sense green activism is largely an emancipatory project and EGGs provide a strong emancipatory contribution to environmental governance. Conversely, organisations without an emphasis on social justice, such as those that maintain a wilderness focus, are likely to be part of the EGS. Due to often oppressive political environments and precarious living conditions environment movements in the South are far more likely to engage with issues of injustice and maintain an emancipatory focus. It is certainly a commitment to social and environmental justice that has driven the formation of EGGs such as ERI.[5] In local and transnational courts and formal settings, however, ERI and other EGGs often appeal to universal human rights norms as avenues for achieving broader environmental justice and security goals due to the power afforded to legal precedents.[6] A decade after Die Grunen's program launched the four pillars a UN report entitled *Human Rights and the Environment* communicated the links between social and environmental concerns to a global audience with the first guiding principle stating that '[h]uman rights, an ecologically sound environment, sustainable development and peace are interdependent and indivisible' (Ksentini 1994: 74). There have been various subsequent attempts to link human rights and the environment through rights-based approaches (Boyle and Anderson 1996; Eckersley 1996; Hancock 2003; 2005; Miller 2002; Woods 2010). Jed Greer and Tyler Giannini, a co-founder of ERI and now Harvard Professor of Law, proposed a new concept, 'earth rights', which are 'those rights that demonstrate the connection between human well-being and a sound environment, and include the right to a healthy environment, the right to speak out and act to protect the environment, and the right to participate in development decisions' (1999: 20).

Although adding some potentially useful analytical tools to environmental campaigns, the problem with all these rights-based approaches is that they are likely to exacerbate problems that already exist within the human rights regime (Woods 2006: 588). A central contestation of human rights is their universality or applicability in various cultural or historical contexts. As Steans notes in her analysis on human rights and women:

> Critics of universal doctrines like human rights raise valid objections that cannot be dismissed easily. However, cultural relativism – in its various guises – is equally problematic. It is no easy task to distinguish between legitimate expressions of identity, community and culture and the (ab)use of 'culture' and 'tradition' to legitimise the exercise of power by authoritarian governments over

---

5   K. Redford, interview with author, Chiang Mai, Thailand, 15 January 2004.
6   T. Giannini, interview with author, Chiang Mai, Thailand, 21 January 2004.

their subjects, or indeed the arbitrary exercise of power by men over women. (2007: 13)

Despite a variety of perspectives on what exactly comprises these rights, however, there is little doubt that their violation often results in greater insecurity for marginalised communities in the South where the interrelationships between environmental protection and human rights are most acute.

The explicit use of a rights-based approach is, however, only one, albeit important, strategy in the pursuit of global and environmental justice. Environmental justice is concerned first and foremost with 'the distribution of environmental risks and harms across individuals and social groups' (Dryzek 1999: 266). Justice advocates suggest there exists an unequal distribution of negative environmental effects flowing from areas of high social mobility and political power to the relatively disadvantaged at both a national and international level (Newell 2005; Walker and Bulkeley 2006). While some theories of environmental justice have focused on distributive justice (Dobson 1998), others have placed increasing emphasis on both the political recognition of diversity within communities and participation in the political processes that govern development projects (Anderson 1996; Fraser and Honneth 2003; Schlosberg 2004; 2007). While most environmental justice analysis, as with environmental politics in general, has been focused on the North (Schlosberg 1999; Walker 2012b), these attempts to empower marginalised communities can also be applied to the South, although the political hurdles faced are often greater. In the South movements for environmental justice have emerged in response to carrying not only the environmental burden of the elites from their own country but that of the North's 'wasteful consumer-driven throw-away society' as well (Bollard 1999: 33). One of the most clearly espoused justice-oriented approaches is Laura Pulido's 'subaltern environmentalism':

> Subaltern environmental struggles are not strictly environmental. Instead, they are about challenging the various lines of domination that produce the environmental conflict or problem experienced by the oppressed group in the first place. Since they must confront multiple sources of domination that include economic marginalisation, patriarchy, nationalism and racism, it is difficult to discern where the environmental part of the struggle begins and where it ends. Indeed, trying to do so may misrepresent the very nature of struggle, as it suggests that environmental encounters are not coloured by political economic structures. (1997: 193)

In the South, while identity politics plays a part, either radical democratic politics, incorporating legal-political rights, or socialist politics, incorporating the shared material interests of the working classes, have dominated environmental justice discourses (Faber 2005: 46). Indeed, in the South there is often little distinction between environmentalists and human rights or environmental justice campaigners as most campaigns directly link environmental sustainability with justice (Williams and Mawdsley 2006), although conservation NGOs do exist (Di Gregorio 2012). As Doyle states in his analysis of the Filipino environment movement, 'its green agenda is not centred on luxury and higher-order ideals … but rather … in the struggle to survive' (2005: 52). People are unlikely to consider 'post-materialist' concerns, such as wilderness preservation, when degradation of 'the environment' in which they live has a direct and instantaneous impact on their health, livelihood and well-being.

Conversely, as I demonstrate later in this book, Northern environmental organisations active in the South may well pursue Northern wilderness aims at the expense of Southern justice and can be considered part of the EGS. As Doyle and I argue elsewhere in our comparison of environment movements in Iran and Myanmar, 'the existence of a larger middle class in Iran, when compared with Burma, may also have contributed to the increasing visibility of … post-materialist banner issues' (2006: 761). In addition, more affluent areas can compartmentalise environmental problems away from much of the population or employ expensive technological solutions, although the poor or marginalised in affluent countries – the South in the North – face similar issues to those in Southern countries, both in terms of recognition and distribution (Doherty and Doyle 2006: 706).

There is often also a clear relationship, particularly in the South, between the struggle for environmental justice and political violence. In her analysis of poor black and indigenous communities in Brazil, Christen Smith (2009) argues that the overt police violence that leaves a trail of dead young men makes the dumping of toxins into their local environment permissible, and vice versa. There is a clear connection in Myanmar between the environmental destruction of ethnic areas and the violence meted out to ethnic communities. This relationship is also apparent in Thailand where struggles for environmental justice have resulted in the assassination of environmental activists by representatives of the state and associated business networks.

An important focus within environmental justice movements is the rights of indigenous peoples. Article 3 of the 2007 UN Declaration on the Rights of Indigenous Peoples provides that 'indigenous peoples have the right to self-determination. By virtue of that right they freely determine their political status and freely pursue their economic, social and cultural development' (UN 2008: 4). While the category of indigenous peoples can be ambiguous or problematic (Sawyer and Gomez 2012), anthropologists are nevertheless reluctant to deny it to 'local communities' who claim it (Dove 2006). It should be recognised, therefore, that the divisions between categories of indigenous peoples and ethnic minorities, as with so many other social or political labels, are somewhat fuzzy. Many of the marginalised ethnic minority communities of Myanmar who have suffered injustices due to the energy projects in this book could justifiably lay claim to indigenous status. Indigenous rights become particularly significant when indigenous peoples are not recognised as such due to the greater protections or concessions that would be afforded them under international agreements (Hirsch 2002: 159). Conflict can also occur between indigenous or ethnic groups over the contested ownership of resource rights, which can turn on the question of the validity of an ethnic identity itself (Dittmer 2010). The Muslim Rohingya minority of Rakhine State in Myanmar are considered 'Bengalis' – interlopers from the west – by the vast majority of the Myanmar population and are therefore denied citizenship. It is a particularly explosive issue because 'if they are recognised as a legitimate identity then resources will follow'.[7] When the resources of indigenous or ethnic minorities are exploited with little recompense, it undermines their ability to manage their own affairs. Through fighting for recognition and resource re-distribution, however, indigenous peoples can have their own consciousness raised in much the same way that other activists are radicalised through their activism (Maddison and Scalmer 2006: 73). As NGOs note in a report prepared for the UN, 'protection of indigenous peoples' cultural integrity is closely linked to recognition of and respect for rights to lands, territories and resources and to protection and preservation of their physical environment' (Forest Peoples Programme and Tebtebba Foundation 2006: 61). Campaigns for environmental justice and indigenous rights can be considered part of a broader global justice movement that uses the tools of globalisation to challenge oppression, rapacious capitalism and neoliberalism on a global scale (Curran 2006: 52; della Porta et al. 2006: 9; Epstein 2001; Eschle 2005: 27; Rootes 2005: 692).

In the 1980s Melucci argued that nation-states were losing their authority from 'above', due to global political and economic interdependence, and from 'below', due to rapidly expanding transnational 'civil societies' (Melucci 1989: 87). Activists in the global justice movement generally oppose the neoliberal project of 'globalisation from above' but they are enthusiastic about a 'globalisation from below' (Brecher, Costello and Smith 2000; Cox 1999: 13; della Porta et al. 2006; Falk 2000: 49). Neoliberal economic globalisation is seen by activists within the movement as further marginalising vulnerable communities across the world and entrenching power within transnational capital. This approach contrasts with conservation groups within the EGS, such as the Wildlife Conservation Society, which generally have a more amenable relationship to business. As I argue elsewhere, to achieve emancipatory outcomes

> the influence of business must always be taken into account by the environment movement when developing a strategy for winning environmental disputes … [This analysis allows activists and scholars] to explore how and why environments continue to suffer in the face of continued acceptance of political practices which are, at their root, antithetical to environmental protection and social justice. (Simpson 1998: 33)

Vandana Shiva argues that this entrenching of business interests 'is not a natural, evolutionary or inevitable phenomenon … [i]t is a political process which has been forced on the weak by the powerful' (1999: 47). The 'negative essence' of neoliberal

---

7   Aung Naing Oo, interview with author, Yangon, Myanmar, 2 May 2013.

globalisation that justice movement activists oppose is the global economic restructuring that imposes on governments 'the discipline of global capital in a manner that promotes economistic policy making in national arenas of decision, subjugating the outlook of governments, political parties, leaders and elites and often accentuating distress to vulnerable and disadvantaged regions and peoples' (Falk 2000: 46). There have been numerous financial crises linked to neoliberal policies across the South. The Asian financial crisis of 1997–98, which was triggered by Thailand's floating of the baht, was a significant backdrop to the protests over the Yadana Gas Pipeline, the first transnational energy project in the region.

The global justice movement includes organisations with an environmental focus but only if it is immersed within a wider agenda for social justice at both local and transnational levels. These organisations are generally EGGs, if their organisational structures match their emancipatory aims, and include ERI and Friends of the Earth (FoE), which joined the campaign against the Thai-Malaysian Pipeline. In recent years, however, even wilderness-oriented NGOs, such as WWF, have in some part increased their focus on social justice issues. The result, according to Rootes, is that 'the environment' has been sidelined within the movement (2006: 779), with, for instance, climate change being a 'bolt-on' to an agenda focused upon 'debt relief, trade and aid, and a useful stick with which to beat' the US government under former president George W Bush (2005: 693). Rather than seeing these emphases as simply a result of a focus on the South, Carter argues that it is more likely due to the important role in the justice movement of 'left wing activists [who] retain lingering suspicions of environmentalism' (Carter 2007: 164). These perspectives are characteristic of attempts to make clear distinctions between 'the environment' and everyday existence in the South, when no such clear distinction exists. Activists in the South believe that Northern priorities such as climate change receive excessive attention when other pressing atmospheric issues, such as air pollution and its impacts on health, are of more immediate concern (Doyle and Doherty 2006: 890; Doyle and Risely 2008; Dyer 2011; O'Neill 2009: 52). For analyses of environmental issues in the global South, therefore, the global justice perspective is a welcome relief from the formerly post-materialist dominated agendas of Northern environmentalists.

As well as facilitating the development of the movement as a whole, the communications technologies of globalisation have revolutionised the movement's tactics, with the internet and email increasingly relied upon to coordinate transnational actions (Curran 2006: 75; Diamond and Plattner 2012; Doherty 2002: 172; Eschle 2005: 21; Klein 2001; Reitan 2007: 80). Despite participation in these transnational protest events, however, most organisations' activities remain strongly entrenched at the local level (della Porta et al. 2006: 234–35; Doherty 2002: 56). This is particularly true in the South. As the case studies in this book demonstrate, activists in the South are now proficient in electronic activism. Social media such as Facebook and websites such as YouTube are used as campaign tools in Myanmar to document oppression from the Myanmar military (Aye Mi San 2007). Indeed, limited political openings in Myanmar, as under the authoritarian regime in Iran, may have actually stimulated these less risky forms of activism (Kelly and Etling 2008). It should be noted, however, that, despite this increased use of real time technologies and the sense of 'time-space compression' (Harvey 1989) in the North and urban areas in the South, much of the world's population 'continues to live a predominantly agrarian existence that is connected to wider global processes only in intangible and attenuated ways [that are] predominantly political-economic and invariably negative' (Beeson and Bellamy 2003: 344).

In the North justice movement protests since the late 1990s have been characterised by carnivalesque visual rhetorics and a revival of neo-Situationist protests (Curran 2006: 192; Szerszynski 2007: 346). A festive atmosphere is a key element of the character of the protest, as is a sense of the absurd, which Torgerson and Doyle both see as essential to green political activism (Doyle 2000: 39–44; Torgerson 1999: 93–94). In the rather more repressive, and culturally conservative, countries that this book examines, there are understandable limitations to the extent of absurd or humorous protest but there are examples. The Moustache Brothers in Myanmar used irony as both rhetorical form and philosophical content (Szerszynski 2007: 348). Their sense of the absurd applied to their traditional a-nyeint comedy and performance, which included improvisation and biting satire about the former military regime, which resulted in the performers being tortured and sentenced to six years' incarceration and hard labour (Amnesty International 2001; Aung San Suu Kyi 1997: 39).[8]

Nonviolence

As the case of the Moustache Brothers above suggests, the consequences for nonviolent forms of dissent can be far more extreme in the South than in the North. Nevertheless, the fourth pillar of green politics, nonviolence, is well established within environment movements of the South. It was, after all, in the South where the most famous advocate of nonviolence, Mahatma Gandhi, used the concept as a powerful and effective form of civil disobedience against imperial Britain. Protest plays a central role in all environment movements as social movements and these are generally undertaken with a commitment to nonviolence (Carter 2007: 54; Doherty 2002: 7). Violence against inanimate objects as part of militant direct action has sometimes been considered a useful way for movements to attract attention in the North, particularly from the media which would otherwise ignore protests or actions (Doherty 2002: 178; Foreman 1991: 113). Nevertheless, even this activity is limited, and nonviolence as a protest tactic, strategy and philosophy has thoroughly permeated environment movements in both the North and South.

Despite its near universal appeal there can be very different consequences of nonviolent protest for activists in the North and the South. Northern activists rarely face life-threatening situations, although there are exceptions. Southern activists,

---

8   Par Par Lay and Lu Maw, interview with author, Mandalay, Myanmar, 15 December 2003.

on the other hand, can face brutal repression by security forces or assassinations that are never investigated. Despite the risks, nonviolent protest remains dominant. Sulak Sivaraksa, a prominent Engaged Buddhist from Thailand, argues that as well as the philosophical rationale there are also practical reasons for nonviolent approaches.[9] A study by Stephan and Chenoweth (2008) supports this position; they suggest that in non-state conflict over the last century, nonviolent resistance has been almost twice as effective as violent campaigns in achieving the stated goals. As noted above, collective nonviolent protest also appears to magnify the emancipatory values of the individuals present, both the individuals' and the group's as a whole (Welzel and Deutsch 2012). Nevertheless, in terms of achieving their goals the Myanmar uprisings of 1988 and 2007 can be seen as failures of nonviolent resistance largely due, according to Williams (2011), to the inability of activists to compel loyalty shifts among the security forces. Debates over the effectiveness of nonviolence in all situations therefore remain, with even the Dalai Lama, one of the most famous contemporary advocates of nonviolence, arguing that the political conditions must be ripe for nonviolence to be successful.[10]

Despite the prevalence of nonviolent approaches amongst activists the existence of severe repression and authoritarian governance in Myanmar bestowed widely regarded legitimacy on ethnic insurgents engaged in civil conflict against the Myanmar military. Ethnic minorities saw insurgent groups, such as the KNU, as their protectors against human rights abuses by the military. As a result, support for marginalised ethnic minorities in Myanmar often translated into at least indirect support for armed insurgent groups. Indeed, the Director of ERI's Southeast Asia Office, Chana Maung, spent six years in the Karen National Liberation Army (KNLA), the armed wing of the KNU, prior to joining ERI. Despite adopting the nonviolent path he still argued that this military response reduced oppression by the Tatmadaw.[11] Other organisations, such as the Kayin-centred KESAN, were quite open about their relationship with the KNU due to its role in promoting security for the Kayin people.[12] Similarly, while the exiled Burma Lawyers' Council (BLC) produced reports and ran seminars on improving basic human rights, democratic rights and the rule of law in Myanmar, individuals within the organisation were also linked to the KNU.[13] While their philosophical emphasis was on nonviolence and basic human rights, these individuals felt little option but to support their ethnic brethren engaged in a battle for their survival. As prominent Engaged Buddhist and founder of Sri Lanka's Sarvodaya Shramadana Movement, A.T.

---

9   Sulak Sivaraksa, interview with author, Schumacher College, Devon, UK, 25 January 1998.

10   Tenzin Gyatso (The Fourteenth Dalai Lama), interview with author, Dharamsala, India, 18 July 1997.

11   Chana Maung, interview with author, Chiang Mai, Thailand, 10 December 2010.

12   Alex Shwe, interview with author, Chiang Mai, Thailand, 8 January 2009.

13   Myint Thein, interview with author, Mae Sot, Thailand, 19 January 2004.

Ariyaratne, argued, 'you have to fight injustice, with or without violence'.[14] The political reality in these areas therefore raised existential issues that activists in the North simply do not need to consider, which can cause conflicts over strategies and tactics between groups in the North and South. Some Northern activists who work closely with people in the South do, however, recognise their privileged position, with one commenting that 'living as a Westerner with a "cushy" background [I] can't comment on armed struggle'.[15]

The differences between nonviolent approaches to protest in the North and South can be demonstrated through comparative studies on the Franklin and Narmada Dam disputes, in Australia and India respectively. In the former the personal risks and costs associated with the campaign pale in comparison with those faced by campaigners in the Narmada Bachao Andolan, who faced the destruction of their homes and threatened to drown themselves in the rising waters if construction was not halted (Doyle 2004; 2005: 133). Franklin Dam activists faced arrest and fines as the price to be paid for 'manufacturing vulnerability' (Doherty 1999) during the protest while Williams and Mawdsley argue that 'the full force of authoritarian state power' was periodically unleashed on the Narmada protestors (2006: 667). For activists in the South there is often no need to manufacture vulnerability; the vulnerability is ever present. Despite the repression of protesters, the employment of Gandhian-influenced tactics, such as satyagrahas or nonviolent resistance (Richards 1991: 48–63), is a common approach throughout the South and is promoted by global justice campaigners such as Vandana Shiva.[16] While these approaches have also been employed by activists in the North, approaches from the North have also been adopted in the South, with the forest occupation in Thailand during the Yadana Pipeline echoing the Franklin and Wet Tropics forest occupations in Australia (Cohen 1997). Even when this transnational cross-fertilisation of philosophies and repertoires of action occur, however, they are often adapted for local conditions (della Porta and Mosca 2007; Doherty and Doyle 2006: 701).

North–South Differences and Conflict

While there have been significant successes in cooperation between movements in the North and South (Tarrow 2005), there is also the potential for conflict. Organisations that traverse the North and South can experience stark differences in foci between Northern and Southern activists due to their vastly different political and cultural environments. There is enormous diversity within the environment movement, even within a single country (Hutton and Connors 1999), so it is not

---

14   A.T. Ariyaratne, interview with author, Moratuwa, Sri Lanka, October 1998.

15   ERI Assistant Team Leader (name withheld), interview with author, Chiang Mai, Thailand, 10 January 2005.

16   V. Shiva, interview with author, Dehra Dun, India, December 2004.

surprising that even greater diversity appears amongst countries with extremely diverse political, economic, cultural and environmental systems which can sometimes cause tensions within the movement and within or between groups.

Within the environment movement in the South the frames of postcolonialism and post-structuralism, rather than those of post-materialism found in the US and Australia and post-industrialism in Europe, usually dominate (Catney and Doyle 2011). Struggles for environmental justice in the South are often fought in the context of a 'postcolonial state, [that] like its colonial precursor ... has been consistently willing to sacrifice both the environment and the poor to a longer-term vision of commercial growth and industrial modernity' (Williams and Mawdsley 2006: 662). As a result environmentalism 'often has its base in the livelihood struggles of the rural poor rather than the aesthetics of emerging middle classes' (Hirsch 1997: 4).

Although environmental campaigns in the South are often conflicts over who should use and benefit from natural resources, the struggles are also more complex and played out over issues of 'interests, knowledge, values and meanings in local as well as national and global arenas' (Dwivedi 2001: 238). Despite a multiplicity of issues that are contested, Hirsch notes in his analysis of Southeast Asia that

> [w]ithin the framework of environmentalism, focused on struggle over control of resources and social or spatial inequities in development, the politics of environment have crystallised most clearly around large-scale resource projects, particularly those that involve appropriation of the local resource base by interests of state, capital, and dominant social groups in the name of national development. (1998: 55)

This accords with research by Kalland and Persoon (1997: 12) that the dominant environment movements in Asia – much of which is considered the South – are movements that 'oppose something'. The locus of these struggles is evident within this book where the central goal of each campaign was to halt a transnational energy project, at least until issues of injustice and inequity were addressed.

Women have played important roles in the informal sector of environment movements in the North (Doherty 2002: 201; Doyle 2000: 33), but women are also central players in struggles throughout the South (Desai 2002: 32–33; Doyle and Simpson 2006: 762; Mies and Shiva 1993; Shiva 1989). In the South men are more likely to work on cash crops while women carry the major burden of supporting the household and doing subsistence agricultural work. Cultural norms and the increasing privatisation of common property resources may increase the difficulty for women in providing for their households so they join or form environmental organisations in order to gain a voice (Miller 1995: 42). Despite the involvement of women in the environment movements of both the North and South, the differing priorities and perspectives echo conflicts that have occurred in the women's movement. In the 1970s and 1980s many white middle-class Northern feminists argued that gender oppression and sisterhood was global and universal, while Southern feminists argued from a postcolonial perspective that much of their oppression related to race, ethnicity, nationality and class (Desai 2002: 28–29). Being at the periphery of an unequal global economy together with the visceral nature of life in the South left many women with a starkly different view compared to those in the North.

An illuminating example of North–South tensions within the environment movement appears in the study by Doherty of a split that occurred in Friends of the Earth International (FoEI) when FoE Ecuador, Accion Eologica, resigned from FoEI in 2002 (Doherty 2006). FoEI, the international federation of FoE groups, is a genuinely grassroots organisation compared with the other large environment NGOs such as WWF and Greenpeace and individual groups from both the South and the North have significant autonomy and independence in their activities (Carter 2007: 150; Wapner 1996: 122). It is also the most non-hierarchical and introspective of these organisations, adopting a critique of neoliberalism with a focus on social and environmental justice (Doherty 2006: 862; Rootes 2006: 769), which included taking part in the campaign against the Thai–Malaysian Pipeline discussed in this book. Nevertheless, Accion Eologica claimed that the Northern groups within FoEI had failed to understand the different contexts faced by groups in the South, including political violence and human rights abuses. Accion Eologica also argued that FoEI's participation in the World Summit on Sustainable Development in South Africa in 2002 had given priority to Northern agendas. This conflict was little different to the first Earth Summit in Rio a decade earlier where Northern agendas included climate change, population and the ozone hole, while the South's agendas were poverty, hunger and desertification (Calvert and Calvert 1999: 189; Chatterjee and Finger 1994). In responses that may be hard for Northern groups to grasp fully, some Southern FoE groups 'stressed the centuries of looting of their national resources by colonial powers, and now by corporations, whether national or transnational, and aligned their struggles with those of indigenous peoples against a global model of development' (Doherty 2006: 869).

Northern imperialism is therefore seen as a precursor to the current exploitation and power imbalances that are often only exacerbated by contemporary 'eco-imperialism', global environmental governance 'from above' (Dyer 2011). In 2002 Accion Eologica argued that Northern groups were supporting a reformist position within a system that required resistance from outside. It saw corporations as 'beyond accountability and argued that seeking to encourage reforms and regulation simply reinforces existing structures' (Doherty 2006: 868). These Southern perspectives, due primarily to precarious political and economic environments, can sometimes result in desperate tactics by those in the South that can be confronting for Northern activists. In 1990 Myanmar activist and journalist Soe Myint used a soap container to hijack a plane from Thailand to Myanmar and redirected it to Calcutta (Kolkata) to hold a press conference about military oppression in Myanmar. These tactics would be unacceptable in the North, particularly since 9/11, but due to the political conditions in Myanmar an Indian court finally acquitted him of all charges.[17]

Conclusion

This chapter provided the theoretical and philosophical foundations for the model of activist environmental governance deployed in analysing the case study environmental campaigns to follow. The emancipatory actors within this activist environmental governance provide alternative perspectives to the 'top-down' eco-imperial or neoliberal approaches that often dominate discussions on environmental governance. These actors are part of emancipatory environmental movements that challenge existing power relations in society, are linked by networks, share a collective identity and engage in protest. These characteristics apply to most environment movements in the South, which frequently face very different social, political and environmental issues from the North. In particular, widespread poverty, environmental insecurity and authoritarian governance mean that Southern movements prioritise human rights and social justice over purely ecological concerns.

Environmental groups can play a positive and emancipatory role in a process of environmental governance, regardless of their formality, by engaging in activities that are informed by more localised, interactive and bottom-up modalities. The model used to analyse these activities, adapted from Doyle and Doherty (2006), proposes a typology of activist environmental governance comprising emancipatory governance groups (EGGs), compromise governance groups (CGGs) and the environmental governance state (EGS). EGGs from both the North and the South value the four core green pillars of nonviolence, sustainability, participatory democracy and social justice in both their aims and organisational structure; they consider the process of activism itself as important as the goals to be achieved. CGGs have emancipatory aims but are constrained by conservative structures that may impact on their ability to achieve truly emancipatory outcomes while organisations of the EGS have no emancipatory characteristics. Although the green pillars provide ideals for emancipatory movements, conflicts often emerge both within and between organisations that traverse the North and South as a result of different cultural and political environments, even within introspective organisations.

The ability of activists and communities within these movements to effectively oppose inappropriate development projects or policies is, however, heavily dependent upon the political regime under which they operate. Transnational capital plays a central role in sustaining illiberal regimes across the South while promoting large-scale energy projects that exacerbate human and environmental insecurity in marginalised communities. In response, civil society actors have

---

17   Soe Myint, interview with author, Delhi, India, 24 December 2004.

engaged in emancipatory activist environmental governance by promoting justice and emancipatory values. The remainder of this book therefore develops this model further through the analysis of local and transnational campaigns against transnational energy projects in Myanmar and Thailand.

Chapter 3
Environmental Politics in Thailand and Myanmar

Introduction

In this chapter I analyse the diverse aspects of environmental politics that were played out between state and non-state actors in the two core countries of this book, Thailand and Myanmar. As I set out in Chapter 1, this book focuses on the Thai case study campaigns in the period from the mid-1990s to the fall of the Thaksin government in 2006, while the Myanmar case study campaigns are examined from their beginnings, in the mid-1990s, until the installation of the new 'civilian' Myanmar government in 2011. Throughout these periods Thailand adopted several roles, not only as an importer, exporter and consumer of energy, but also as sometimes-reluctant host to much of the Myanmar activist diaspora campaigning against the projects being pursued at home. Despite the growing influence of transnational actors the ability to engage in activist environmental governance in these countries was still highly dependent on the nature of the domestic political regimes. The overriding factor that determines domestic political openings for activism is the extent to which democratic principles are applied and respected within a society. The degree to which basic human rights – such as freedom of speech and freedom of association – are permitted or curtailed determines the openness of environmental protest and the forms in which it is most efficacious. In addition to these civil rights there are now widely recognised human rights related to a healthy environment (Ksentini 1994), which, if ignored by the regime in power, can exacerbate environmental insecurity. This chapter therefore develops a political regime model, based on the level of political competition and authoritarianism, which can then be used to assess the regime types of Thailand and Myanmar.

Throughout the chapter I also examine the role of business in either helping or hindering authoritarian governance in each country. Business interests in both countries were powerful and had great influence over the formation of public policy, but the mechanisms through which this power operates varied once again with the nature of the regime. In Thailand large capital interests gained unprecedented political power through the electoral success of Thai billionaire Thaksin Shinawatra and his Thai Rak Thai (TRT) party. This success challenged the existing power structures that dominated Thai society: the monarchy, the military and provincial business interests. In Myanmar the military dominated the economy as it dominated the political sphere, and privatisations from the early

1990s only served to widen the economic gulf between those with political access to the military elites and the rest of the population.

In both Thailand and Myanmar communities faced environmental insecurities that were often exacerbated by large-scale energy projects. Environmental security in Myanmar was particularly tenuous with the country often inhabiting the lowest ranks in measures of human development, corruption and democracy, and the suffering caused by natural disasters was further aggravated by the brutality and neglect of the military regime. Thailand was more affluent and, in general, more democratic than Myanmar, but marginalised communities were still exposed to environmental insecurities and human rights abuses. The repression faced by environmental activists who campaigned against energy projects represented a symbiotic convergence of interests between large business interests and the most powerful political class. Personal, economic and political linkages and reciprocity between elites in government and business therefore facilitated both the causes of insecurity and the resultant attempts to silence voices of dissent. The attempts to close down political debate over these energy projects were representative of wider disjunctures in society between the interests of large capital and those more focused on democracy, justice and sustainability. The situation faced by activists and marginalised communities in Thailand and Myanmar is not uncommon throughout the South where illiberal and authoritarian regimes, characterised by opaque governance, are prevalent. The countries examined here therefore provide important lessons on the implications of the nature of political regimes and business power for environmental politics in the South and provide the context for analysis of the specific environmental campaigns against the cross-border energy projects in the following three chapters.

Assessing Competition and Authoritarianism in Political Regimes

Democracy is one of the four green pillars and therefore a central value that underpins emancipatory activist environmental governance. Democratic regimes are likely to allow more open activism, with activists free to voice dissent and critique government decisions. Communities under less democratic regimes may also be more susceptible to environmental insecurity, as citizens are less able to protest or register opposition to protect their interests. The nature of political regimes is, therefore, likely to have a significant impact on the nature and extent of environmental activism and may also determine to what extent transnational influences are able to traverse borders and influence domestic environmental policy and debate. Historically, however, there have been few in-depth analyses of the impact of political regimes on environmental activism (Doyle and McEachern 2008: 22–30; Doyle and Simpson 2006). Part of the reason for this lacuna appears, once again, to be the traditional focus on the North, where states tend to be more democratic. In this section I outline a theoretical model for political regimes while the rest of the chapter provides more detailed assessments of Thailand and Myanmar based on sites of competition and authoritarianism.

The dominant political regimes in Thailand during the case study campaigns are quite difficult to categorise; there were greater political freedoms than in Myanmar but also limitations. It could therefore be considered one of a growing number of hybrid regimes 'that are neither clearly democratic nor conventionally authoritarian' (Diamond 2002: 25). Some of these regimes reflect the 'new forms of authoritarianism' that Huntington cautioned in the early 1990s could emerge that suited 'wealthy, information-dominated, technology-based societies' (Huntington 1991: 316). Due to the increasing prevalence of these 'new' regimes, a useful typology for exploring environmental politics is to group regimes as liberal democracies, traditional authoritarian regimes and hybrid regimes (or amalgam regimes), where certain sub-categories may also exist.

Some of the specific terms used to describe hybrid regimes, such as 'transitional democracies', can be misleading, however, suggesting unidirectionality, while others gloss over important differences. It can be tempting to lump all regimes with authoritarian tendencies in the same group but there are certain qualitative differences that make activism in unequivocally non-democratic countries a very different proposition to activism under more liberal regimes that also have authoritarian tendencies, such as Thailand. In recent years models of semi-authoritarianism (Ottaway 2003), limited multiparty regimes (Hadenius and Teorell 2007: 147), competitive authoritarianism (Levitsky and Way 2002; 2010), and a proliferation of other hybrid regimes (Brownlee 2007: 25–26; Diamond 2002), have been proposed to analyse regimes that fall neither squarely into the authoritarian nor liberal democratic camps.

Many of these categorisations are, however, problematic. They tend to overlook the dominant role of capital in the establishment of many of these regimes, which is evident in the cases of Singapore and Malaysia (Rodan 2004). This issue is part of a broader problem that Jayasuriya and Rodan identify in these models which overlooks the causes of the formation of hybrid regimes in favour of 'descriptions of institutional performance' (2007: 774). In another critique Sim (2006: 146) argues that Diamond's uncritical use of Freedom House's numerical rankings of authoritarianism can result in 'ideologically driven' categorisations that preference economic freedoms. Likewise, Hadenius and Teorell (2007: 144) argue that the 'degree of competitiveness' that is used to determine Diamond's categories is problematic and understates qualitative differences between authoritarian regimes.

Despite the shortcomings of these models, Levitsky and Way's criteria for competitive authoritarianism provides a useful starting point for the analysis of environmental activism under the political regimes of this book. Throughout this chapter I address the concerns over these models expressed by Jayasuriya and Rodan and others by examining the dominant role of capital in the formation and maintenance of these regimes to avoid preferencing economic over political freedoms. Levitsky and Way define competitive authoritarian regimes as those in which 'formal democratic institutions are widely viewed as the principal means of obtaining and exercising political authority. Incumbents violate those rules so often and to such an extent, however, that the regime fails to meet conventional minimum standards for democracy' (2002: 52).

One of the most important aspects of competitive authoritarianism, therefore, is that, despite facing authoritarian obstacles, opposition forces may periodically 'challenge, weaken, and occasionally even defeat autocratic incumbents' through formal democratic institutions (2002: 54). Indeed, this is the main point of difference from full-blown traditional authoritarian regimes. Levitsky and Way argue that there are four important areas of democratic contestation where this competitiveness reveals itself: the electoral arena, the legislature, the judiciary and the media (2002: 54–58). First, under competitive authoritarian regimes elections are regularly held, seriously contested and often bitterly fought. Although the electoral process is generally free of massive fraud, it may also be characterised by 'large-scale abuses of state power, biased media coverage, (often violent) harassment of opposition candidates and activists, and an overall lack of transparency' (2002: 55).

Second, legislatures in competitive authoritarian regimes are relatively weak, but they can occasionally become focal points for opposition. Third, governments routinely attempt to subordinate the judiciary, either blatantly or through more subtle techniques such as bribery, extortion and other mechanisms of co-option. The combination, however, of formal judicial independence and incomplete control by the executive allows 'maverick' or strongly independent judges to make decisions that can contain the scope of authoritarian control. Fourth, the media is formally free and journalists may emerge as important opposition figures. Nevertheless, governments often seek to suppress the independent media, using relatively subtle mechanisms of repression such as 'bribery, the selective allocation of state advertising, the manipulation of debts and taxes owed by media outlets ... and restrictive press laws that facilitate the prosecution of independent and opposition journalists' (2002: 57–58).

As useful as these four criteria are, a major omission in the Levitsky and Way model is the independence and accountability of the various uniformed security and law enforcement agencies, including the military and police. It is extremely difficult for an independent judiciary to achieve just outcomes if law enforcement agencies are corrupt and free to manipulate and manufacture evidence without adequate institutional oversight and review. In traditional authoritarian regimes complicity by law enforcement agencies is crucial to a state's ability to undermine democratic participation through compromised court proceedings, which may extend, in the worst cases, to extrajudicial killings by the security services. Under liberal regimes security services are highly accountable to elected officials with high levels of transparency in decision making. Competitive authoritarian regimes sit somewhere in between these two extremes; security services do not operate with a culture of complete impunity but elected governments and the judiciary are similarly constrained in their ability to provide rigorous oversight due to entrenched privileges and power. These structural advantages are often left over after the transition from traditional authoritarian regimes under which security services enjoyed undisputed and unconstrained power. Under competitive authoritarian regimes there are, therefore, elements of contestation and competition in this fifth arena.

In this chapter I argue that the Thai government of Chuan Leekpai exhibited more features of a liberal democratic regime while, using the five criteria above, the tenure of Thaksin Shinawatra was characterised by competitive authoritarianism, although these characterisations represent tendencies rather than absolutes. It should also be noted that under Thaksin the government also qualified as a dominant-party regime, a subset of Hadenius and Teorel's (2007: 148) limited multiparty regimes in which the dominant party achieves two thirds of the vote. In discussing opportunities for democratic change, Hadenius and Teorel (2007: 152) also argue that the demise of an authoritarian regime does not necessarily signify a more liberal replacement, as in most cases one authoritarian regime simply gives way to another. This prognosis appeared to have been borne out in Thailand, at least in the short term, as it was ruled by a military dictatorship for fifteen months following a coup in September 2006 in which Thaksin was deposed (Connors and Hewison 2008). More sites of competition have, however, returned to Thailand since military rule with the return of national elections and more liberal democratic governance, although the military's new constitution remains in place.

Under traditional authoritarian regimes activism is more constrained than in those defined as competitive authoritarian. Under these politically closed regimes (Diamond 2002: 31) there is extremely limited competition and contestation in the five areas discussed above and Myanmar, particularly during the period of direct military rule between 1988 and 2011, fulfilled this requirement. Myanmar could have been classified as a single-party authoritarian regime under the Burma Socialist Programme Party (BSPP) from 1962 to 1988, but I have followed Geddes' convention by using the subsequent period as the basis for classification, primarily because the military has effectively ruled since 1962 (Geddes 1999: 122–23). Military regimes are states in which military officers are the major or predominant political actors by virtue of their actual or threatened use of force (Hadenius and Teorell 2007: 146). Geddes argues, however, that despite

> a consensus in the literature that most professional soldiers place a higher value on the survival and efficacy of the military itself than on anything else ... in countries in which joining the military has become a standard path to personal enrichment acquisitive motives can be assumed to rank high in most officers preferences. (1999: 126)

As a result, of all authoritarian regimes Geddes (1999: 122) argues that military regimes tend to be the most unstable and most susceptible to internal disintegration. Hadenius and Teorell (2007: 150) also found that military regimes are more short-lived than one-party regimes but that they outlast limited multiparty regimes, or competitive authoritarian regimes. In this context the Myanmar military regime was extremely enduring, with the military effectively in power between 1962 and

2011. The political reforms since the 2010 elections and the presence of Aung San Suu Kyi and the NLD in parliament have resulted in a shift away from this traditional authoritarianism. With the structural political advantages allocated to the military in the constitution, however, it is still too early to definitively allocate it a new regime model although a shift towards a more competitive authoritarian regime appears underway.

Under traditional authoritarianism, civic engagement and activism does not necessarily represent a threat to the ruling class (Brownlee 2007: 217–18), and sometimes it may actually represent a consolidation of power (Doyle and Simpson 2006: 764; Jamal 2007). Emancipatory environmental activism and governance, however, is a radical form of civic engagement which promotes justice by challenging existing inequitable political structures and it therefore represents a threat to politically closed regimes.

In addition to Myanmar and Thailand the campaigns against the energy projects in this book also spilled over into other countries, although this activism was more peripheral and I will therefore not examine their political regimes in detail. Nevertheless, elements of their regimes still influenced the level and type of activism associated with each project. In the one-party state of China, activism has increased but it is still constrained by a traditional authoritarian regime (Mol and Carter 2006; Tang and Zhan 2008; Xie 2009). In the liberal democracy of South Korea activism has flourished since the demise of its military dictatorship (Lee 1999) and there is similarly vibrant political engagement in India's liberal democracy, despite authoritarian elements (Williams and Mawdsley 2006).

Applying the label liberal democratic does not imply that the country is free of oppression with equal access to the political process. Rights and freedoms exist on a continuum and even within nominal liberal democracies may ebb and flow (Hadiz 2006). The countries that are included in this category, however, have institutions with democratic characteristics, including a representative parliament elected by the people (Beetham 1992: 40). Nevertheless, highly concentrated media ownership and powerful vested political interests can undermine these elements of democracy, even in countries of the North. Voting systems for these regimes also vary enormously, and some are far more representative than others (Reilly 2004). In essence, whatever political regime activists operate under, green democratic ideals are a long way from fruition but the existence, or otherwise, of basic democratic rights in Thailand and Myanmar played an important role in the extent and effectiveness of activist environmental governance in each country.

Political Regimes and Environmental Politics in Thailand

The ability of social activists and movements to protest openly in Thailand has been largely determined by the nature of the contemporaneous political regime. The ability to substantially influence policy and political outcomes has been tenuous and, even under its most democratic governments, tended to reflect the extent of accommodation by existing political power structures. These power structures, often allied to the monarchy, run deeply through Thai society and stretch back to its earliest history.

Thailand's nominally modern and democratic political era began in 1932 when a constitutional monarchy replaced absolute rule, although this event simply formalised a longer term process that was already under way (Connors 2007: 40). For many of the subsequent years the military played a significant role in Thai politics. In fact the prime minister was a military officer for all but eight years over the period 1938–88 (Pasuk Phongpaichit and Baker 2009). This militaristic authoritarian rule generally constrained public dissent and criticism of the government. During the 1970s workers and the rural majority acquired a political voice for the first time and popular protests resulted in the bloody crackdown of October 1976 (Chaiwat Satha-Anand 2007; Connors and Hewison 2008), in which various authors have argued that the long-serving King Bhumibol was somewhat complicit (Handley 2006: 235–38; McCargo 2005: 504). Following a military coup in 1991 and a violent crackdown on unarmed protesters, massive street demonstrations in May 1992 and a carefully orchestrated intervention by the king squeezed the military from power. There followed a rapid expansion of social activism throughout the 1990s and 2000s in which there was a dramatic increase in NGO activism and increased public debate by academics and intellectuals, although many of the country's powerful and conservative bureaucratic and military structures remained. During this time of increased civil activism environmental organisations began to target the energy projects examined in this book.

The progress of cross-border projects between Thailand and Myanmar tended to reflect the Thai government's relationship with Myanmar's military and was therefore often dependent on the government of the day. In the lead-up to the protests over the Yadana Gas Pipeline in 1997 the prime minister was General Chavalit Yongchaiyudh, who had a long history of business interests in Myanmar (Chang Noi 2009: 23–25; Fahn 2003: 129; Pavin Chachavalpongpun 2005; Piya Pangsapa and Smith 2008). When the government floated the Thai baht on 2 July 1997, its rapid devaluation triggered the Asian financial crisis and cost the Chavalit-led coalition government power in November of that year (Baker and Pasuk Phongpaichit 2005: 254).

Chuan Leekpai of the Democrat Party then became prime minister for a second time. A new constitution was promulgated that attempted to increase public participation in government decision making and established several independent bodies to oversee and regulate formal politics in Thailand. The new constitution resulted in the 'judicialisation' of politics, with a more active and central role for the judiciary (Dressel 2009; 2010; Hewison 2010; Hicken 2007; Thitinan Pongsudhirak 2012). It was a significant attempt to undermine the paternalistic attitude and 'money politics' of previous governments and allowed the public greater input into political discourse (Laird 2000). Nevertheless, Beeson and Bellamy (2008: 126) argue that many of the democratic changes since 1992, including the 1997 constitution, belied the underlying structural power of the military, and Thailand's militarist and authoritarian 'strategic culture' was not challenged.

Chuan's government did not handle the Asian financial crisis well and in its wake Thaksin Shinawatra and his Thai Rak Thai (TRT) party dominated Thai national politics, winning comprehensive election victories in 2001 and 2005 (Ufen 2008: 342). Most of the activism against the Thai–Malaysian Pipeline in Thailand's south occurred during the lead-up to the 2001 election and in the succeeding years. In both elections there was evidence of TRT vote buying throughout the country but not so much as to seem worse than previous governments (Somchai Phatharathananunth 2008). It was also obvious that the great financial resources of TRT had been deployed in far more effective ways, even if not necessarily to the benefit of the democratic process, than other parties had historically achieved. Thaksin's rise also represented a shift in Thailand's socioeconomic dynamics with Thai peasants becoming a more vocal and powerful political force (Walker 2012a). The challenge to the established order resulted in his tenure being curtailed by a royal-backed military coup in September 2006 (Albritton and Thawilwadee Bureekul 2007; Connors 2008; Heiduk 2011; Farrelly 2013b; Ockey 2007). A new military junta-sponsored constitution, designed to reduce Thaksin's influence, was promulgated in 2007 that on the surface held to its predecessor's liberal constitutional content but actually eroded its democratic core (Dressel 2009; Hicken 2007).

Despite further setbacks, including the TRT party being dissolved by the Constitutional Court, many of the TRT politicians returned to power in December of 2007 in the newly formed People's Power Party (PPP). The PPP adopted most of Thaksin's policy agenda until it too was dissolved in December 2008 and was succeeded by the Pheu Thai Party (PTP). The PTP, led by Thaksin's younger sister Yingluck, in turn achieved a landslide win in the following elections in July 2011. For over a decade, therefore, Thaksin or his proxies have won every competitive national election (Thitinan Pongsudhirak 2012). Since the formation of TRT in 1998 'Thailand's politics has revolved around Thaksin' (Hewison 2010: 120), and his national influence during the Thai environmental campaigns in this book warrants further examination.

From the outset TRT was essentially a vehicle to carry Thaksin to the prime ministership, but it was also a vehicle for domestic capital to regain the edge over foreign companies by expelling the Chuan government, which domestic capital saw as implementing the IMF's neoliberal agenda that gave little or no benefit to local companies (Connors 2004: 2; Hewison 2002: 244; 2004: 504; Pasuk Phongpaichit and Baker 2008; Rodan and Hewison 2006: 114). Following the initial victory, however, his government bullied the media and used state violence in ways reminiscent of Thailand's previous military regimes (Pasuk Phongpaichit and Baker 2009; Thak Chaloemtiarana 2007a). His strong mandates and 'high levels of responsiveness were offset by executive abuses, corrupt practices, limits on civil liberties and gross violations of human rights' (Case 2007: 622). In earlier periods military dictatorships had rapidly lost legitimacy when violence was unleashed on demonstrators but Thaksin remained a popular democratically elected leader despite engaging in repression (Chaiwat Satha-Anand 2006a: 186–7). He politicised the bureaucracy (Painter 2006), and his approach to organised labour, which his private companies also adopted, failed to protect basic rights and weakened trade unions (Glassman, Park and Choi 2008: 368). This approach was simply 'part of a wider process through which the government sought to subvert forms of representative politics by curbing the demands of civil society' (Brown 2007: 827). His approach to the Muslims in southern Thailand reignited an insurgency that cost over 5,000 lives (Askew 2008; Harish and Liow 2007; McCargo 2009; 2012; Srisompob Jitpiromrsi and McCargo 2008). Like General Chavalit before him, Thaksin's influence on breaches of human rights in the region was particularly significant due to his willingness to do business with the military regime in Myanmar, both in his capacity as prime minister and through his family-owned company, Shin Corp (Pavin Chachavalpongpun 2010).

From the launch of TRT Thaksin used nationalism and the discourse of national economic growth to attract popular support and, when in power, stifle dissent. According to Pasuk and Baker (2009) the 'new nationalism' promoted by Thaksin and TRT was not based on an imagined ethnic identity; rather the interests of the people 'bundled together' in Thailand were paramount and 'economic sovereignty' was the key to their interests. Glassman (2004: 199) demonstrates that the support of wealthy elites for the TRT was based largely on the twin policies of specific state interventions to shore up their own industries and support for Thailand's export-oriented economic profile, resulting in little more than 'neo-mercantilist opportunism'. Thaksin's nationalist rhetoric was, therefore, largely a cover for TRT's hidden agenda of self-enrichment and empowerment (McCargo and Ukrist Pathmanand 2005: 181). His time in office also saw an unprecedented increase in foreign capital, but only in industries in which the TRT hierarchy were not dominant (Pasuk Phongpaichit and Baker 2008: 12–13). Thaksin and TRT were, however, just the last in a long line of elite groups manipulating nationalism for their own ends. The nationalist tradition of the strong authoritarian state always provided a cloak for rapid capital accumulation by the political elite (Baker and Pasuk Phongpaichit 2005: 263–65).

Thaksin's nationalist rhetoric also suggested that individual liberties, such as freedom of speech, should be surrendered in exchange for protection by an economically strong state in a social contract. This was not the first form of social contract in Thailand; many of Thailand's leaders since the late 1950s had established a developmental social contract whereby top-down paternalistic governance was offset by industrialisation and an expansion of the middle class (Hewison 2005: 323–26). In this case, however, it became a social contract to participate in the market economy. This increasing symbiosis between capital and authoritarian governance was epitomised in Thaksin's encouragement of 'loyal' sectors of the military to participate in lucrative economic activities (Beeson and Bellamy 2008: 124). Despite an overtly populist approach (Jayasuriya and Hewison 2004), the effect of Thaksin's more significant economic policies was to deepen capitalism and the reach of the market in Thailand, towards what Cerny (2000) describes as a 'competition state'. In exchange for the public accepting economic reforms, Thaksin needed to deliver benefits 'beyond the domestic capitalist class in order to deliver legitimacy for the government of the rich while local business was restructured' (Hewison 2005: 324). His response was to introduce the million baht village fund and 30 baht health schemes, which were particularly popular in rural areas (Selway 2011).

Thaksin's approach had implications for the energy sector, however, as he financed his promises to the poor by neoliberal privatisations of state assets. His view of neoliberalism was complex as economic nationalism had been a cornerstone of the TRT election platform and his program was explicitly designed to protect domestic capital from the effects of the IMF package, yet he also had policies of utility privatisation. The first privatisation was the sale of around 32 per cent of the Petroleum Authority of Thailand (PTT), which was a partner in both the Yadana and Thai–Malaysian Pipeline projects, in late 2001. PTT was Asia's third largest oil and gas firm after China-based Sinpec and PetroChina. The sale, however, seemed to have been 'managed' as large holdings were issued to government ministers' families and friends. The issue price was also undervalued as it quintupled over two years. Five other smaller privatisation projects followed the same pattern (Pasuk Phongpaichit and Baker 2009: 13). Privatisation was clearly used to enrich the already rich and powerful. As a result, a movement against privatisation in Thailand formed with over one hundred civic organisations, development groups and trade unions joining together to oppose both privatisation and the politics behind policy corruption (Connors 2004).

Between 2004 and 2006 the government proposed twelve further privatisation projects. The first and biggest was the Electricity Generating Authority of Thailand (EGAT), which was the central customer of both the Yadana Pipeline and Salween Dam projects. When the privatisation was announced in January 2004, the proposal was strongly opposed by EGAT's union, its former governors and activist groups (Connors 2004: 12). Nevertheless, in 2005 the government enacted two Royal Decrees that dissolved the EGAT state enterprise and created the charter of EGAT Plc. The Supreme Administrative Court later revoked these decrees, however, and the 2006 coup effectively ended the privatisation push. While subsequent governments did not abandon all of the neoliberal philosophies underpinning the privatisation process, the ability of activists, in conjunction with independent judges, to at least temporarily undermine the privatisation process provided some evidence of the capacity of local social and environmental movements to influence important policy outcomes in Thailand.

Thaksin's growing power between 2001 and 2005 challenged not only environmental activists but also the power of the king and what McCargo defined as the 'network monarchy', replacing it with 'a network based on insider dealing and structural corruption' (McCargo 2005: 512). In response to the Thaksin challenge, monarchist elites joined forces with environmental and human rights activists to form the People's Alliance for Democracy (PAD) – the 'yellow shirts'

– and demand Thaksin's resignation (Pye and Schaffar 2008). On 19 September 2006, however, with a TRT victory looking likely in the forthcoming election, Thailand experienced its eighteenth coup since 1932 (Beeson and Bellamy 2008: 97) and the 1997 constitution was abolished within a few hours (Hewison 2007: 931). Ironically, part of the support for the coup came from the perception by monarchists, the intelligentsia and Bangkok's middle class that Thaksin was a 'proxy of global capitalism' (Thongchai Winichakul 2008: 588), precisely the perception that plagued the second Chuan government. Most environmental activists left the PAD soon after its emergence as it morphed into an ultra-nationalist monarchist movement that displayed little interest in genuine democracy (Hewison 2010; Kengkij Kitirianglarp and Hewison 2009; Pavin Chachavalpongpun 2009).[1]

Despite Thaksin remaining in self-imposed exile, the popularity of Thaksin and TRT in rural and poor areas remained high, as demonstrated by his proxies' continued electoral success (Funston 2009; Hewison 2008: 207). It was only after PPP, the second incarnation of TRT, was dissolved in December 2008 that the Democrat Party was able to form government for a single term under Abhisit Vejjajiva after key Thaksin allies joined a Democrat coalition in exchange for ministries and the adoption of many of Thaksin's policies. In response to the activism of the PAD and the 2006 coup Thaksin's supporters formed the United Front for Democracy Against Dictatorship (UDD) – the 'red shirts' – and in the years after Abhisit became prime minister the UDD undertook regular protests, often occupying large areas of central Bangkok, that were easily identifiable by the sea of red shirts and flags that characterised the protests. These protests were largely peaceful although sporadic violence did erupt, and a final crackdown in the central shopping district of Bangkok resulted in over 85 dead and 2,000 injured between March and May 2010 (Connors 2011; Forsyth 2010; Glassman 2010). In addition to this crackdown the Abhisit government, despite adopting a less aggressive stance to most NGOs, took a harder line on potential anti-monarchy influences in society, with the introduction of harsh internet laws and prolific use of article 112 of the Criminal Code, the archaic *lèse majesté* law, to silence dissent (Streckfuss 2011). The PTP government, led to victory by Yingluck Shinawatra in July 2011, continued to face a divided country but the Thaksin era undoubtedly changed the nature of politics in Thailand forever (Hewison 2010; Thitinan Pongsudhirak 2012).

Assessing Competitive Authoritarianism in Thailand

To evaluate the conditions faced by activists under the governments led by Chuan (1997–2001) and Thaksin (2001–6) – when the Yadana and Thai–Malaysian Pipeline campaigns were in full swing – it is necessary to appraise both the competitive and authoritarian tendencies that existed at the time. I focus particularly on the Thaksin regime, however, as it was the most dominant over this period

---

1   Pipob Udomittipong, interview with author, Chiang Mai, Thailand, 5 April 2009.

and the most significant in terms of establishing new coalitions between large capital interests and political elites that undermined the activities of environmental movements.

Handley describes Chuan as, unlike the prime ministers that preceded and succeeded him, a 'modest and incorruptible lawyer-politician and lifetime advocate of stronger democratic institutions and the rule of law' (2006: 364), although he maintained a traditional disregard for public protest. The Chuan regime was relatively liberal for Thailand and became more democratic when the independent institutions of the 1997 constitution were created to provide checks and balances within the state. The new institutions included an Election Commission, a National Human Rights Commission, a Constitutional Court, an Administrative Court and Parliamentary Ombudsmen (Thitinan Pongsudhirak 2012). Unfortunately the creation of these new democratic institutions coincided with the financial crisis that persisted throughout Chuan's tenure, resulting in increasing rural protests, which were met with a mixture of repression and selective concessions (Pasuk Phongpaichit and Baker 2000: 147). Protesters received less public sympathy because of the atmosphere of crisis, and the government took a hard line on acts of civil disobedience, arguing that protests ruined the nation's image and scared away much needed tourists (Missingham 2003: 201–3). This approach to dissent was in evidence through restrictions on public participation during public hearings for the Thai–Malaysian Pipeline in 2000. Nevertheless, as a result of the new constitution, and despite structural impediments and lingering corrupt practices, under the Chuan government the country unmistakably moved towards greater democratic accountability and greater competition in most political spheres.

Prior to his election Thaksin appeared to present a different style of politics to Chuan, cultivating the support of activists and NGOs and offering them a political legitimacy that had been hitherto denied (Missingham 2003: 201–3). Soon after his election, however, Thaksin demonstrated little compunction about quashing public dissent by subverting the rule of law and democratic processes. He progressively undermined the new institutions of the new constitution; his approach was to 'penetrate them, politicise them and subordinate them to his own will and purpose' (McCargo and Ukrist Pathmanand 2005: 16). While some analysts argue that the quality of democracy under Thaksin was simply 'low' (Case 2007), I argue that Thailand, particularly under the corporate agenda of Thaksin and TRT, satisfied the five criteria for a hybrid competitive authoritarian regime discussed above, namely: the electoral arena; the legislature; the judiciary; the media; and the uniformed security services. First, despite Thailand holding regular elections, they were often characterised by corruption and a lack of transparency. With the financial backing of Thaksin and other leading capitalists TRT took these practices to new heights with rampant vote buying, the use of police and military officers to threaten voters and the 'logrolling' of MPs from other parties into the TRT camp with cash inducements (Connors 2004: 6–7; Pasuk Phongpaichit and Baker 2009). The fragility of elections as a democratic process and the entrenchment of this culture in Thailand was emphasised by an ABAC poll conducted by Bangkok's

Assumption University in October 2007 which showed that nearly two out of three Thais were quite ready to accept gifts or money in exchange for their votes and over 80 per cent would not report election corruption (*Bangkok Post* 2007).

Second, the parliament under Thaksin was rendered almost obsolete. While the 1997 constitution had attempted to empower the prime minister, it was not intended to completely sideline the legislative arm of government. Following the 2001 and 2005 elections, however, the House of Representatives was dominated by TRT and, as Thaksin consolidated his power, he gradually built up an effective majority in the Senate, draining it of its supposed monitoring power (Dressel 2009: 307–9). Legislation required little debate but parliament was effectively boycotted by Thaksin and thereafter by much of his party. For the first two years House sessions were halted five times because they were inquorate (McCargo and Ukrist Pathmanand 2005: 106), but by 2004 this was occurring approximately once every week (Pasuk Phongpaichit and Baker 2009).

Third, although elements of the judiciary were severely compromised, some judges and courts remained relatively independent. The decision by the Constitutional Court to acquit Thaksin of concealing assets, just after his election in 2001 and despite overwhelming evidence to the contrary, raised questions about the court's independence and integrity (Connors 2007: 172). On the other hand, throughout Thaksin's premiership the Administrative Court made several rulings that went against the government including the nullifying of the EGAT privatisation decrees. The judiciary was far more active in politics throughout this period although this 'judicialisation' failed to resolve Thailand's political schism (Thitinan Pongsudhirak 2012).

Fourth, much of the Thai media was cowed during Thaksin's tenure. Thaksin controlled television stations either through ownership by the government and his family companies or through government and corporate advertising revenue (Freedman 2006: 53; Nelson 2005: 3; Pasuk Phongpaichit and Baker 2005: 66). In 2000 Shin Corp acquired Thailand's only independent television station, iTV, and in the lead-up to the election in the following year more than twenty journalists were summarily sacked after complaining that Thaksin was interfering in election reporting. Following the election, contracts with critical production companies were revoked, entertainment was increased at the expense of current affairs, the iTV chairman resigned and there were regular reports of journalists being punished for being too critical of the government or Thaksin (McCargo and Ukrist Pathmanand 2005: 48). While Thaksin effectively closed down the one possible critical outlet on the television, the print media in Thailand at the turn of the century was considered the most vibrant in the region, being 'an island of outspokenness in a tight-lipped ocean' (McCargo 2000: 1). In Thaksin's period in office, however, he managed to suppress a large section of the press, and even the independent-minded Nation media group and *Bangkok Post* faced various forms of physical and legal intimidation, including attempted hostile takeovers. Following the military coup in September 2006 freedom in the media was curtailed even further, although these restrictions were eased following the 2007 election.

Fifth, the security services in Thailand were notorious for corruption and facilitating repressive regimes. Thaksin had attended the Armed Forces Academy Preparatory School and later entered the police force (Thak Chaloemtiarana 2007a). He married into a prestigious police family and many in the Shinawatra clan had entered military service. He was, therefore, well placed to use the uniformed services to further his business dealings and underwrite his political ambitions; his first business venture was a government concession selling computers to the police force (McCargo and Ukrist Pathmanand 2005: 127). Chuan's two governments in the 1990s had attempted to challenge the centrality of the military in political and economic life (Chaiwat Satha-Anand 1999: 152) but under Thaksin it was fully rehabilitated with its funding and prestige restored. Thirty-five of Thaksin's former classmates were appointed to key military positions and his cousin rose from a lowly inactive post to commander of the entire army in less than two years. The police and military were granted immunity from prosecution in the south of the country and were complicit in human rights abuses both there and during the 'War on Drugs'. Despite Thaksin alliance-building, however, much of the military remained aligned to the king and the 'network monarchy', which undermined the government's legitimacy and facilitated the 2006 coup (Heiduk 2011). With the resumption of elections after the coup Jon Ungphakorn, a former senator, argued that '[h]uman rights in Thailand will not improve with an elected government back in power as there are structural flaws embedded with authoritarianism [with] police and army officers ... at the core of human rights problems' (Achara Ashayagachat 2008).

While periods of military rule effectively closed down opportunities for dissent, under Thaksin's competitive authoritarianism environmental activists also found it difficult to voice dissent and influence political discourse. His administration was significant in pursuing authoritarianism within an existing competitive system (Case 2009; Connors 2009). Although subsequent governments increased the use of *lèse majesté* and associated laws to restrict freedom of speech, they, and Thaksin's predecessor, were unwilling or unable to suppress political activism to the same extent as occurred under Thaksin. In the next section I investigate how environmental politics, in particular, played out in Thailand over this period.

Environmental Politics and Security in Thailand

Although Thailand has a larger middle class than most of its neighbours, including Myanmar, many communities have also faced environmental insecurity due to the adoption of the industrial development model in the 1950s. Rapid industrialisation increased pressures on ecosystems, particularly after the start of the 'boom' decade in the mid-1980s when incentives by Thailand's Board of Investments increased the pollution intensity of industry at the same time that the economy shifted to export-oriented industrialisation driven by low labour costs (Fahn 2003: 4; Forsyth 1997: 183; Mounier and Voravidh Charoenloet 2010; Simpson 2014b). By 1991 hazardous waste-generating industries accounted for 58 per cent of Thailand's industrial GDP (Zarsky 2002: 42). Forest cover declined from

54 per cent in 1961 to around 25 per cent in 2005, which both contributed to climate change and magnified the impacts of climate change variations such as the increased severity and frequency of droughts and floods (Marks 2011: 243). There was dramatic growth in the number of large dams and their adverse environmental impacts, driven primarily by EGAT (Bello, Cunningham and Poh 1998: 206). Although there is debate over the extent of an 'environmental crisis' in Thailand, with Forsyth and Walker arguing that a simplified 'crisis' narrative adopted by some activists has been unhelpful (Forsyth and Walker 2008), there is little doubt that marginalised communities are subjected to various types of environmental insecurity (Hirsch 1997; Piya Pangsapa and Smith 2008). Thailand's electrification rate was 99.3 per cent (International Energy Agency 2010b), which was much higher than Myanmar, but this energy security was often achieved at the cost of other environmental insecurities. While the health and well-being of Thailand's middle classes may have improved due to rising living standards, the toxic by-product of this affluence has been most acutely felt by less fortunate communities, such as those near the Map Ta Phut industrial estate on the eastern seaboard. These villagers have had their land encroached upon by toxic industrial factories and, in addition to facing long-term pollution that causes serious illness, more immediate threats to their security also occur in the form of industrial accidents (*Bangkok Post* 2012; Supara Janchitfah 2004).

Despite increasing pressures on ecosystems, Thailand's environment movement also experienced some success in this period achieving an official ban on logging in 1989 and the blocking of World Bank–backed Nam Choan Dam in Kanchanaburi Province in 1988 (Forsyth 2001: 5; Rigg 1991: 46). Unfortunately, illegal logging continued and the ban saw logging expand unchecked in neighbouring countries such as Myanmar, Laos and Cambodia (Hirsch 2001: 241; Mukherjee and Chakraborty 2014). These impacts had parallels in the energy sector, particularly in the wake of the Nam Choan Dam cancellation, with EGAT focusing on cross-border energy projects to import energy from its more authoritarian neighbours through projects such as the Yadana Gas Pipeline and Salween Dams. These projects, nominally for the pursuit of national energy security, have had adverse impacts on the environmental security of local communities (Giannini et al. 2003; Hirsch 1998: 68; KRW 2004; Piya Pangsapa and Smith 2008; Simpson 2007). Despite some success the opportunities for activists to genuinely contribute to Thailand's environmental policy development have been strictly limited (Ungera and Patcharee Sirorosb 2011).

The main developmental focus of most recent Thai governments has been on ensuring sufficient electricity for unrestricted domestic industrial development and acting as a regional hub of an ASEAN Power Grid (Chuenchom Sangarasri Greacen and Greacen 2004: 538). As prime minister, Thaksin stated that his aim was for Thailand to become a regional energy-exporting hub facilitating the energy security of Asia as a whole (Moses 2003). About 70 per cent of Thailand's electricity is generated using natural gas and approximately one third of that gas comes from Myanmar through the Yadana and Yetagun Pipelines (International

Energy Agency 2010a; Kate 2005). Approximately 5 per cent of electricity generation capacity is derived from large-scale hydropower. Despite assertions about the necessity of these projects, the actual electricity needs of Southeast Asian countries are often overstated, with Thailand's energy industry continually overestimating its projected electricity requirements. In 2004 the government's National Economic and Social Advisory Council examined projections by EGAT over the previous decade. It found that in the utility's previous eleven forecasts, ten had overestimated demand, sometimes by as much as 40 per cent. In addition, Thailand's use of energy has been quite inefficient; it uses three times more energy per dollar of GDP than Japan (Imhof 2005). Improved energy efficiency measures in conjunction with smaller scale decentralised renewable energy projects could have made some the large-scale energy projects pursued by EGAT redundant.

Hirsch argues in his analysis of environmental politics in Southeast Asia that these large-scale resource projects attract the most environmental activism, 'particularly those that involve appropriation of the local resource base by interests of state, capital, and dominant social groups in the name of national development' (1998: 55). Hamburg (2008: 108–9) suggests that in Thailand women are most likely to be involved in these natural resource disputes due to their reliance on natural resources for cooking and domestic tasks, and because land inheritance is passed down through daughters. Indeed, women in Thailand have found that membership of organisations that address issues within women's traditional domain is an empowering outlet within a society where there are barriers to women's participation in formal politics (Iwanaga 2008: 7). Women have been particularly active in the Thai campaigns examined in this book.

The borders of Thailand are, unlike those of Myanmar, relatively open to the free exchange of information and communications and readily permeable by transnational activists and influences. There is, therefore, significant cross-fertilisation or 'contamination' (della Porta and Mosca 2007) of transnational activist techniques and philosophies through transnational NGOs such as ERI, which, although based in Thailand, tends to steer clear of Thai politics to minimise political attention. Much of the activism in Thailand is associated with an environment movement readily comparable with those in the North. Social class can influence the environmental discourse adopted (Forsyth 2001), but many of the environmental campaigns undertaken in Thailand are based in marginalised communities including the poor Muslim villages in the south of Thailand that campaigned against the Thai–Malaysian Pipeline. There are also peculiarities in the Thai movement. Buddhism plays a central role in the social and political life of most Thais and, while this can sometimes have the effect of marginalising religious minorities such as Malay Muslims in the south of the country, it is often a source of inspiration for other activists. In recent years the concept of Engaged Buddhism has been used by Thai activists trying to domesticate the discourses of human rights and democracy by finding parallels within Thai culture. Islam has also begun to play an important role in environmental activism in southern Thailand, with activists using it as a symbol of resistance during the campaign against the Thai–Malaysian Pipeline. This was the first time Thai environmental activists had used Islam as a significant symbol.

Engaged Buddhism, however, is well entrenched in Thailand through activism within the Thai *sangha* (monkhood). In response to rampant deforestation monks in forest monasteries began to 'take an active role in ecosystem and environmental management while simultaneously maintaining the monastic tradition of the meditative recluse' (Taylor 1991: 111). Although Engaged Buddhist monks have been involved with various aspects of 'development' since the 1930s (Walter 2007: 334), one of the first cases in which Buddhist monks took an environmentalist position involved the 1985 proposal to build a cable car on Doi Suthep mountain in Chiang Mai (Darlington 2003: 103). Other forest monks became active in response to similar development proposals that seemed to conflict with various Engaged Buddhist tenets including the ethos of protection of all living beings (Taylor 1993; 1997). One response to deforestation was the ordination of trees which has been carried out by environmentalist monks since the late 1980s (Darlington 2003: 96; Swearer 1999: 220; Walter 2007: 335). These tactics were used extensively, in conjunction with the Kanchanaburi Conservation Group (KCG) and other environment groups, during the forest protests against the Yadana Gas Pipeline in the late 1990s.

Perhaps the most prominent environmental activist in Thailand is Sulak Sivaraksa, whom Donald Swearer describes as 'the one person justifiably singled out as the progenitor of contemporary Thai Buddhist social activism' (1999: 219). Originally from the aristocracy, he was a trainee of Prince Dhani in the 1960s before turning against the ruling elites to became a social critic (Handley 2006: 185). He founded his first NGO, the Sathirakoses-Nagapradeepa Foundation, in 1968. In 1989 he co-founded the International Network of Engaged Buddhists in Bangkok which dealt with 'alternative education and spiritual training, gender issues, human rights, ecology, alternative concepts of development, and activism' (Sathirakoses-Nagapradeepa Foundation 2008). His assistant in many of these projects from the early 1990s was Pipob Udomittipong who also participated in the Yadana protests and later worked with ERI and Salween Watch.

Sulak's outspoken criticism of the Thai government over several decades resulted in repressive reprisals. He was arrested in 1984, but released after four months. In September 1991 the military government issued a warrant for his arrest on charges of *lèse majesté* and he was forced to live in exile for a year until the return of democracy. He was cleared of all charges in 1995 under international pressure (Handley 2006: 447; Sulak Sivaraksa 1998; Swearer 1999: 220), but was later arrested during the forest protests against the Yadana Pipeline. His criticism spread beyond the Thai governing elites. He argued that collusion between authoritarian regimes and corporations was central to undermining the social fabric of communities and the exploitation of people. He also regarded institutionalised Buddhism as complicit, arguing that 'Buddhism, as practiced in most Asia countries today, serves mainly to legitimise dictatorial regimes and multinational corporations' (Sulak Sivaraksa 1992: 68). While critiquing globalisation he supported the 'localism' movement in Thailand. The localism movement combined traditional Thai approaches to self-sufficiency with the global environment movement's emphasis on decentralisation, although in Thailand this approach could be construed as perpetuating the paternalistic relationship between the king and peasantry (Connors 2005: 267).

The governing elites' response to this opposition was often repressive. According to one activist 'Thai bureaucracies see the environment as political' and activists were often under surveillance from the intelligence services during study tours or similarly innocuous activities.[2] Although corruption was endemic within the military and particularly the police, the overarching attitude of the government of the day also influenced the actions of the rank and file. Chuan was generally more inclined to enforce the rule of law during his tenure but Thaksin's rhetoric against various sectors of society provided cover for repressive crackdowns by the security services.

Environmental activists were often targeted under the cover of these crackdowns. The most significant repression during Thaksin's tenure occurred during the 'War on Drugs', with the government releasing a daily body count of alleged drug traffickers killed; approximately 2,500 in total between February and May 2003 (Amnesty International 2004; Subhatra Bhumiprabhas 2003). Police claimed traffickers were killing each other to protect their networks but local and international human rights groups accused the authorities of summary execution of suspects, arguing that many innocent people were killed on the basis of hearsay or 'to settle local disputes and, at the same time, score political points with the government' (Human Rights Watch 2007a: 30). UN Special Envoy for Human Rights, Hina Jilani, noted in the report of her mission to Thailand soon after that many Thai activists, including those opposing the Thai–Malaysian Pipeline, had reported that they were afraid to highlight human rights violations for fear of retaliation by local authorities, 'including possibly being killed under cover of the anti-drugs campaign' (Jilani 2004: 18). The threat to environmentalists at this time was highlighted by the murder of prominent activist Charoen Wataksorn in June 2004 (Somchai Phatharathananunth 2006: 222). There were additional concerns that activists were being targeted during Thaksin's heavy-handed approach to the Muslim insurgency in the three southernmost provinces of Yala, Narathiwat and Pattani, which bordered Songkhla Province and the Thai-Malaysian Pipeline. These actions included declaring martial law and an emergency decree in July 2005 that granted police and military officers immunity from prosecution for their actions, despite evidence of involvement and complicity in abductions and assassinations (Chaiwat Satha-Anand 2006a; Funston 2006; Human Rights Watch 2007a; ICG 2005; 2007; McCargo 2007; UN 2008).

In Thaksin's bellicose nationalist rhetoric he opposed not only the IMF intervention in the country but also foreign journalists, UN agencies, foreign NGOs and international sponsors of Thai NGOs. This approach provided convenient

---

2   Varaporn Chamsanit, interview with author, Canberra, Australia, 3 February 2006.

cover for Thaksin's attacks on activists who opposed major development projects; he argued that protests were organised simply to secure foreign funds (Supara Janchitfah 2004: 120). Despite hopes in activist circles prior to his election that Thaksin would look favourably on community activists, he demonstrated, once his prime ministership was secure, that his initial sympathetic overtures to NGOs such as the Assembly of the Poor had been entirely strategic (Missingham 2003: 211). In the years following his election he argued that NGOs and activists were inhibitors of economic growth that should be suppressed by the uniformed services in the national economic interest. His stated goal was to make the police a tool of the state to help increase national income (Pasuk Phongpaichit and Baker 2009). He held a similar goal for the military, and this often resulted in the violent repression or intimidation of NGOs who opposed his economic agenda. He also seemed to target two types of NGOs in particular: those that received foreign funding and those that worked on Myanmar issues, as this also clashed with his business interests in Myanmar.[3]

Apart from the overt harassment of activists and NGOs by the military and police, there was also an increasing threat of violence perpetrated by non-uniformed assassins. Between 2001 and 2005 at least twenty environmentalists, human rights activists and community leaders, including monks, were killed in separate incidents, most of them shot (Amnesty International 2004; Biel, Hicks and McClintock 2006: 22). The issues addressed by the activists who were killed included opposing dams, quarries, plantations, rubbish dumps, illegal logging and waste treatment plants and protecting mangroves, forests and river catchments (AITPN 2005; Jilani 2004). While there was generally little evidence to link the government to the assassinations directly, the pattern followed in most cases indicated some form of collusion between local authorities and capital interests. In almost every case the victim was opposing a development project on social or environmental grounds or attempting to protect a public area from encroaching private activities. The lack of successful prosecutions in almost all cases suggested that the government was unwilling to pursue the perpetrators, and the granting of immunity from prosecution for police and military officers in the south only exacerbated the propensity for authoritarian responses. This climate of impunity, in addition to Thaksin's attacks on the legitimacy of NGOs and environmental activists, did nothing to discourage the assassination of those who spoke out against development projects. While this atmosphere made activist environmental governance and the pursuit of environmental security in Thailand challenging, the opportunities for political activism were still much greater than those available to activists in neighbouring Myanmar.

---

3   Wandee Suntivutimetee, interview with author, Chiang Mai, Thailand, 11 January 2005.

Political Regimes and Environmental Politics in Myanmar

In all societies, but particularly in the countries of the South, achieving environmental security is linked to democratic governance and the protection of human rights. There are few countries in the world where the evidence for such links is more compelling than Myanmar. As I set out below, Myanmar's government from 1962 to 2011 was, unlike Thailand's hybrid political regime, unequivocally undemocratic and traditionally authoritarian. The possibilities for genuine public dialogue and dissent over environmental issues were particularly constrained, and public participation in both informal and formal politics was strictly limited. Other indicators of authoritarianism during this period included an estimated one soldier for every hundred citizens, a pervasive surveillance apparatus and pre-censorship for all literature (Philp and Mercer 2002: 1588).

Despite Myanmar's dismal history of authoritarian governance incremental openings emerged in the domestic political space following the first national elections for two decades in November 2010 (Cheesman, Skidmore and Wilson 2012; Farrelly 2013a; Jones 2013; Kyaw Yin Hlaing 2012; Macdonald 2012). There is little doubt that in the decade prior to the elections, and particularly since Cyclone Nargis in 2008, domestic environment movements had become more active, but restrictions on freedom of speech, of association and of the media still made the collection and dissemination of information on energy projects extremely difficult. The political and economic reforms under the new 'civilian' government under the former general, President Thein Sein, may, however, result in an expansion of domestic environmental activism with echoes of that which accompanied the fall of socialism in Eastern Europe (Carmin and Fagan 2010).

There remain limitations on political and economic freedoms in Myanmar but from 2011 the political and security atmosphere in much of Myanmar was transformed as media restrictions were eased and Aung San Suu Kyi, the National League for Democracy (NLD) General Secretary, entered the national parliament in April 2012 along with forty two of her colleagues (Simpson and Park 2013). These reforms were rewarded with a gradual easing of Western sanctions and the granting of the ASEAN chair in 2014, despite the historically contrasting approaches between the two blocs (Egreteau and Jagan 2013; Haacke 2006; 2010b; Jones 2010; 2012; Roberts 2010). Although it is possible that some of the promising reforms could be reversed at any time, by a military coup or a reassertion of political power by more conservative forces within the government, the momentum of these reforms is significant. Given the historic and enduring dominance of the military in all sectors of the country (Maung Aung Myoe 2009; Selth 2001), there are unlikely to be any more significant shifts in the domestic political system than the military's 'managed' or 'disciplined' democracy in the near future so the best prospects for improved governance and development are for the international community and local activists to leverage change through the present political openings (Holliday 2011; Kyaw Yin Hlaing 2009; Pedersen 2011: 64–65; Simpson 2012).

Regardless of broader political changes the local environmental movement remains, despite recent progress, embryonic. There are limits to both its experience and expertise. Attempts by governments of the North to use state-to-state leverage to promote democracy and the development of civil society in Myanmar met with little success for many years as there were few linkages – economic, social, communication, intergovernmental and transnational civil society relationships – tying Myanmar to the North, which Levitsky and Way (2006: 396) argue would be the more potent contributors to reform. By focusing on the connections between domestic and Northern actors, however, Levitsky and Way tend to underestimate the significant role played by the exiled activist community in neighbouring countries. Under the authoritarian military regime that existed for the period examined in this book the activist diaspora occupying Myanmar's borderlands provided the most fertile and important environmental governance of energy projects in Myanmar. To establish the context for this activism, the rest of this chapter examines the historical development of the domestic political environment and the consequences for environmental politics.

Myanmar's history has been entwined with the issues of postcolonial state building since it gained independence from Britain in January 1948 (Tin Maung Maung Than 2005a: 65). Throughout the colonial era, due to British policies and administration, contact and cooperation between the diverse ethnic peoples of Myanmar declined while ethnic conflict increased (Holliday 2011; Naw 2001: 195; Walton 2008: 893). Since independence ethnic minorities have struggled for recognition and autonomy from the Bamar (Burman) majority, which comprises two thirds of the total population. These struggles have been particularly significant for environmental politics as most large-scale energy projects have been based in the mountainous, resource-rich ethnic minority borderlands that surround Myanmar's central Ayeyarwady (Irrawaddy) plain (Pedersen 2008; Scott 2009; South 2009; Tin Maung Maung Than 2005a: 67; Walton 2008).

Following the assassination of General Aung San, the independence hero and father of Aung San Suu Kyi, on the eve of independence, the first years of the state were characterised by unstable but occasionally democratic governments punctuated by interventions by the Tatmadaw. Until recently, the last attempt at anything remotely resembling democratic rule ended following a military coup by General Ne Win on 2 March 1962 (Smith 1999: 196). As leader of the Revolutionary Council and then the BSPP he led the country into a 26-year era of isolation following his 'Burmese way to socialism', an admixture of Buddhist, Marxist and nationalist principles that ethnic minorities in the country interpreted as 'Burmanisation' (Callahan 2003: 202–4; Charney 2009; Myat Thein 2004: 52; Steinberg 2010; Taylor 2009; Thant Myint-U 2006: 291; Tin Maung Maung Than 2007). This process saw Myanmar decline 'from a country once regarded as amongst the most fertile and mineral rich in Asia to one of the world's 10 poorest nations' (Smith 1999: 24). The BSPP remained in power until 1988 when student-led protests over shortages of essential goods and spiralling rice prices in March led to a brutal crackdown that left up to 100 civilians dead. In the wake of the worsening security situation Ne Win resigned in July, prompting an escalation of protests across the country demanding a return to multiparty democracy. A general strike and a mass demonstration was called for the auspicious date of 8-8-88, resulting in a brutal crackdown (Cumming-Bruce 1988; Smith 1999: 2–5).

Ne Win loyalists, under the increasing influence of his protégé Lt Gen (and later General) Than Shwe, ended the BSPP era on 18 September 1988 by seizing power to establish the State Law and Order Restoration Council (SLORC). There followed a country-wide crackdown on any form of protest in which up to 1,000 people, including schoolgirls, monks and students, were killed in Yangon (Rangoon) in the following three days alone. The death toll from the year's violence may well have exceeded 10,000 – almost all unarmed civilians – although the final death toll will never be known (Lintner 1990; Smith 1994; 1999: 15–16; Wellner 1994).

In the lead-up to the first multiparty national elections in three decades in 1990 there was no relaxation of SLORC's martial law and many of the opposition's major leaders were kept under arrest (Smith 1999: 412). After assuming that a year-long propaganda campaign had been effective, and underestimating the public hatred towards the military, SLORC allowed a surprising degree of openness on election day, 27 May 1990. According to Bertil Lintner, 'after months of repression, severely restricted campaigning and harassment of candidates and political activists, during the election itself even foreign journalists were invited to cover the event and there were no reports of tampering with the voting registers' (1999: 382). The turnout was almost 73 per cent and in the final result the NLD, led by Aung San Suu Kyi who was under house arrest, captured 60 per cent of the votes and 392 of the 485 seats contested. Most of the rest went to NLD allies from the various ethnic minorities while the military-backed National Unity Party captured a mere 10 seats, 2 per cent of the total (Lintner 1999: 382; Smith 1999: 414; Steinberg 2007: 115).

Despite their resounding rejection by the electorate, SLORC refused to acknowledge the results and arrested many more political opponents, including elected parliamentarians. Aung San Suu Kyi was kept under house arrest for 15 of the next 21 years, and suffered attempts on her life by the military or its proxy, including the 2003 Depayin massacre, during brief periods of relative freedom (Ad Hoc Commission on the Depayin Massacre 2003; Aung Htoo 2003; Davis and Kumar 2003). A rapid expansion of the Tatmadaw following the 1988 protests laid the foundations for the perpetuation of military rule (Maung Aung Myoe 2009; Selth 1996: 145; 2001). On 15 November 1997 SLORC reorganised itself, shifting some of its personnel and changing its name to the State Peace and Development Council (SPDC) following advice from a Washington-based public relations company, although these changes were largely cosmetic (Charney 2009: 179; Silverstein 2001: 119; Tin Maung Maung Than 2005a: 78). At this time the SPDC also used the 'Asian values' debate to justify their autocratic rule, arguing that Western conceptions of democracy and human rights were foreign to Myanmar (Steinberg 2007: 117). Ne Win died in 2002 but General Than Shwe remained in firm control as chairman of the SPDC. In May 2008 a new constitution was accepted by referendum, in the midst of Cyclone Nargis, and national and regional elections were held in 2010 for the first time in two decades (Taylor 2009; Tin Maung Maung Than 2011). Numerous provisions included in the constitution, however, gave the military a central role in running the country.

To increase its perceived legitimacy throughout the period of military rule the Tatmadaw drew on religion, continuing Ne Win's portrayal of Myanmar's leaders as devout Buddhists. Despite being a dictatorship, the military still saw political benefits in presenting itself as the country's patrons of Buddhism, even though urban residents were sceptical of these gestures (Ardeth Maung Thawnghmung 2004: 22; Fink 2009: 230; McCarthy 2008). While the early military government of the 1960s kept a calculated distance from religious life, from the 1980s the military increasingly invoked Buddhism as a unifying national force to achieve its political aims (Matthews 1999: 38). Having appropriated Buddhism, the military projected a vision through the state-run media of a united state, *sangha* and laity as a way of disciplining the population (Philp and Mercer 2002: 1591).

The Tatmadaw have long understood that building new monasteries and supporting pliant monks is a shrewd investment. In 1999 the military sponsored a major restoration of the Shwedagon Pagoda in Yangon (Fink 2009), the most revered pagoda in the country. The Shwedagon's walls were adorned not only with photos of devout military leaders in their uniforms making offerings with pliant monks, but also with illustrated frescos of the leaders in the form of parables of the Buddha in harmony with nature. This focus may have gained some kudos with the Buddhist Bamar and Shan peoples, but led to a further deterioration of relations with predominantly Christian ethnicities such as the Kayin and Kachin.

Even after the new government was formed in 2011 the UN Secretary General's Special Rapporteur on the situation of human rights in Myanmar, Tomás Ojea Quintana, reported that ongoing tensions in ethnic border areas and armed conflict, 'particularly in Kachin, Shan and Kayin States, continue to engender serious human rights violations, including attacks against civilian populations, extrajudicial killings, sexual violence, arbitrary arrest and detention, internal displacement, land confiscations, the recruitment of child soldiers and forced labour and portering' (Quintana 2011: 9).

The systematic sexual assault of ethnic minorities by the military in association with energy projects and military offensives has been documented in numerous NGO reports (see Apple and Martin 2003; KWO 2007; SHRF and SWAN 2002). This took on new legal significance with the unanimous resolution by the UN Security Council in June 2008 that 'rape and other forms of sexual violence can constitute war crimes, crimes against humanity or a constitutive act with respect to genocide' (UNSC 2008). The Tatmadaw's actions also led to Myanmar being repeatedly sanctioned by the International Labour Organization (ILO) for using forced labour and for providing 'safe haven' to drug traffickers. In 2004 the ILO expressed 'grave concern' over the continuing scale and scope of forced labour in Myanmar (ILO 2004), and in November 2006 its governing body, citing 'great frustration' with the country's authorities, took the rare step of moving towards legal action in the International Court of Justice (ILO 2006). Despite this record the ILO was one of the few UN organisations that made some progress in gaining cooperation from the regime in the decade prior to the new government (Horsey 2011). In 2011 an ILO labour specialist from Geneva oversaw the development of the country's new labour law, which came into force in March 2012 and, according to the ILO's Liaison Officer in Yangon, was 'pretty close' to meeting the government's requirements under international agreements.[4]

After its second catastrophic demonetisation in two years (Turnell 2009: 252–54), Myanmar was designated by the UN as a 'least developed country' in 1987 (Smith 1999: 24) and by 2011 it still endured endemic rent seeking and corruption, with Transparency International (2011) ranking Myanmar as third worst for corruption out of 182 countries. So in the same year that the United Nations Development Programme ranked Myanmar last in public health expenditure in the world (UNDP 2006), the youngest daughter of the SPDC's leader, Senior General Than Shwe, was married in a wedding of extreme opulence that instigated outrage both at home and abroad (*The Irrawaddy* 2006). Some estimates suggested that the total per capita expenditure on health under military rule was approximately $4 with the military regime itself contributing approximately 9 cents (Vicary 2007: 4). Oehlers argued that under military rule it was 'simply impossible to imagine any meaningful progress in health or any other socially important measure' (2005: 204), although the new government after 2011 delivered increases to the health and education budgets.

Despite the new regime the military is unlikely to give up its privileged position of political and economic power voluntarily as under a truly democratic government it may be forced to face the consequences of its harsh rule and it might also lose the large share of the national budget – up to 40 per cent – that it has traditionally enjoyed (Kyaw Yin Hlaing 2009; Selth 1998: 102–3; 2000: 62; Simpson 2007: 546; Smith 1999: 415). Apart from their income through the national budget, the higher echelons of the military and their business colleagues also increased their wealth through the private sector after the dubious privatisations of state assets after 1988 that accelerated to fever pitch in the lead-up to the 2010 elections (Ba Kaung 2010; Charney 2009: 180; Turnell 2010: 35). Through both state and individual interests, therefore, the 'iron glove of the military [envelops] the invisible hand of the private sector' (Steinberg 2005a: 61).[5] In Thailand separating the political regime from business was problematic, particularly under Thaksin, but in Myanmar it was virtually impossible. The lack of transparency and depth of corruption in the country is symptomatic of the domination of both these areas by the military.

---

4   S. Marshall, interview with author, Yangon, Myanmar, 9 May 2011 and 17 February 2012.

5   This wealth was not transferred to the lower echelons of the military, however. Desertion rates were high due to mistreatment by superior officers, low pay and poor living conditions (Human Rights Watch 2007b: 30).

Given the almost complete grip of the Myanmar military on the formal economy during military rule it was 'clear that sanctions [did] affect the generals' income, the generals' families' income, as well as the generals' families' friends' income' (ICFTU 2005: 2). For this reason, the Federation of Trade Unions–Burma and most other organisations opposing military rule supported sanctions against foreign investment in Myanmar (FTUB 2005). Despite debates over the impacts of sanctions and corporate engagement with Myanmar (Holliday 2005a; 2005b; Kudo 2008; Kudo and Mieno 2009), Myanmar was thus in the unique position of being the only country in the world for which international trade unions were calling for disinvestment and sanctions. As in other military regimes (Geddes 1999: 126), the senior military rulers relied very much on the distribution of the country's wealth to maintain their rule. As a result, Williams (2011) argued, the reduction in rents to key players as a result of political and economic changes in the country meant that support for the regime was fracturing, even before the elections. In this context the energy projects in this book were of central importance to senior military figures in maintaining their positions of power and continuing the profitable process of 'primitive accumulation' during this period (Jones 2013).

Assessing Competitive Authoritarianism in Myanmar

To analyse the opportunities for political participation and dissent in Myanmar I use the same criteria used to determine Thailand's competitive and authoritarian characteristics, namely: the electoral arena; the legislature; the judiciary; the media; and the uniformed security services. With the new government in 2011 there have been improvements in some of these areas, which are briefly examined, but the case study campaigns predominantly cover the period of military rule so this is the period focused upon here.

Addressing the first criterion, the multiparty elections held in 1990, which gave the NLD and its allies a resounding win with over 80 per cent of the seats, were ignored by the military regime and many opposition candidates were imprisoned or exiled. Of the NLD's original 22-person Central Executive Committee, 18 were detained and 40 elected MPs were arrested, with two dying shortly after amidst allegations of torture (Smith 1999: 412–19). The only subsequent elections have been those held in November 2010 under the new constitution, which was clearly designed to consolidate the power of the Myanmar military. Members of religious orders are now ineligible to vote (s. 392), meaning that the thousands of Buddhist monks who protested in September 2007 were excluded from the electoral process. The NLD Central Executive Committee effectively chose to boycott the election, as Aung San Suu Kyi was precluded from participating, leaving the military-supported Union Solidarity and Development Party (USDP), which had replaced the regime-sponsored NGO, the Union Solidarity and Development Association (USDA), as a virtually uncontested winner. Although the existence of the USDP as a party may eventually improve the prospects for democracy (Wright and Escriba-Folch 2012), the absence of the NLD and any effective pro-democracy opposition meant that in this election it was almost impossible for new figures to 'enter the political arena and make their voices heard' (Holliday 2008: 1047). Of the ethnic parties only the Rakhine Nationalities Development Party (RNDP) and the Shan Nationalities Democratic Party gained significant representation as many others aligned with the NLD also boycotted. Once the new government was formed, however, with the military and the military-backed USDP controlling over 80 per cent of the seats in the national parliament, President Thein Sein and other 'reformers' in the government pushed for changes in the constitution and electoral laws that increased the legitimacy of the process by allowing Aung San Suu Kyi and the NLD to contest elections. The NLD re-registered and subsequently swept the by-elections of April 2012. The real test for this criterion will come in the 2015 election, which may result in an electoral landslide for the NLD and its allies, stripping the military of any residual legitimacy. If the military allows a free and fair election and changes to the constitution then elections in Myanmar can be considered a genuine site of competition once more for the first time in over five decades.

On the second criterion, there were extremely limited opportunities for legislative activity after Ne Win took power in 1962. Until 1988 the BSPP was the only legal political party and after SLORC came to power in 1988 it annulled the 1974 constitution and dissolved the legislature. Between 1988 and 2011 the military reserved all legislative and executive power for itself (Taylor 2005: 22–23). There was, therefore, no legislative check on the executive. Following the 2010 elections there is now a national parliament (Pyidaungsu Hluttaw) with a lower house (Pyithu Hluttaw) and an upper house (Amyotha Hluttaw), as well as provincial parliaments, but under the current constitutional provisions these remain dominated by the military with 25 per cent of the seats in the national parliament reserved for serving military personnel. While the NLD gained representation following the 2011 by-elections its influence will remain marginal, at least until its next significant opportunity at the 2015 general election.

Third, there was a similar lack of judicial independence in Myanmar. In the BSPP era the judicial system included courts with judges from the military and BSPP (Fink 2009), but during direct military rule the independence of these courts was limited even further with the military holding all judicial authority (Taylor 2005: 23). Although court cases did take place, there was little room for lawyers to manoeuvre; three defence lawyers representing dissidents of the September 2007 protests were imprisoned for between four and six months for complaining of unfair treatment (Saw Yan Naing 2008b). Little change occurred in the initial months after the new government was formed with the Special Rapporteur noting that 'there do not appear to be any major structural transformations within the judiciary' (Quintana 2011) but improved independence was expected to eventually accompany the entrenching of more democratic modes of governance.

Fourth, under military rule there were restrictive media regulations in Myanmar which resulted in a severe lack of media freedom and freedom of speech (Lewis 2006: 51–52). All editorial and advertising material was censored. Every publication in Myanmar was required to submit a draft of its final layout with photographs and captions for prior approval by the Press Scrutiny and Registration Board. As a result of this time-consuming and often arbitrary process, the limited number of private newspapers and periodicals that existed in Myanmar only published weekly or monthly and were never critical of the ruling military. Order 5/96 from 1996 allowed for 20-year jail terms for anyone airing views or issuing statements critical of the regime (Fink 2009; Philp and Mercer 2002). In an easing of print censorship in June 2011 the new government announced that it was moving from a 'pre-censorship' to a 'post-censorship' mode, initially on topics such as sport and children's literature. In April 2013 private daily newspapers appeared on Myanmar's streets for the first since 1964. Under the military regime the internet had also been targeted: a young blogger who was a major source of information on the September 2007 protests was sentenced to over 20 years' imprisonment for posting a cartoon of Than Shwe on his website (Saw Yan Naing 2008b). In 2010 Myanmar ranked 174 out of 178 countries in the Worldwide Press Freedom Index by Reporters Without Borders (RSF 2010), while the Committee to Protect Journalists (2010) ranked it the worst country in the world in which to be a blogger. The websites of Yahoo, Hotmail, Google, the *Bangkok Post*, the Australian Broadcasting Corporation and all exiled media such as *The Irrawaddy* were banned until restrictions were eased in September 2011.

Fifth, since 1962 the military has been the dominant social and political institution in Myanmar and under military rule the associated intelligence services maintained close surveillance of its citizens. Regulations were changed at the whim of the military leadership and corruption within the military was rife but unpredictable, making everyday existence difficult (Steinberg 2005b: 93). Systematic and ad-hoc human rights abuses were common, particularly in ethnic minority areas. Military interests controlled 'the executive, judiciary, administrative, legislative and economic branches of the state' (Ardeth Maung Thawnghmung 2004: 17), which resulted in a culture of impunity and a lack of rule of law. While the new government is nominally civilian the structural position of the military in society through its entrenched political and economic power ensures that it is unlikely to be subservient to a truly democratic and civilian government for some time to come. There is a lack of accountability and transparency in the dealings of the military, which continues to this day, but during military rule this aspect of political life was a strong contributor to traditional authoritarianism.

Military-ruled Myanmar, therefore, evidently lacked competition in the five areas and failed the criteria for competitive authoritarianism or any other hybrid regime, but was traditionally authoritarian. The military systematically closed down all the usual avenues for competition, resulting in a politically closed society that lacked legitimate opportunities for environmental activists, or any political activists, to voice dissent. At the time of writing it was still too early to determine what regime type would emerge from the reforms being undertaken under the new government and it is unlikely that such an assessment could be made until the dust settles from the 2015 general elections. Nevertheless, it was clear that compared with military rule the opportunities for political discourse and engagement were dramatically increasing with more competition likely to emerge in the five areas discussed above.

Environmental Politics and Security in Myanmar

As a result of protracted authoritarianism Myanmar is a country in dire need of effective environmental governance (Simpson 2014a; Smith 2007; Tun Myint 2007). Rapacious exploitation of the country's environment and natural resources by the military and its business associates has had adverse implications for the human and environmental security of the country's population. Throughout military rule a lack of coherent environmental governance and policy making resulted in widespread and ad hoc mining, logging and energy projects that were undertaken without regard for the adverse environmental consequences, a situation compounded by civil conflict between the central government and ethnic minorities in the mountainous and resource-rich border regions. Upland regions, such as Kachin State in the north and Kayin (Karen) State in the east, face a plethora of environmental problems with unchecked logging being a central concern for local communities and often blamed for increased flood events (MacLean 2003; Tint Lwin Thaung 2007). Likewise large-scale, artisanal and small-scale mining put together have had an enormous environmental impact due to a lack of environmental regulations, resulting in deforestation and the pollution of rivers from mine tailings. Mines are spread throughout Myanmar and produce zinc, lead, silver, tin, gold, iron, coal and gemstones although one of the biggest is the Letpadaung (Monywa) Copper Mine in Sagaing Region, which under the new Thein Sein government was the site for protest and conflict (AFP 2013; Smith 2007).

In the early decades of military rule, during the period of 'the Burmese road to socialism', state authoritarianism and incompetence depleted ecosystems while running down the economy and a precipitous fall in foreign aid following the crackdown in 1988 left the economy on the verge of collapse. As SLORC came to power in 1988 they offered attractive incentives for foreign investment through the Union of Myanmar Foreign Investment Law. This created a market economy that opened the door to joint ventures with foreign companies that were interested in exploiting Myanmar's natural resources and resulted in the various transnational energy projects discussed in this book (Callahan 2009: 47–48; Carroll and Sovacool 2010: 638–42; Chenyang 2010; Hughes 2011: 195–96; Lintner 1990: 165; MacLean 2003: 16; McCarthy 2000: 235; Myat Thein 2004: 123; Simpson 2008; Turnell 2007). In the subsequent two decades the energy sector, including hydropower and oil and gas, was the primary recipient of FDI and in the 2006–7 fiscal year it accounted for more than 98 per cent of all foreign investment. By 2008 Myanmar Oil and Gas Enterprise (MOGE), Myanmar's state-owned and sole oil and gas operator, had entered 21 long-term contracts with oil and gas TNCs from 13 countries (ITUC 2008: 17; *The Myanmar Times* 2008; Ye Lwin 2008).

Revenues from the energy sector continue to increase with natural gas being the major source of foreign currency (James 2006: 112; Turnell 2008). Gas revenues, being approximately $3 billion gross and $1.5 billion net in 2010–11 (Turnell 2012: 146), constituted about 45 per cent of the country's total exports, with most of the gas delivered to Thailand through the Yadana and Yetagun Pipelines, although this income was likely to increase dramatically after the Shwe Pipeline came online in July 2013. In addition, PTT's Zawtika Gas Pipeline from the Gulf of Mottama (Martaban) to Thailand was also expected to be completed in 2014. Gas income had increased substantially since the first royalties started flowing at the turn of the century and evidence suggests that these revenues supported the expansion of the military and its ability to harass its opponents, while little was spent on health and education. In 2001 the *Bangkok Post* reported that the military regime was buying 10 MiG-29 jet fighters from Russia for US$130 million with the funds received from Thai gas sales. The $40 million down-payment for the fighters was transferred in the same week that Myanmar received its initial share of royalties, of approximately $100 million, from PTT (*Bangkok Post* 2001). Once these gas revenues started flowing the Myanmar military went on an accelerated arms-buying spree, upgrading navy and air force weapons and increasing the size of its army (IISS 2005).

China and its TNCs are playing an increasingly prominent role in the South through FDI in hydropower and oil and gas and Myanmar is an important ally in the region both for its domestic energy supplies and its geographic location (Haacke 2010a; Kyaw Yin Hlaing 2009; Steinberg and Fan 2012; Thant Myint-U 2011). As it shares a border with China, Myanmar offers a conduit for Middle East oil that bypasses the Straits of Malacca (Brewer 2008; Clarke and Dalliwall 2008; ERI 2008b). As well as vast investments in Myanmar's energy resources China has invested in Myanmar's infrastructure such as the Kyauk Phyu port facilities in Rakhine State, but China's FDI requirement to employ largely Chinese labour, whose pay is typically deposited into bank accounts in China, means the local development impact is limited (Miller 2010: 105–6). Russia is also increasing its presence both through the oil and gas industry and the supply of arms, which are bought with the proceeds of energy sales. In January 2007 Russia and China jointly vetoed a UN Security Council resolution for the first time since 1972 to scuttle a US and UK-sponsored resolution calling on Myanmar's military regime to stop the persecution of minority and opposition groups (Reuters 2007b; Simpson 2008: 224; Tisdall 2007). India could not provide enough inducements to compete with China's strategic Security Council veto, which resulted in Myanmar selling the gas from the Shwe Project to China rather than India, despite the latter offering a higher price (Alamgir 2008: 981; Clarke and Dalliwall 2008; Egreteau 2008: 953).

These energy exports were made in the context of extreme energy shortages and widespread energy poverty throughout the country itself (Sovacool 2013). Although exact measures are difficult to come by the electrification rate was estimated at 13 per cent in 2009 by the International Energy Agency (International Energy Agency 2010b) while the ADB suggested a rate of 26 per cent in 2011 (Asian Development Bank 2012: 23). This ADB figure included an average of 16 per cent across rural areas while Yangon recorded the highest rate of 67 per cent.

Table 3.1   Electricity consumption in Myanmar

Area	Population	Total electricity consumption (GWh)	Per capita consumption (kWh)
Yangon (Rangoon) Region	5,420,000	3,084.0	569.0
Tanintharyi (Tenasserim) Region	1,327,000	6.6	5.0
Rakhine (Arakan) State	2,698,000	9.4	3.5

Source: Author's calculations (Modins.net 2004; Thiha Aung 2005a; 2005b; 2005c)

Outside of the major centres electrification rates are paltry and in some remote ethnic minority regions it is non-existent. Figures extracted from the government mouthpiece, the *New Light of Myanmar*, demonstrated that the two ethnic minority provinces that hosted the Yadana and Shwe Gas Pipelines, Tanintharyi (Tenasserim) Region and Rakhine (Arakan) State respectively, had the two lowest per capita levels of electricity usage in the country. Six years after the gas in the Yadana Pipeline started flowing the promises of improved access to electricity for local communities remained unfulfilled with usage in Yangon 114 times higher (see Table 3.1). Environmental groups from across the region, such as the NGOs TERRA and MEENet in Thailand, argued that instead of pursuing these large-scale developments Myanmar's energy needs could be substantially fulfilled by small-scale electricity projects such as micro-hydro, biomass and solar (Kyaw Hsu Mon 2012).[6] As with many other countries of the global South simple biomass technologies such as fuelwood, charcoal, agricultural residue and animal waste have historically provided the dominant fuel source, supplying almost 70 per cent of the country's primary energy (Asian Development Bank 2012: 3).

As an indicator of the increasing responsiveness of the new government to environmental concerns over large-scale development projects and the export of energy, in September 2011 it announced the suspension of the $3.6 billion Myitsone Dam in Kachin State and the cancellation of the 4,000MW coal-fired power station at the Dawei Development Project in Tanintharyi (Tenasserim) Region the following January (Simpson 2014c); 90 per cent of the electricity from the projects was to be exported to China and Thailand respectively. It was the first time that a Myanmar government had responded to environmental activism and gave some hope to Myanmar activists campaigning on other energy projects. Nevertheless, such decisions were still subject to the arbitrary whims of government ministers. A new Environmental Law to establish environmental institutions and standards was passed in 2012 but by late 2013 the regulations were still to be released. The new Foreign Investment Law of 2012 required EIAs for all major development projects but until the final details of the rules are released, including the method of application, it will be difficult to assess their

---

6   Premrudee Daoroung, interview with author, Bangkok, Thailand, 13 May 2011.

rigour (Simpson 2014a). In addition, the efficacy of the rules will be dependent on the ability and integrity of responsible ministers and civil servants who often lack the relevant skills and capacity according to Win Myo Thu, an influential domestic environmentalist and the founder of the NGO Economically Progressive Ecosystem Development (ECODEV).[7]

The new government was no doubt aware that energy shortages had played a key role in precipitating the national protests in 2007. In August of that year the SPDC had announced enormous increases in fuel prices, which increased the cost of living and transport, further reducing already precarious human security for many people in Myanmar. The price of compressed natural gas, which was used by public buses, reportedly increased by 500 per cent. Myanmar exports vast amounts of natural gas to Thailand through the Yadana and Yetagun pipelines, which could have otherwise been used to maintain lower gas prices (Carroll and Sovacool 2010). The state's security apparatus usually ensures that public displays of dissent are quickly extinguished but in September 2007, for the first time since 1988, widespread opposition to the regime, led by Engaged Buddhist monks, overflowed onto the streets across the country before being brutally suppressed by security forces and pro-military militias (AFP 2007b; *BBC News* 2007; ICG 2008a; McCarthy 2008; Selth 2008; Skidmore and Wilson 2008).

Apart from these very occasional public displays, and the ethnic conflicts in the remote border regions, dissent in Myanmar during almost five decades of authoritarian rule was private and unseen and an independent domestic civil society critical of the state largely absent. By the end of the 1990s Steinberg contended that the regime had 'attempted to divide the opposition, both ethnic and political, and ... eliminated all vestiges of civil society in Myanmar ... independent NGOs do not exist beyond village temple societies' (Steinberg 1998: 275). South (2009: 180) later argued that this view was overly pessimistic, suggesting that in the subsequent decade vast numbers of local groups or community-based organisations (CBOs) had emerged throughout Myanmar although the majority were centred on religion and worked at the 'primary' level of welfare activities. A few CBOs had expanded to the 'secondary' level working on community development, such as Metta Development Foundation which trained farmers in the absence of education by the state (MDF 2006), although this type of social mobilisation was unlikely to represent a threat to the ruling class (Brownlee 2007: 217–18). There were, however, few CBOs or NGOs that worked at the 'tertiary' level of rights-based activities and conflict resolution and those that existed found it difficult to cooperate openly with international groups as many domestic activists received three-year prison sentences simply for contacting 'unlawful' (exiled) associations.[8] ECODEV was one of the few environmental organisations that undertook both secondary and tertiary-level activities, but even these were

---

7   Win Myo Thu, interview with author, Yangon, Myanmar, 10 May 2011.

8   Phyo Phyo (pseudonym), interview with author, Yangon, Myanmar, 18 February 2012.

relatively non-confrontational with the struggle for land and natural resource rights undertaken through Myanmar's compromised court system. The absence, from the 1980s, of many international organisations, such as the World Bank and Asian Development Bank, exacerbated the international isolation and stifled the development of domestic civil society (Simpson 2014c; Simpson and Park 2013). Despite widespread criticism of the role of these organisations in the South their projects could be used, according to Win Myo Thu, to increase the involvement of local people in decision making and management: 'there are still benefits for governance and participation even if the project isn't as effective as it could be'.[9] These barriers to the free flow of information into Myanmar and a muzzled media ensured that its borders were more than mere 'speed bumps' for the transnational sharing of activist strategies, tactics and philosophies. In recent years further growth in service-oriented civil society has centred on humanitarian assistance as domestic NGOs such as Network Activities Group delivered emergency relief following natural disasters.[10]

Natural disasters in Myanmar have been exacerbated by the widespread destruction of forests, mangroves and ecosystems. Throughout the SLORC/SPDC era McCarthy (2000: 260–61) argues that there was a 'hard sell' of the country's natural resources with no evidence that this wealth was redistributed among Myanmar's people nor any evidence of long-term planning guiding the foreign investment projects being approved by the Myanmar Investment Commission. Ethnic minorities were particularly poorly rewarded from major development projects, even though they were often located in ethnic areas (Skidmore and Wilson 2007). Despite the daily restrictions on the general population, it was these ethnic minorities in Myanmar's mountainous border regions, including the Kayin, Shan, Mon and Rakhine (Arakanese), who particularly suffered from repression and environmental insecurity (Lintner 1999; Smith 1999; South 2009).

In the multiplicity of individual security challenges facing the people of Myanmar rampant logging and environmental destruction together with a lack of legislated EIA were intrinsically linked to non-democratic governance and authoritarian military rule. The role of the state was central to ethnic minorities' insecurity, both through assaults on their person and on their environment through the four cuts campaign that aimed to restrict insurgents' access to food, funds, intelligence and recruits (Smith 1999). These peoples therefore faced challenges to their human security, whether considered from a narrow (political violence by the state) or broad (freedom from want) perspective (Kerr 2007: 95). The military's contribution to environmental insecurity was highlighted by their response to Cyclone Nargis in early 2008 that killed more than 140,000 people, destroyed 800,000 houses and left millions of Ayeyarwady (Irrawaddy) Delta residents homeless and facing disease and malnutrition. The government's immediate response following the cyclone was to hold up visa applications for foreign

---

9   Win Myo Thu, interview with author, Yangon, Myanmar, 5 January 2011.
10   Bobby Maung, interview with author, Yangon, Myanmar, 7 January 2011.

journalists and aid workers and to deny entry to Western aid deliveries, leading to a massive build-up of food, medicine and disaster response expertise in Bangkok in the crucial days following the event (Fink 2009: 108–10; Larkin 2010: 8–10; Paik 2011; South 2009: 227; Vicary 2010: 214–18). Extensive coastal mangrove destruction over previous years, which was partially caused by the military's shrimp and fish farms on confiscated land, exacerbated the damage (*Independent Bangladesh* 2005; Kinver 2008).

Environmental politics in Myanmar therefore merges with the politics of survival, with active and sometimes militant opposition to the state. The historical legacies of colonial rule, constant marginalisation and military repression have resulted in widespread insurgent activities by ethnic minorities. Despite negotiating 17 ceasefires with various armed ethnic groups between 1989 and 1996 civil conflict still reigns in northern and eastern Myanmar (Fink 2008; Maung Aung Myoe 2009; South 2011). Although in the east the KNU signed a ceasefire with the government in 2012, sporadic clashes with the KLNA still occurred and in 2011 renewed fighting broke out between the Tatmadaw and the Kachin Independence Army in the north. Conflict in these regions resulted in internally displaced peoples (IDPs) and refugee flows into Thailand. Ethnic communities were forced from their homes by the military and their crops either burned or expropriated. In 2005 Human Rights Watch found an average of 30 'displacement episodes' per person among the Kayin they interviewed in Kayin State (Human Rights Watch 2005). Tatmadaw offensives in eastern Myanmar, in the region of the Yadana Pipeline and Salween Dams, also resulted in the systematic rape and sexual assault of women in ethnic minority communities.

As a result of these multiple insecurities, Myanmar activists often removed themselves from the Myanmar government's sphere of influence, either to the border liberated areas independent of Tatmadaw control such as Ei Tu Hta IDP camp on the Salween River or, where possible, to less authoritarian neighbouring countries to facilitate their operations. These activists become the transnational agents who engaged in the campaigns against the energy projects. These activists may have been in the environmental movement but their concerns were related directly to human rights abuses and they experienced a parallel process to one described by O'Kane for women in the area:

> for those trapped in the unsettled and ambiguous Burma-Thailand borderland space, distinctions between public/private, politics/survival, mother/activist, freedom fighter/illegal alien collapsed and become inseparable experiences. The collapse and/or significant restructuring of how these binary categories of relations are lived in the transversal spaces resulted in transformations in [their] political awareness. (O'Kane 2005: 15)

Hsiplopo, the Ei Tu Hta camp leader, epitomised this complexity. Having grown up in Yangon he joined the KNU in 1973 and thereafter lived in the forest and the camps along the Thai–Myanmar border. His multiple identities included KNU member, father, camp leader, and IDP and anti-dam activist. O'Kane's analysis echoes the arguments of Kaiser and Nikiforova that '[b]orderlands are not marginalised spaces … but rather … central nodes of power where place and identity across a multiplicity of geographical scales are made and unmade' (2006: 940).

These displaced and rebellious borderland communities were therefore the most fertile spaces for the development of environment movements against transnational energy projects in Myanmar. Despite the dangers these activists often re-entered Myanmar incognito to undertake research for NGOs such as ERI that were based outside Myanmar's borders.[11] These activists developed networks between displaced communities inside Myanmar and the exiles across the border (South 2004). The indeterminate nature of the zones of operation that environmental activists inhabited led them to undertake a broad range of activities. No border-based EGGs were simply lobby groups; their various other activities included undertaking covert field research, writing reports and providing clandestine education and training in Myanmar.

I argue that this exodus from authoritarian-ruled Myanmar to engage in activism in its borderlands and neighbouring countries resulted in what can be considered an 'activist diaspora', which provided activist environmental governance of policies and projects in Myanmar. The inclusive domain of the term 'diaspora' has, at times, been stretched to render it almost meaningless, with academic literature on, for example, liberal or queer diaspora leading Brubaker to argue that '[i]f everyone is diasporic, then no one is distinctively so' (2005: 3). Myanmar's expatriate activists, however, fulfil not only traditional aspects of the term based on dislocation or 'the dispersal of a people from [their] original homeland' (Butler 2001: 189), but also what Sökefeld argues are 'imagined transnational communities' with a 'transnationally dispersed collectivity that distinguishes itself by clear self-imaginations as community' (2006: 267). His focus on social movement theory and forms of mobilisation dovetails with the concept of an activist diasporic community. This concept, deriving as it does from a largely progressive and democratic activist community, also avoids the pitfalls that afflict some diasporic communities, such as a lack of trans-ethnic solidarity and gender awareness (Anthias 1998).

Despite friction existing between some ethnic communities of Myanmar, even in exile, this friction was largely absent in the environmental activist communities examined here who made significant attempts to develop multi-ethnic collaboration for both strategic and ideological reasons. An additional stimulus to cooperation was the geographic reach of both the Shwe Gas Pipeline and the Salween Dams, which crossed several ethnic states and regions, precipitating a high level of cross-ethnic cooperation in the development of the campaigns. This collaboration, while widespread, did not necessarily extend to all groups. The Muslim Rohingya in Rakhine State are the most marginalised of all ethnic groups in Myanmar and in 2012 they faced what Human Rights Watch argue was a campaign of ethnic

---

11   Ka Hsaw Wa, interview with author, Chiang Mai, Thailand, 14 January 2004.

cleansing due to the pervasive view in Myanmar that they were not a legitimate Myanmar ethnicity (Human Rights Watch 2013; Simpson 2004: 33).[12] Local Buddhist Rakhine politicians, however, saw the Rakhine community as the oppressed indigenous ethnicity of Rakhine State – with the natural resource rights that this entailed – facing resource and population pressures from Bangladeshi immigrants from the west and Bamar labourers from the east. While arguing that the human rights of the Rohingya should be respected it was clear that they also considered the Rohingya as interlopers from the west who should therefore be denied citizenship.[13] Exiled Rakhine Shwe activists shied away from entering this ethnic debate while groups such as EarthRights International argued that they had not worked with Rohingya communities because they were not located along the pipeline route.[14] While any ambivalence from Rakhine Shwe activists on this issue could be problematic, after five decades of marginalisation under military rule their desire to promote their own ethnic rights is understandable. Their experience and the cultural and political milieu in which the environmental groups operated may have influenced their perspectives but their cooperation with a range of ethnicities still represented an emancipatory approach unmatched by the general exiled community. This multi-ethnic collaboration by exiled activists is therefore the first defining characteristic of an activist diaspora.

The second characteristic is that an activist diaspora actively establishes linkages and provides specialist knowledge and otherwise unavailable expertise to broader campaign networks or coalitions in foreign or 'outsider' communities, which contrasts with the often isolationist approach of other members of diasporic communities. The importance of exiled activists to the outside groups lies in their specific linguistic and cultural knowledge together with a local political understanding, often sharpened by intense personal experiences, which outsiders, in this case from the North or Thailand, are unlikely ever to acquire or fully comprehend. This 'insider' knowledge of Myanmar exiles allowed them to undertake covert research back in Myanmar proper that would otherwise be unachievable and to provide expertise to outside or transnational civil society groups that was inaccessible for other activists.

The third aspect is that through this contribution of expertise to broader campaigns and engagement with activists outside the exile community the activist diaspora plays an important role in bolstering the confidence and skills, including English language proficiency, of exiles who may otherwise feel isolated in a foreign country. The delicate balance of emotions that exists in exile communities between 'hope [and] resistance' on the one hand and 'helplessness, suffering and apathy' on the other suggests that those who establish strong links both to their own diasporic

---

12   M. Smith, interview with author, Yangon, Myanmar, 26 June 2012.

13   Aye Tha Aung, interview with author, Yangon, Myanmar, 20 February 2012; Oo Hla Saw, interview with author, Yangon, Myanmar, 20 February 2012.

14   P. Donowitz, interview with author, Chiang Mai, Thailand, 18 April 2013 and email to author, 6 September 2013.

community and the outsider community are those most likely to adapt successfully to their new lives (Mavroudi 2008: 70). The networks within Myanmar's activist diaspora facilitated linkages and communication with other transnational activists, based predominantly in Thailand, that successfully conveyed their campaigns, on both cross-border energy projects and democratisation in Myanmar, to a more spatially dispersed transnational audience.

Conclusions

Through an analysis of environmental politics in Thailand and Myanmar this chapter has demonstrated the intrinsic relationship between the democratic qualities of a political regime and the ability of local environmental movements to undertake activism. A lack of democratic openings at home tends to stimulate transnational activism through the appearance of an activist diaspora. I also revealed the close linkages between large business interests and authoritarian governance in both competitive authoritarian regimes and more traditional authoritarian military regimes, which exacerbates pressures on environmental security for marginalised communities. There are six broad conclusions to be drawn from this analysis. First, while political regimes in Thailand were more liberal than the authoritarian military regime of Myanmar, there were powerful anti-democratic forces in both countries.

Second, despite some success in opposing major development projects in Thailand, environmentalists in both Thailand and Myanmar faced a variety of difficulties in their campaigns to highlight concerns over development practices. The strictures faced by Thai activists were generally much less severe than those in Myanmar although activist assassinations occurred and police repression was not uncommon. In response to successful campaigns, however, Thai governments and businesses attempted to export the negative impacts of energy production by pursuing energy projects in neighbouring authoritarian states, particularly Myanmar. Until 2011 the Myanmar government crushed domestic public dissent and environmental activists therefore removed themselves from the military's sphere of influence, resulting in an activist diaspora that facilitated transnational activist environmental governance of policies and projects in Myanmar.

Third, while there was some transnational transfer of environmental strategies, tactics and philosophies into both Thailand and Myanmar, the more restricted media space and communication technologies of Myanmar meant that the country's borders were not, unlike Thailand's, mere speed bumps to the free transfer of information. Under military rule Myanmar activists communicated their messages to the international community primarily through the conduit of the activist diaspora while Myanmar's porous and fuzzy borderlands, particularly along the Thai border, allowed activists direct access to some ethnic minority communities, although at great personal risk.

Fourth, while in both countries environment movements demonstrated characteristics of transnational activist philosophy and techniques, both also had strong local cultural influences, particularly those related to Engaged Buddhism.

Fifth, in both Thailand and Myanmar there was a strong symbiotic relationship between large business interests and political elites. In Thailand this relationship reached its apogee with the election of Thaksin and Thai Rak Thai (TRT), Thailand's most significant consolidation of large capital in a single party. In Myanmar large business interests were intimately intertwined with the upper echelons of the military, with state privatisations only entrenching the country's wealth in the hands of the military and its supporters. The convergence of interests between large business interests and political elites resulted in the dominant development philosophies in both countries closing down debate on development issues and undermining attempts at genuine public participation in development processes.

Sixth, as a result of this convergence of interests, elites within both business and government in Thailand and Myanmar pursued development practices and energy projects that increased human and environmental insecurity for marginalised communities, particularly in the vicinity of energy projects. In Thailand, with a more open political system and environmental laws in place, there were generally avenues available to challenge environmental and human rights issues related to development proposals. Nevertheless, projects were pursued and undertaken without regard for legal requirements and opportunities for redress were dependent on the nature of the contemporaneous political leadership. In Myanmar environmental destruction relating to energy and development projects was rampant. EIAs were not required, were rarely undertaken and never released publicly. Egregious human rights abuses, forced labour and systematic sexual assault also often accompanied projects.

While these conclusions are drawn from the analysis of environmental politics in Thailand and Myanmar, they also provide important lessons for environmental politics within the South in general. Political regimes in the South exhibit various degrees of authoritarianism, with often opaque relationships between large business interests and government and widespread environmental insecurity. Authoritarian governance can persist despite the existence of formal institutions of democracy. The following chapters, which examine the campaigns against the energy projects in more detail, provide important insights into the dynamics of local and transnational activist environmental governance in the South and its impacts on environmental security.

Chapter 4
Local Activism

Introduction

Activist environmental governance can occur at a variety of scales and, in the era of globalisation, transnational activism is becoming a crucial component of environmental campaigns throughout the South. Nonetheless, local activism remains important (Rootes 2008), and its significance to the case study campaigns is analysed in this chapter. Although often stimulated by local issues it can provide activists with their first exposure to environmental activism and, through a process of activist radicalisation and transformation, influence broader transnational campaigns. As discussed in Chapter 1, defining local activism in these campaigns is somewhat difficult due to the fuzzy nature of the border areas and their associated populations (see Chaturvedi 2003; Christiansen, Petito and Tonra 2000). In the context of these campaigns, however, I define local activism as activism undertaken within borders primarily directed at a domestic audience in the home country, which is the physical location of the project.

A principal consideration for this chapter is the significance of local conditions in encouraging the extent and types of local activism within the broader transnational environmental disputes. I therefore address whether local living conditions, local cultural or religious influences, local business interests and authoritarian governance impacted on the involvement of local people in activism and the resultant organisational structures, philosophies, strategies and tactics in the campaigns. This analysis will also provide some insight into the effectiveness of local actors in forming EGGs and engaging in activist environmental governance.

To analyse the forms and effectiveness of local activist environmental governance I examine two separate, although related, components. The first component is public protest, which is a central feature of any social movement (della Porta et al. 2006: 28–29; Doherty 2002: 7; Doherty and Doyle 2006: 702–3). The likelihood of protest in an environmental campaign can, however, vary greatly depending on the cultural norms and political regime involved, especially in the South. These factors play a pivotal role in the appearance or otherwise of public opposition: very little local opposition was voiced under authoritarian rule in Myanmar while protest was a common occurrence under the relatively competitive political system in Thailand.

The second component is engagement with the state and business through public participation in state-sanctioned fora, although, as I will demonstrate throughout this chapter, the two forms of activism sometimes converged due to the hollow participation regimes on offer by the state. This component is generally much more available in domestic, rather than transnational, environments as local governance structures are more hierarchical and well defined than in the international system. State-sanctioned fora in this context include consultative processes whereby participation from the public is sought in development decisions, primarily through processes linked to EIA. As I noted in Chapter 2, within the environmental movement there is a broad consensus that environmental justice and an overarching ecological rationality governing our societies are best pursued by strengthening democratic forces and increasing community involvement in decision making. Despite some progress in embedding public participation in policy development, truly effective participatory governance remains elusive (Reddel and Woolcock 2004). In theory, EIA could be a step on society's path towards ecological rationality (Bartlett 2005; Dryzek 1987), but it could also be employed as a strategic state response to neutralise protests and other forms of activism and dissent. The measure of a particular EIA process on the continuum between these two poles is its transparency and the ability of the public to participate freely and influence the state's decisions. As I shall demonstrate, however, on the few occasions where EIA processes were followed the gap between the theory of participation and what occurred in practice was significant.

In examining local activism some campaigns necessarily received more attention than others. Local activism, as defined above, was extremely limited under the authoritarian military regime in Myanmar, although it has increased significantly under the new government. Given the time periods examined the campaigns that experienced local activism in Thailand, such as the Yadana and TTM campaigns, constitute the majority of the analysis here. Although these two campaigns were not successful in stopping the projects, at least Thais were able to voice their opposition; in Myanmar this sort of activism was 'impossible' under military rule (Lewis 2006: 54–55). Most of the local activism discussed in this chapter was therefore undertaken in Thailand by Thais where the target of the activism was often the Thai government, Thai media or Thai corporations, although there were, of course, complex linkages between these local campaigns and those directed at the international community.

Yadana Gas Pipeline Project

The key local organisation in the campaign against the Yadana Pipeline Project in Thailand was the informal Kanchanaburi Conservation Group (KCG), which was initially formed to oppose the Nam Choan Dam project in Kanchanaburi Province in the 1980s (Fahn 2003: 88; Supara Janchitfah 1998). The Nam Choan dispute politicised people in the area who were concerned about the local environmental consequences of the dam. Phinan Chotirosseranee (also Bhinand Jotiroseranee), a shop-owning woman in Kanchanaburi township, was a central voice in opposition to the dam and co-founded the KCG to direct the local campaign. Once the dam was cancelled in 1989, however, the group lay dormant until the campaign against the Yadana Pipeline emerged in the mid-1990s. Phinan was co-president of KCG – the group ensured equal gender representation in the leadership – but despite a nominal hierarchy it was largely consensus-based and the leadership roles were essentially those of spokespeople.[1] Leadership roles for women and a concern for both gender equality and a lack of hierarchy are characteristic of environment groups in the North (Desai 2002: 32–33; Doyle 2000: 33), but the organisational structure and composition of KCG demonstrated that environment groups in the South also attempt to address these issues.

Despite its dormancy between the Nam Choan and Yadana disputes, the KCG ultimately endured, supporting the proposition that once formed by a particular issue environment groups tend to survive while informal networks are more likely to disintegrate (Doyle and Kellow 1995: 110). While the KCG was initially a not-in-my-back-yard (NIMBY) group – Phinan later conceded the group had a 'selfish', parochial outlook in the early stages of the Yadana campaign[2] – the group eventually broadened its interests beyond Kanchanaburi Province. Phinan adopted an emancipatory outlook and later co-authored a report with Sulak Sivaraksa for the UN Human Rights Committee (Pibhop Dhongchai et al. 2005). As Doherty (2002: 185) notes, radical transformation through the process of activism itself means that even NIMBY local environment groups, such as the KCG, may eventually become social movement actors, and both KCG's organisation and its emerging focus adhered to the green pillars, qualifying it as an EGG.

Phinan had already, however, demonstrated a growing interest in broader social concerns in the early 1990s by founding the Kanchanaburi Women's Group, which also participated in the Yadana campaign. This group introduced 'women's issues' into the campaign and provided symbols of eco-femininity – siding with the forest against the brute masculinity of the mechanised excavators, which they blocked with their bodies – evoking images of the women of the Chipko movement in India; a movement which Shiva argued had been 'fuelled by the ecological insights and political and moral strengths of women' (1989: 67). Inferences of feminine essentialism have been found in other women's movements of the South (Blondet 2002), but as Porter (2003) argues women's concerns are more likely to simply derive from their traditional roles within the family. Nevertheless, there exist many parallels between women's and environmental movements as they both tend to be emancipatory in nature.

After travelling across Tanintharyi Region (formerly Tenasserim Division) in southern Myanmar the Yadana Pipeline route crossed the Thai-Myanmar border at Nat-E-Taung (see Figure 4.1), then passed through Huay Kayeng (Huai Khayeng) Forest Reserve, now part of Thong Pha Phum National Park, in Thailand's Kanchanaburi Province on its path to the Ratchaburi power station (Giannini et al.

---

1   Phinan Chotirosseranee, interview with author (translator: Ellen Cowhey), Kanchanaburi Province, Thailand, 5 October 1998.

2   Ibid.

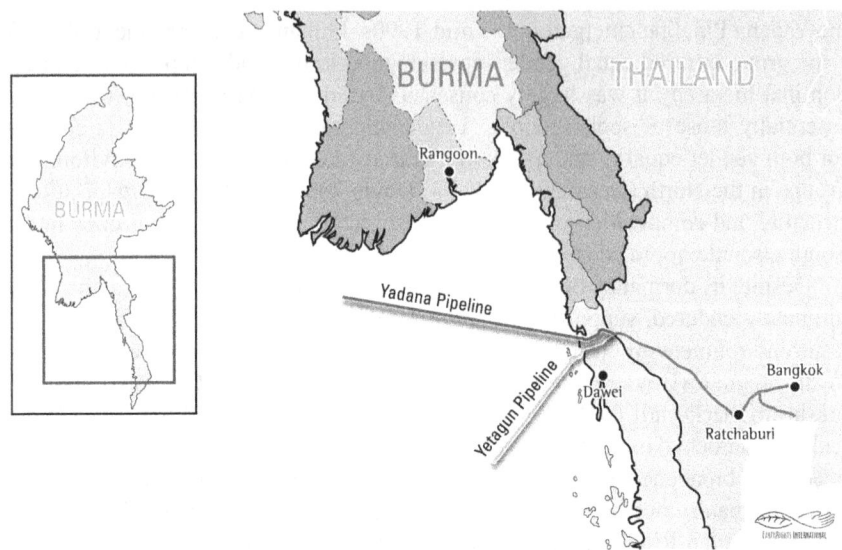

Figure 4.1   The Yadana and Yetagun Gas Pipelines
Source: EarthRights International

2003: 142).[3] Initially, it was the proposed destruction of such pristine environments and related impacts on water resources that instigated the protests from the KCG and other community members (Fahn 2003: 199). Doherty (2002: 184–85) notes that, although direct action is generally a last resort for local environment groups, they are often willing to undertake protest action if it is likely to be effective and not alienate other locals. In this case, despite a demonstration in favour of the pipeline, which activists accused of being staged by the Petroleum Authority of Thailand (PTT) (Nantiya Tangwisutijit 1998), most of the local population supported the protesters. This support reflected the KCG's standing in the community through its transparent operations and its youth training projects, which had the support of the United Nations Development Programme small grant program (Supara Janchitfah 1998). In addition a concert was held to promote awareness of their campaign, which increased local knowledge and involvement.

Despite the formation of the KCG in the 1980s the localised response to the initial pipeline proposal was rather slow to emerge in Kanchanaburi itself. In Bangkok the NGO TERRA had expressed concern about the project's impact on Thailand's western forest complex in early 1995 (Fahn 2003: 199). In July 1996 EarthRights International (ERI), which was formed partly in response to the project,[4] and the Southeast Asian Information Network (SAIN) published their

---

3   Although the Yadana Pipeline was eventually joined at Nat-E-Taung by the Yetagun Pipeline to form a single pipeline in Thailand the campaigns focused on the Yadana, as the original project, which is reflected in the analysis here.

4   K. Redford, interview with author, Chiang Mai, Thailand, 15 January 2004.

seminal report on the project, *Total Denial*, which set the standard in reporting for many activist groups in the region (ERI and SAIN 1996; Giannini et al. 2003). Although the report considered the environmental issues in both Thailand and Myanmar at length, most of the report was devoted to the human rights abuses meted out to ethnic minorities in the pipeline area on the Myanmar side of the border and the complicity of the TNCs involved in the project: Unocal (US), Total (France), PTT and EGAT (Thailand). This focus highlighted an initial difference between groups that were formed with a distinct justice focus and the local Kanchanaburi groups that, when they first joined the campaign, adopted a broadly NIMBY approach focused on issues of local environmental amenity. The Kanchanaburi activists were concerned about environmental issues, but they faced no existential threats. This contrasted with the mostly ethnic Kayin along the pipeline route on the Myanmar side of the border who were engaged in a constant fight for their survival against both entrenched poverty and the Tatmadaw. In these Kayin areas environmental insecurity was pervasive, as the Kayin had no access to electricity and precarious supplies of food and water. These insecurities were exacerbated by forced relocations associated with the pipeline, with some villages burnt to the ground (Giannini et al. 2003: 43).

ERI followed their *Total Denial* report with a class action for Kayin villagers against Unocal in the US courts in October 1996. At this time, despite international activism and national media attention, the NIMBY Kanchanaburi groups were still relatively unconcerned about the project as they had been assured by local politicians that the pipeline would bypass Kanchanaburi, entering Thailand through Ratchaburi Province to the south (Fahn 2003: 203). By the end of 1996, however, it was established beyond doubt that the project was to pass through Kanchanaburi and local groups began to mobilise, with a national campaign of 88 organisations consisting of 'NGOs, environmentalists, conservationists, human rights groups and local affected people' eventually joining the protest (Ogunlana, Yotsinsak and Yisa 2001: 208). Although groups such as KCG transformed into emancipatory actors the muted initial reaction suggests that it can sometimes take local environment groups more than one major campaign – in this case both the Nam Choan and the Yadana – to be transformed from parochial NIMBY groups to radicalised social movement actors.

Public hearings for major development projects in Thailand were introduced as a way of including public input in 1996, but there were no laws or regulations at that time for mandatory public participation within the EIA process. As a result, an initial EIA for the Yadana Project on the Thai side was prepared for PTT but only a limited technical hearing and questionnaire survey were conducted, with local issues inadequately considered (Ogunlana, Yotsinsak and Yisa 2001: 209). The Office of Environmental Policy and Planning rejected both this EIA in January 1996 and another in February 1997, arguing that they were unsatisfactory. Nevertheless, the National Environment Board approved the project in March 1997 (ERI and SAIN 1996: 76; Fahn 2003: 200–201; Giannini et al. 2003: 149–50). In Thailand an approved EIA had been mandatory prior to government assent for major projects since 1992 but this directive was ignored in this case (Simpson 1999: 507). The EIAs were also not made publicly available but were eventually leaked to activists. They were written in highly technical English, however, which made them inaccessible for most of the Thai-speaking communities in Kanchanaburi Province.[5]

Throughout 1997 the local Kanchanaburi activists attempted to influence PTT and the government over the project and its proposed route, but with little success. By the end of the year it was apparent that in the absence of a more formal and influential participatory process their input was being ignored so a more direct action strategy evolved. This response accorded with Doherty's (2002: 191) study of environmental activism in the North, which suggests that disputes only become confrontational when political opportunities are restricted or authorities act unfairly. From 7 December the activists undertook occasional protests at the sites in Huay Kayeng Forest Reserve where the trees were being cleared and on 21 December 1997 approximately 50 activists, including the KCG, set up a permanent encampment in the forest, adopting a *satyagraha* strategy (Richards 1991: 48–63), in an attempt to bring the laying of the pipeline to a halt (Fahn 2003: 199–202; Ogunlana, Yotsinsak and Yisa 2001: 212). These forest protest camps continued into February of the new year. Engaged Buddhist Sulak Sivaraksa left the camp in January to fly to the UK to teach a course on Buddhist Economics at Schumacher College where he placed the protests in the context of local-global campaigns against neoliberal globalisation.[6] During the protests numerous tactics and strategies were used, but from the beginning nonviolence was self-consciously the overarching governing principle.[7] Activists emphasised the Buddhist heritage of their chosen forms of activism, and Sulak noted the importance of individuals training themselves to act nonviolently· 'from a Buddhist perspective, from the Gandhian perspective; each individual must train himself/herself to be nonviolent'.[8]

Both Sulak and Phinan, and many others who have adopted Buddhist-inspired nonviolent activism, recognise the philosophical debt to Gandhi due to his pursuit of truth through nonviolence or *ahimsa* (Bond 2003: 128; Cabezon 1996: 305; Sulak Sivaraksa 1988: 94). Nonetheless, this activism resulted in tactics that were also reminiscent of Northern actions, such as forest protests in Australia or road protests in the UK (Doyle 2005). In one instance activists linked arms and placed their bodies in the path of excavators. While this approach did not demonstrate the technical innovations to engender manufactured vulnerability found in the affluent countries of the North (Doherty 1999), in a remote forest of Thailand it did, nonetheless, represent a similar philosophy. Thailand's past is littered with the

---

5   Phinan Chotirosseranee, interview with author (translator: Ellen Cowhey), Kanchanaburi Province, Thailand, 5 October 1998.

6   Sulak Sivaraksa, interview with author, Devon, UK, 25 January 1998.

7   Phinan Chotirosseranee, interview with author (translator: Ellen Cowhey), Kanchanaburi Province, Thailand, 5 October 1998.

8   Sulak Sivaraksa, interview with author, Devon, UK, 25 January 1998.

broken bodies of dissenters and activists, so this sort of nonviolent activism was certainly potentially dangerous (Baker and Pasuk Phongpaichit 2005; Handley 2006; Thak Chaloemtiarana 2007b; Wyatt 2004). While in Thailand under the Chuan-led government this tactic achieved at least a pause in proceedings, particularly due to the presence of local media, on the Myanmar side of the border this sort of tactic was not even considered; in the remote forests of Tanintharyi Region the excavators would have been unlikely to halt.

A young activist from the KCG at these protests, Pipob Udomittipong, also worked with Sulak as coordinator of the Buddhist-inspired Spirit in Education Movement and on the working and executive committees of the International Network of Engaged Buddhists. At the forest protest site in February 1998 he argued that the two main reasons for opposing the pipeline were the attendant environmental destruction and the human rights abuses in Myanmar (Aung Hla 1998). He later worked with ERI and Salween Watch and acknowledged that justice issues were intertwined with spiritual development, which, in much of the Thai context, translated into Buddhist practice.[9] As with Sulak, in the Thai context he saw a personal practice of Engaged Buddhism as central to the success of nonviolent action.

Apart from supplying the philosophical bedrock for the broad strategy of nonviolence, there were also other, more tangible, examples of Buddhist influence on the campaign. Since the mid-1980s tree ordinations had been carried out by environmentalist monks to save forested areas (Darlington 1998; Darlington 2003: 103; Taylor 1993: 11), and during the Yadana campaign Phinan and the KCG enlisted the support of Thai environmentalist monks who conducted five tree ordination ceremonies during the protest (*Bangkok Post* 1997).[10] Bringing in the *sangha* to support a protest was an important statement in Thailand but it did not always achieve the desired effect. While Buddhist monks were revered throughout the country, forest monks have also been seen by those in the urban centres of power, like the forest itself, as 'untamed' and 'uncivilised' (Taylor 1991: 107). Nevertheless, the symbolic act by monks of wrapping a tree with saffron robes in an ordination ceremony added a level of gravitas to the protest that would otherwise have been missing.

Another peculiarly Thai tactic, but one that had its roots in Gandhian approaches, was erecting a poster of Thailand's queen over the pipeline route. Queen Sirikit became Thailand's queen when King Bhumibol ascended the throne in 1946 and was highly revered despite her previous role in supporting the rightist repression of the 1970s (Handley 2006: 238). This choice was significant given that the KCG had been in direct opposition to the king and his government in the 1980s over the Nam Choan Dam, which Bhumibol had promoted as part of his extensive dam building plans for Thailand. While some of the activists were

---

9   Pipob Udomittipong, interview with author, Chiang Mai, Thailand, 3 January 2005.

10   Phinan Chotirosseranee, interview with author (translator: Ellen Cowhey), Kanchanaburi Province, Thailand, 5 October 1998.

republicans who saw the monarchy as an impediment to democracy,[11] others were less radical, but the choice of the queen, rather than the king, provided a potent symbol that again linked the feminine to the forest. As a tactic it was temporarily successful in blocking the excavators although after negotiations it was removed as it was seen as potentially 'politicising' the monarchy; as Phinan recalled, neither the excavator operators nor their managers wanted to be seen crashing through a poster of the queen.[12]

While the Thai government was largely hostile to the forest protesters they garnered sympathy from some journalists, with whom they were acquainted, in Thailand's English language newspapers, the *Bangkok Post* and the *Nation*, although the circulation was limited to expatriates and educated Thais (Lewis 2006: 93).[13] There was also support from academics and students around the country, and 16 chapters of the Confederation of Students for Conservation staged a nonviolent protest in front of Government House on 1 February 1998 (AHRC 1998).

Ultimately, however, the forest protest and its supporters met with at least temporary success. The KCG and most other activists had agreed to leave the forest site if a public inquiry was set up and on 12 February 1998 Prime Minister Chuan Leekpai put the project on hold and appointed a Central Advisory Committee to undertake public hearings to resolve the conflict (Warasak Phuangcharoen 2005: 15). Phinan and some other activists, who were by this time feeling the physical and emotional strain after over two months in the camps, willingly left the forest on the day the committee was announced in anticipation of venting their frustrations in more formal public settings (Fahn 2003: 201).[14]

The committee staged an 'open public forum' in Bangkok, first at Government House then at Chulalongkorn University, to review the conflicts caused by the construction of the pipeline, although some affected people complained that these fora were too far away from their communities for them to be adequately represented. Nevertheless, on 23 February 1998 the committee concluded that the PTT had produced a flawed EIA report and that the EIA process lacked transparency (Warasak Phuangcharoen 2005: 18). They suggested that PTT had failed to accommodate public opinion and genuine concerns on environmental impacts (Ogunlana, Yotsinsak and Yisa 2001: 222). The committee also critiqued the EIA process for not including public participation, arguing that it was only 'public relations' (Fahn 2003: 204).

---

11   Pipob Udomittipong, interview with author, Chiang Mai, Thailand, 3 January 2005.

12   Phinan Chotirosseranee, interview with author (translator: Ellen Cowhey), Kanchanaburi Province, Thailand, 5 October 1998.

13   Pipob Udomittipong, interview with author, Chiang Mai, Thailand, 3 January 2005; Supara Janchitfah, email to author, Bangkok, Thailand, 19 September 2008.

14   Phinan Chotirosseranee, interview with author (translator: Ellen Cowhey), Kanchanaburi Province, Thailand, 5 October 1998.

Despite a rather harsh assessment of the whole process the committee failed to recommend that the project be delayed or that the route be altered and on 28 February 1998 Chuan decided that the the the pipeline would proceed (AHRC 1998; Fahn 2003: 204; Warasak Phuangcharoen 2005: 11). The project's opponents argued that the hearings were held too late in the development process and did not adequately address their key issues (Ogunlana, Yotsinsak and Yisa 2001: 222–23).

Following Chuan's announcement Sulak, who had not made any commitment to leave the forest, continued what he termed a 'fierce campaign' at the Huay Kayeng forest site on 3 March 1998 with fellow students and activists, arguing that doing business with the Myanmar regime 'was unconscionable'.[15] On 6 March 1998 Sulak and 40 other protesters were arrested under the *Petroleum Act*, a statute that protected PTT's energy operations, for allegedly obstructing the pipeline construction (Lewis 2006: 55). In a statement made on the day of his arrest, Sulak questioned the local benefit of the project.

> My friends and I may not be able to protect the forest but we want to demonstrate that development without consideration for human rights, consideration for the environment issues and without consideration of the local participation is fundamentally wrong. Development must benefit the poor, the grass roots, animals and trees ... Most grand schemes of economic development and technical advances benefit multinational corporations and the super rich, but harm the majority of people. (Kalayanamitra Council 1998)

He also emphasised the human rights aspect, linking the gas revenues to military spending and resultant repression: 'Is it ... just, to pay money to Burma for her to buy arms to kill Burmese people?' (Kalayanamitra Council 1998). The next day a further 20 villagers and student activists were temporarily detained by police in Dan Makham Tia District in Kanchanaburi for allegedly obstructing the construction work (AHRC 1998). Despite the eventual completion of the pipeline and several changes in government, Sulak remained on bail for eight years. The case was only dropped in August 2006 because the law under which Sulak was charged related to the state-owned Petroleum Authority of Thailand, which had since been partially privatised (Sai Silp 2006).

While Sulak was prosecuted in court on numerous occasions and was forced to leave Thailand in the early 1990s due to charges of *lèse majesté* (Handley 2006: 298),[16] the penalties for local activists speaking out in Myanmar were generally far more immediate and visceral. A prominent example was that of the Moustache Brothers, an *a-nyeint* comedy troupe who used nonviolent forms of performance protest such as comedy and satire, and employed irony as both rhetorical form and philosophical content (Szerszynski 2007: 348). They had a close friendship with Aung San Suu Kyi and supported the NLD's opposition at this time to foreign

---

15   Sulak Sivaraksa, interview with author, Bangkok, Thailand, 20 September 1998.
16   Sulak Sivaraksa, interview with author, Devon, UK, 25 January 1998.

investment in projects such as the Yadana Pipeline. For an innocuous joke within their performance two of the performers, Par Par Lay and Lu Zaw, were tortured and received six years' incarceration with hard labour for their sense of the absurd. They knew they would probably be arrested soon after the performance and yet they continued anyway.[17] This was manufactured vulnerability, Myanmar style.

Although there were limitations to public participation in Thailand, the opportunities were bountiful compared with the Myanmar side of the border where genuine public participation in the development process was non-existent, despite more than 1,000 villagers being forced by the military to attend an 'opening ceremony' of the pipeline on 1 July 1998 (Giannini 1999: 15). As Lambrecht notes, development projects that occur in Myanmar's borderlands 'are only participatory inasmuch as they are financed predominantly through forced labour and the taxation of the rural populace' (2004: 172).

The local activism against the Yadana Project therefore entailed a wide variety of tactics and philosophies, with many activists influenced by their various ethnic or cultural backgrounds, but the opportunities for public dissent were largely defined by the openness of the governing political regime. In Thailand there was a relatively liberal climate under Chuan but it became obvious that, at the very least, the consultation process was severely constrained by the 'range of policy process norms to which governments adhere' (Holland 2002: 76). In contrast, public opposition in Myanmar was quickly quashed by the military and, as a result, there was little public campaigning. Activism did occur within Myanmar in the form of research and education undertaken in the pipeline conflict zone by cross-border groups but as these groups were based outside of Myanmar proper they are classified as transnational actors and are discussed more fully in Chapters 5 and 6. While the local Thai activists opposing the Yadana Project were rural or provincial they were still relatively affluent. In the south of the country a similar project posed a more existential threat to the livelihoods of poor Muslim fishing communities.

Trans Thai–Malaysia (TTM) Gas Pipeline Project

The Trans Thai–Malaysia (TTM) Pipeline Project in the largely Muslim Songkhla Province in the south of Thailand was punctuated by a variety of local protests and actions from its announcement in 1998. The project required offshore drilling in the Malaysia–Thailand Joint Development Area in the Gulf of Thailand, the construction of two gas separation plants (GSPs) in Chana District on the coast and the laying of a gas pipeline, with a parallel liquefied petroleum gas (LPG) pipeline, from the GSPs to the border with Malaysia in the west (TTM 2012). As the pipeline fed directly from Thailand into the pre-existing Malaysian pipeline network, the protests and activism were focused on the Thai side of the border.

---

17   Par Par Lay and Lu Maw, interview with author, Mandalay, Myanmar, 15 December 2003.

The protests began under the government of Chuan Leekpai but continued after Thaksin Chinawatra became prime minister in January 2001.

Despite some assistance from transnational networks the campaign against the TTM was driven primarily by the local environment movement in Thailand, for three key reasons. First, the project was based in Thailand and, while there were authoritarian responses by the state to public protests, the generalised wide-scale repression that characterised Myanmar was absent and transnational NGOs were therefore less likely to assign resources to it. Second, the absence of Northern TNCs as central players in the project, apart for Barclays Bank's involvement in a financing role, minimised attention from Northern activists within the global justice movement. Third, the environment movement in Thailand was already relatively mature, dynamic and self-sufficient.

Activists began the campaign following the signing of a Memorandum of Understanding (MoU) between Chuan and Malaysian Prime Minister Mahathir Mohamad in April 1998 and the discovery of plans relating to the industrial development of Songkhla Province within the Indonesia–Malaysia–Thailand Growth Triangle (Asian Development Bank 2003). To illustrate the potential impacts of the TTM Project and its associated development Penchom Tang (Saetang) of the NGO Campaign for Alternative Industry Network accompanied 40 Songkhla villagers and other activists, some of whom had been involved with the Yadana campaign, on a three-day study trip to the heavily industrialised eastern seaboard in early March 2000 (Penchom Tang and Pipob Udomittipong 2003: 4; Vasana Chinvarakorn 2000).[18] On a similar fieldtrip Varaporn Chamsanit, a human rights activist and former journalist with *The Nation*, took Bangkok students on a study tour to both these industrial areas and to Chana District and Hat Yai in Songkhla Province for meetings with fisherfolk and local activists as part of the human rights program for the Thailand Research Fund.[19] Following these trips, a seminar on large-scale industrialisation in the south and Thailand's Energy Development Policy was held at Bangkok's Chulalongkorn University on 8 March 2000 (Vasana Chinvarakorn 2000). These contacts and networks of activism broadened the outlook of local villagers from the localised disruptions caused by the construction of the pipeline and GSPs to concerns over a more widespread pattern of industrial development and environmental insecurity driven by the forces of globalisation.[20]

A broad movement against the project emerged, comprising environmental organisations, academics and local fisherfolk, which argued that the project would cause serious deleterious impacts upon local communities and their environments. Local academics at Prince of Songkla University in Hat Yai, the capital of Songkhla Province, also questioned the need for the project, pointing out that Thailand

---

18 Penchom Tang, email to author, Bangkok, Thailand, 17 October 2005.

19 Varaporn Chamsanit, interview with author, Canberra, Australia, 3 February 2006.

20 Ida Aroonwong, interview with author, Bangkok, Thailand, 25 November 2008.

was importing gas from Myanmar while planning to export gas to Malaysia.[21] Reungchai Tansakul, an Associate Professor of Biology and authority on EIA in the area, asked: 'Why don't we reserve our natural resources when we still have excess energy? We are not wise; we use up our own deposits when we still have alternative sources' (cited in Supara Janchitfah 2004: 42–43).

His concerns about the project, including environmental laws not being properly adhered to, coincided with other local inhabitants and fisherfolk.[22] He also placed the project in the wider context of the planned industrialisation of the region, arguing that government moves to revitalise the nearby polluted Songkhla Lake would be meaningless unless the government revised its plan to turn the southern provinces, particularly Songkhla, into an industrial hub: '[The] decision to approve the Thai–Malaysian gas pipeline and gas separation plants in Chana District would bring many more dirty factories to the province. These factories will dump more pollution to the lake' (cited by Kultida Samabuddhi 2002). This support from Thai academics, which was often sought out by the villagers themselves, added credibility to the campaign and was crucial to disseminating its message throughout the broader community. By November 2002, 1,371 academics had signed a statement on Chiang Mai University's Midnight University website urging the government to review the TTM (Midnight University 2002).

Unlike the Yadana campaign, in which local activists were slow to get involved, the TTM campaign was entrenched in local communities from its inception and local villagers were both informed and active, setting up the Lan Hoy Siab protest encampment on the beach near the GSP site in Taling Chan Subdistrict (Penchom Tang and Pipob Udomittipong 2003). Experienced NGO activists had learnt from the Yadana campaign that it was important to establish linkages with local communities early in the campaign. To both demonstrate solidarity and to learn about local issues some Bangkok activists, such as Ida Aroonwong of the NGO Alternative Energy Project for Sustainability, therefore immersed themselves in the local community by living with the Chana villagers for a period of two years during the campaign.[23]

These efforts by outside activists to link up with Songkhla activists and the improved communications technologies available since the Yadana dispute, particularly mobile phones and the internet, allowed activist networks to be established quickly and effectively. One of the most important roles activists outside the villages played was to link the campaign to broader societal issues and inequities. This role was similar to that played by some women's organisations in the South whose 'most significant achievement [has often] been to contribute to an increase in the consciousness and confidence of [women] workers to demand their rights' (Hale and Wills 2007: 458). As a result the Muslim identity of local villagers became a prominent focal point for their actions. Activist Varaporn

---

21   Prasart Meetam, interview with author, Hat Yai, Thailand, 11 February 2005.

22   Reungchai Tansakul, email to author, Hat Yai, Thailand, 8 November 2005.

23   Ida Aroonwong, interview with author, Bangkok, Thailand, 25 November 2008.

Chamsanit noted that throughout the campaign against the TTM, 'Muslim minorities increased their awareness of their Islamic identity. At the ocean in Chana District the villagers essentially said "We must protect the ocean because God gave it to us". I don't think that they would have said that thirty years ago'.[24] This response reflected a wider pattern across southern Thailand where Muslims started placing their local marginalisation in the larger context of the global Orientalist (Falk 2009: 48–51) attack on Muslim identity in the global 'War on Terror' (Funston 2006: 87–88). Varaporn suggested that in the TTM campaign this process of identification was expedited by outside activists who 'brought greater awareness [to local villagers] of international/global aspects of the TTM such as anti-Muslim sentiment after 9/11'.[25]

As with the Yadana Project it was local communities and ethnic minorities who were not only most adversely affected by the project, but were also the most voiceless communities in the decision-making process. Much of the local activism against the TTM project concerned the lack of genuine consultation and participation in the decision-making processes. The project was carried out under the 1997 Thai constitution, which required greater public participation in development processes (s.59 and s.60) and improved checks and balances through the establishment of supervisory bodies such as the National Human Rights Commission.

The constitution was promulgated on 11 October 1997 and its public participation requirements did not apply to the Yadana Project, which had been approved by the National Environment Board the previous March. The TTM Project therefore became one of the first major tests of the public consultation processes under the constitution in which public hearings were required to be held either during a project's feasibility study or EIA, but prior to any decision being made. The EIA process for the TTM initially included some *tambon* (subdistrict) level and *amphoe* (district) level public meetings in 1999 where the public voiced concerns about the project. There was, however, little information available at this time so it was difficult to mount convincing arguments against the project (Chatchai Ratanachai 2000a: 2–22). As a result of complaints over the inadequacy of this process, the Ministry of Industry set up a public hearing in Hat Yai in July 2000, although the EIA had already been completed and published four months earlier (Warasak Phuangcharoen 2005: 27). Like the Yadana EIA, it was originally in English and difficult for locals to read, but a Thai translation eventually became available (Chatchai Ratanachai 2000a; 2000b; Supara Janchitfah 2004: 114).

While the EIA process was underway, four contracts – the Gas Sale Agreement, the Balancing Agreement, the Master Agreement and the Shareholders Agreement – had all been signed by PTT and the Malaysian TNC Petronas on 30 October 1999. As with the Yadana Project this indicated that the main decisions on the project had been made prior to the results of the EIA and public hearings. Compounding this impression was the manipulation of public events to avoid dissenting voices.

---

24 Varaporn Chamsanit, interview with author, Canberra, Australia, 3 February 2006.
25 Ibid.

Prior to the hearing, military officers were employed as public relations officers and they intimidated and harassed project opponents (Penchom Tang and Pipob Udomittipong 2003). The public hearing was held at the Municipality Hall in Hat Yai on 29 July 2000 and academics and university students tried to broaden the discussion to consider the industrialisation program surrounding the proposed Indonesia–Malaysia–Thailand Growth Triangle but the chair of the hearing, General Charan Kulavanija, closed down the discussion. Many of the villagers were also excluded from the hearing and the frustrations resulted in clashes between the opponents of the project and its industry and military supporters and the hearing collapsed (Warasak Phuangcharoen 2005: 27–28).

The Ministry of Industry arranged a second public hearing at Jiranakorn Stadium in Hat Yai on 21 October 2000 but, again, access was restricted with barbed wire barricades preventing opponents from entering the venue, although the chief of police argue that these had been erected to protect the stadium's plants and trees (Supara Janchitfah 2004: 51). The limitations of access led to further clashes between project supporters and opponents and the suspension of that meeting too. In January 2001 the Industry Minister appointed a special committee to review the results of the public hearing and later that month the committee, to the disbelief of activists, concluded that the public hearing results were 'satisfactory'. The new government of Thaksin Shinawatra had promised to approach the activists differently and it commissioned reports from the Senate Committee on Environment, the National Human Rights Commission and Chulalongkorn University while Thaksin visited the protesters at Lan Hoy Siab on 4 January 2002 to promise a fair hearing. By May 2002 the reports had been submitted, all recommending that a final decision on the project be postponed until numerous issues related to human rights, the environment and the future energy needs of the region were resolved. Nevertheless, on 11 May 2002 the government ignored this advice and announced that the pipeline was approved (Warasak Phuangcharoen 2005: 29–33). This result supported Forsyth and Walker's contention in relation to land use planning in Thailand that unless participatory processes 'provide a forum for the expression of genuinely alternative environmental visions, it serves as just another element of the statemaking process' (2008: 246).

For the rest of 2002 villagers in Chana District and environmental organisations, students and academics around the country continued to lobby the government and campaign against the project. On 20 December approximately 1,000 villagers accompanied by students and human rights activists travelled the 50 kilometres from their villages in Chana District to Hat Yai to protest against the TTM project and to hand a petition to the prime minister who was meeting his Malaysian counterpart (*Alexander's Gas and Oil Connections* 2003; Jilani 2004: 15). According to activists Thaksin's aide told them to wait in a specified area and, after they complied, hundreds of policemen surrounded them and attacked them with batons (Pibhop Dhongchai et al. 2005: 44).

The discontent over the TTM Pipeline was widespread and it triggered the largest environmental campaign in the country. Local villagers used their Islamic identity as a cultural symbol of resistance but also adopted the global symbol of human rights, both for the local media and in discussions with the UN Special Envoy for Human Rights, Hina Jilani. Muslim villagers were harassed by both uniformed and plainclothed Border Patrol Police, some carrying M-16 machine guns (Lohmann 2007: 15–19). Police in the neighbouring and predominantly Muslim southern provinces were also granted immunity from prosecution under the executive decree of 2005, inflaming tensions in already marginalised communities (ICG 2007: i).

An editorial in *The Nation* cited the government's crackdown on those opposing the TTM Pipeline as the clearest evidence that the government was crushing dissent, with a 'thick and dark cloud [settling] over the remnants of civil society' (*The Nation* 2004). Repeated human rights violations due to the heavy-handed police and military tactics at Hat Yai and other protests led Hina Jilani to describe the situation facing Thai civil society at this time as a 'climate of fear' (Chimprabha 2003; Human Rights Watch 2007a: 30; Lohmann 2007: 20; SEAPA 2005).

The police prosecuted twenty activists from the Hat Yai protest for encouraging the use of force and causing a public disturbance. In the Provincial Court case the police produced a manipulated video of the event but the defendants played their own video, which showed that the protesters, who were mostly Muslims, had just finished their evening prayers and were sitting down having dinner when the violence broke out (*The Nation* 2002).[26] Several hundred police attacked the protestors with batons, beating people brutally and overturning a truck and other vehicles. Both the National Human Rights Commission and the Senate separately investigated the incident and both implicated the interior minister and the police in the violence (Pasuk Phongpaichit and Baker 2004: 146–47; Sanitsuda Ekachai 2003; Supara Janchitfah 2004: 117).

On 30 December 2004 the court acquitted all twenty defendants (Somroutai Sapsomboon 2004). The judge pronounced that 'getting together to express their collective opinions was lawful' under section 44 of the 1997 constitution (*Provincial Court of Songkhla, Thailand* 2004). This decision was upheld on appeal in 2007 (*The Nation* 2007). The decision, given the video evidence, should never have been in doubt but given Thaksin's influence over the media and parts of the judiciary at this time it was still a significant decision, indicating that the judiciary retained an independence that was completely lacking in Myanmar. On 14 February 2007 Judge Sudawan Riksathien of the Provincial Court noted that the constitution guaranteed that local people be informed of developments in their area and be allowed to give their views and participate in planning. She ruled that Chana District residents who were dissatisfied with the public hearings had the right to demonstrate their opposition to the project and to demand a government review (Lohmann 2007: 148).

Despite the violent clashes that resulted from the authoritarian tactics, the general strategy and tactics employed by the demonstrators was one of

---

26   Pipob Udomittipong, interview with author, Chiang Mai, Thailand, 3 January 2005.

nonviolence. This approach was influenced both by Buddhist activists from across Thailand but also by the Islamic background of the Chana villagers.[27] As Chaiwat Satha-Anand (2006b: 202) demonstrates, Muslim nonviolent action in the face of injustice is common in contemporary societies, despite the demonisation of Muslims following 9/11 in 2001. In Taling Chan in early 2003, to emphasise their nonviolent approach, a bicycle convoy of 300 mostly Muslim local children distributed leaflets and flowers to the 400 police deployed to guard the site of the GSPs (Anchalee Kongrut 2003).

A central argument of the local Islamic community against the GSPs in particular was that some of the land to be used was communal land that had been donated to the community as *wakaf* (or *waqf*) land in an Islamic religious ceremony, which, according to local interpretation, was both non-saleable and also non-exchangeable (Lohmann 2008: 91). The government argued that Thai law did not recognise the donation because it was a Muslim ceremony, further enhancing their feelings of marginalisation.[28] In the early stages of the pipeline dispute, the proposed GSPs did not garner much attention until the area that villagers had been using for generations was closed off to community access. By April 2005 it had become a major issue. Pipeline opponents in Pa Ngam village in Taling Chan employed the politics of identity with a sign that read: 'The gas separation plant company has grabbed Muslim *wakaf* land. Our *wayip* duty as Muslims means we must all take responsibility' (Oilwatch SEA 2005b). In July up to 300 protesters rallied at the Chana District government land office to present a petition signed by 1,563 people opposing permission being given to TTM to use the public land (Oilwatch SEA 2005a).

The reasons behind the campaign against the TTM project over these years were many and varied. Thai activists from outside of Songkhla Province saw the project as the start of an industrialisation drive in the south following in the wake of the eastern seaboard, which would cause widespread environmental security issues and the further marginalisation of local Muslim communities. These fishing villagers viewed the project as a threat to their livelihoods and way of life, but through their own experience and contact with outside activists their perspective evolved to also see it as an attack on their Muslim identity.

Varaporn Chamsanit argued that engaging in environmental activism was itself a transformative process for these these villagers.[29] Although there were differences, in both the Yadana and TTM projects, as with most large-scale projects, remote and largely rural communities were particularly affected. They were affected both detrimentally, due to their proximity to the project, but also positively due to the consciousness raising activities of environmental activism. They become politicised and engaged in political processes while learning to organise and give voice to their interests. The protests in Hat Yai were well attended by NGOs,

---

27   Varaporn Chamsanit, interview with author, Canberra, Australia, 3 February 2006.
28   Ida Aroonwong, interview with author, Bangkok, Thailand, 25 November 2008.
29   Varaporn Chamsanit, interview with author, Canberra, Australia, 3 February 2006.

students and academics – empowered predominantly Buddhist groups in Thailand – but the protests in the villages were dominated by local Muslim villagers. As in the North, the consequences of this engagement led local activists to a more critical view of state authority (Doherty 2002: 199–202). As Somchai Phatharathananunth notes in the Thai context in particular, this activism generated 'counter-hegemonic ideologies that are crucial for democratic development' (2006: 17).

The cooperation between Buddhists and Muslims to promote environmental security during the campaign echoed activism undertaken in the Philippines, where Christian and Islamic groups on the southern island of Mindanao collaborated to oppose exploitative mining practices. Doyle and McEachern conclude that, in this situation,

> [the] religious adherents … find that, at their foundations, the religions have more in common with each other when engaged in protecting both people and the Earth. In this way religions are often deinstitutionalised, removing the staid power structures that usually govern them, and returning to their emancipatory roots. (2008: 81)

This was the same argument that many Engaged Buddhists used in their opposition to destructive developments.

As with the Yadana campaign, women activists featured strongly in the public dissent exhibited throughout the TTM campaign. Ida Aroonwong, who spent two years with the Chana villagers, argued that this was not unexpected as 'projects in Thailand [like the TTM] affect households so women naturally get involved'.[30] Alisa Manla was a local Muslim woman from the fishing village of Ban Nai Rai who took up the opportunity with other villagers to visit the eastern seaboard in 2000 and became an active community voice against the TTM project, despite death threats. She invoked the symbols of her familial responsibilities in the campaign by voicing the threats to her house, the health of her children, and their future (Vasana Chinvarakorn 2002). Her transition from leader of a village housewives' group and daycare centre to community spokesperson was initially resisted by her village, but eventually accepted. As Razavi (2006) demonstrates, despite the conservative reputation of Islam in much of the North, modernist and reformist currents in political Islam can be very open to gender equality.

Other women included Penchom Tang, director of the Campaign for Alternative Industry Network, who helped organise an eastern seaboard fieldtrip and in 2006–7 conducted a study on women leaders in Thailand's environmental movement.[31] These activities demonstrated once again the centrality of women to activism in the region particularly those who, like Phinan in the Yadana campaign, used women's issues in strategic ways. While women's movements in some parts of the world are waning due in part to the 'professionalisation' of feminist movements (Cornwall

---

30   Ida Aroonwong, interview with author, Bangkok, Thailand, 25 November 2008.
31   Penchom Tang, email to author, Bangkok, Thailand, 9 February 2008.

and Molyneux 2006: 1184), the activism in these campaigns suggested that women still play a key role in the dynamism of the environmental movements in the South. While activism against the TTM demonstrated the strength and flexibility of local Thai environmental movements the limited local activism against the Salween Dam Projects illustrated the repressive restrictions on forming similar movements in Myanmar.

Salween Dam Projects

The Salween River passes through Shan, Kayah (Karenni) State and Kayin (Karen) States in Myanmar before emptying into the Gulf of Mottama (Martaban) at Mawlamyine (Moulmein) in Mon State. The series of proposed dams on the Salween River therefore precipitated a multi-ethnic dimension to the campaigns. Environmental groups across the region had been aware of the proposed dams since the 1990s[32] but it was not until 2003 that the campaigns gained significant media coverage outside of Myanmar. In December 2003 the *Bangkok Post*, whose environmental writers were networked with Salween activists such as Pipob Udomittipong,[33] ran a front page story with the prominent headline 'China plans 13 dams on Salween' (Kultida Samabuddhi and Yuthana Praiwan 2003). Although concentrating on the upper Salween in China (the Nujiang or Nu River), it also drew widespread attention to the impending dam projects in Myanmar and became the first of many articles on the topic. There was no coverage of the issue in Myanmar itself, however, and even in Thailand the coverage was restricted to the English-language press as, according to the editor of the Salween News Network, 'the analysis on Myanmar in the Thai language press is often extremely limited'.[34]

Unlike the projects in Thailand activists faced problems of access to communities in the Salween region due to civil conflict and repressive governance under military rule. Under Thein Sein's new quasi-civilian government activists hoped that reforms and increased responsiveness, as demonstrated by his decision to suspend the Myitsone Dam on the Ayeyarwady (Irrawaddy) River in September 2011 (Simpson 2013b), would eventually result in a positive response to the campaigning against the Salween Dams, but for most of the period examined here local Salween activists faced a brutal military regime in an insecure civil conflict zone. Transnationally, the issues relating to the Salween Dams, and indeed dams throughout Myanmar, resulted in a sprawling network of exiled groups and umbrella organisations, predominantly in Thailand, that produced a plethora of websites and detailed reports. After the Yadana experience, these environmental groups were aware of the need to create networks early on between local communities

---

32   T. Giannini, interview with author, Bangkok, Thailand, 22 January 2000.

33   Pipob Udomittipong, interview with author, Chiang Mai, Thailand, 14 January 2004.

34   Wandee Suntivutimetee, interview with author, Chiang Mai, Thailand, 11 January 2005.

and activists, despite the difficulties in access to the area. Technologies such as the internet, mobile phones and desktop publishing were much more accessible to these exiled groups in the years after the Yadana and TTM campaigns and resulted in greater coordination and publicity for a wider international audience. In contrast to this extensive transnational activity, however, the level of local activism under the difficult political conditions of authoritarian rule was extremely low.

Many of the areas along the Salween River on the Myanmar side of the border were nominally in the control of the Myanmar military but its authority did not go unchallenged as these were areas of longstanding civil conflict. The KNU signed a tentative ceasefire with the new government in early 2012 but they had been at war with the Tatmadaw in Kayin State since 1948. In 2007 Aung Ngyeh, an exiled Kayah activist, set out the manifold problems for activists and local communities in this region:

> The Salween River flows through Shan State, Karenni State, Karen State and Mon state. Thailand, China and the Burmese military regime set up plans to build dams across the Salween River ... but the people inside Myanmar, they didn't know about the plans. Therefore, the first thing that we have to do inside Myanmar is to raise awareness of dams construction to our Burmese people ... To organise public action inside Myanmar, it's very hard as Burmese people have subjected to living under oppressive regime. Most dam construction plans are located in ethnic lands ... where long running civil war is found. So, the villagers who are staying in those dam construction sites have suffered from various kinds of human rights abuse for long time and they have to struggle for their [survival]. The people who will be affected from the dams, specially from Karen, Karenni and Shan states have to live in their own lands as internally displace people [IDPs]. Their lives are full of risks and ... their lives can be destroyed [at any time] so they have to hide in deep jungle for their safety. So when, we have tried to deliver the messages of dam construction plans to them, we have also faced difficulties in reaching them ... Because of these difficulties, we can't organise public activities inside Burma yet except for raising awareness on dam construction.[35]

The first large-scale hydropower project in Myanmar was the Mobye Dam and Lawpita Hydropower Project on a tributary of the Salween River in Kayah (Karenni) State in the 1950s which resulted in large-scale relocations when 12,000 villagers were forcibly moved (Burma Rivers Network 2012).[36] As a result of civil conflict, forced relocations around the Salween River increased from 1996, when 212 villages in an area thought to be sympathetic to the Karenni National

---

35 Aung Ngyeh (pseudonym), email to author, Mae Hong Song, Thailand, 13 December 2007.

36 Nge Reh, interview with author, Ban Nai Soi, Mae Hong Son, Thailand, 8 December 2010.

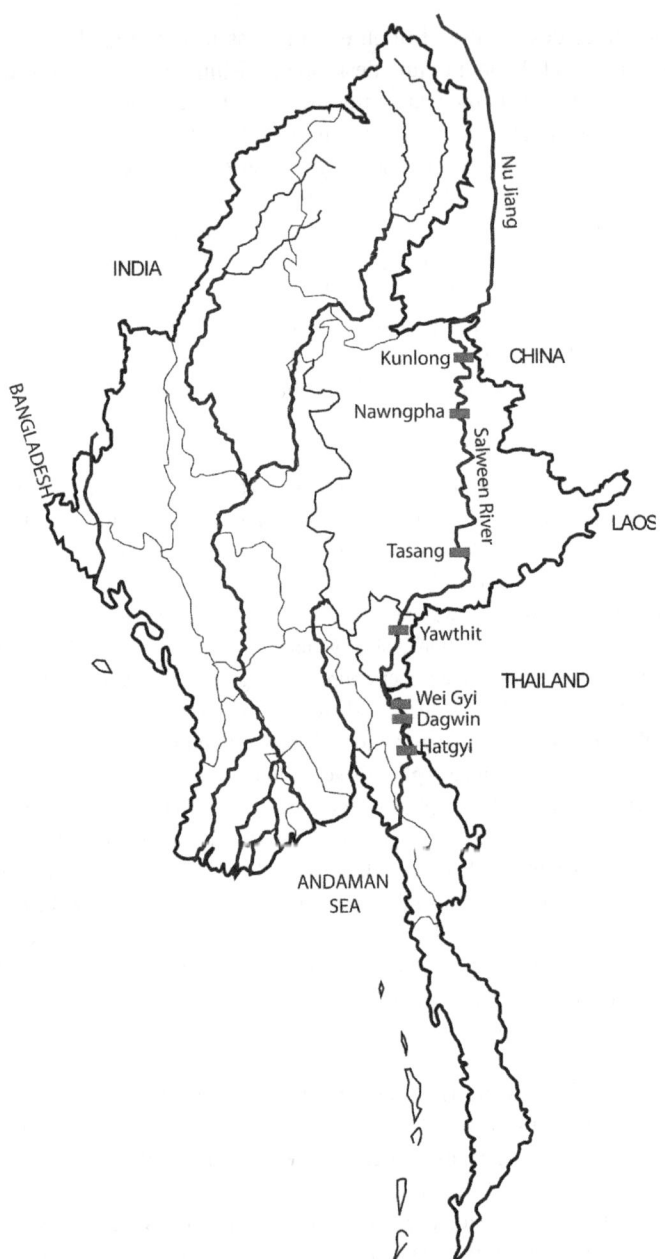

Figure 4.2   Proposed dam sites on the Salween River
Source: Burma Rivers Network

Progressive Party (KNPP) were relocated as the area became progressively militarised (KDRG 2006: 15). Further relocations occurred to clear the area for the Wei Gyi and Ywathit Dams.[37] According to the Shan Sapawa Environmental Organization (Sapawa), this also occurred upstream in Shan State, in a pattern common along the Salween where, as part of a wider anti-insurgency campaign, 60,000 villagers from areas adjoining the Tasang Dam site and flood zone were also relocated in 1996 (Sapawa 2006: 20–24). Sapawa, based in Chiang Mai, was the first Shan environmental organisation formed and they were particularly active over the issues associated with the Tasang Dam Project. Although Sapawa conducted research in Myanmar, their activities were restricted and they were unable to organise significant public activities in the Salween region (Sapawa 2006: 23–24).[38] In contrast, according to Sapawa, the Myanmar military forced over 400 villagers, many of whom had worked on projects as forced labour, to attend the official launch of the Tasang Dam in March 2007 with Thai construction company MDX and high-ranking Burmese military officials (Sapawa 2007).

Downriver from both Shan and Kayah States the Salween River forms part of the border with Thailand in Kayin State, the sites for both the Wei Gyi and Dar Gwin Dams that surround the Ei Tu Hta IDP Camp, with the Hat Gyi Dam entirely within Kayin State further downriver (see Figure 4.2) (Burma Rivers Network 2012). Displacement was also rampant throughout Kayin State. The Kayin founder and director of the Karen Environmental and Social Action Network (KESAN), who grew up near the Wei Gyi Dam site, was repeatedly displaced downriver with his family each time there was an attack by the Tatmadaw until he finally left Myanmar permanently in 1995 after the fall of the Karen base of Manerplaw.[39]

Despite the difficulties, the proximity to Thailand made it possible to organise some local actions on the river. Karen Rivers Watch (KRW), an exiled coalition of Kayin organisations formed in June 2003 including KESAN, the Karen Office of Relief and Development (KORD) and the Karen Women's Organization (KWO), collaborated with villagers from 2005 to organise protests along the river near the dam sites every year (Cho 2008; Saw Karen 2007). The events had transnational elements, with assistance coming from these exiled groups, but there also existed a large local component, with local activists within the villages raising awareness about the projects and the campaign.[40] As one activist from KRW noted, 'we ask local villagers to share their feelings and knowledge; we mobilise the community from the Karen side'.[41] Although not participating in the protests themselves, a growing presence in local environmental activism in Myanmar was the local

---

37   Eddie Mee Reh, interview with author, Ban Nai Soi, Mae Hong Son, Thailand, 7 December 2010.

38   Sai Sai, interview with author, Chiang Mai, Thailand, 9 January 2009.

39   Paul Sein Twa, interview with author, Thai–Myanmar border (location withheld), 6 April 2009.

40   Sai Sai, interview with author, Chiang Mai, Thailand, 9 January 2009.

41   Nay Tha Blay, interview with author, Mae Sariang, Thailand, 7 January 2009.

NGO ECODEV. Based primarily in Yangon the organisation collaborated with ethnic groups, such as KESAN, on dam issues although, as a local group inside Myanmar, it adopted a less openly confrontational approach than the exiled groups and focused on community forestry and development.[42]

Although the protests occurred inside the official borders of Myanmar, the locations where they took place were politically fuzzy to the extent that they were not under the complete control of the Myanmar government. These areas were, like the Ei Tu Hta IDP Camp, largely controlled by the KNU. The ethnic Kayin considered them to be 'liberated areas', and contested the Myanmar state's claim of sovereignty.[43] Absolute control over these areas was fluid, but security considerations were paramount and activists were reticent to discuss the locations of the protests. Chana Maung of ERI suggested that the protests happened at 'safe areas for … activists and villagers',[44] while a KESAN activist later disclosed the location on the understanding that it was not divulged.[45] Control over particular areas changed over time, according to the Tatmadaw's military operations, making it difficult to arrange locations far in advance.

The protests 'inside' Myanmar were not, therefore, undertaken to appeal directly to the Myanmar military regime. Indeed to draw its attention may have resulted in violent suppression. As Sai Sai noted, 'the Salween movement action does not get a lot of attention inside because it's impossible to protest under the ruling military regime'.[46] The main purpose of the protests was to raise awareness of the implications of the dams amongst the local villages and promote empowerment through knowledge, although they contributed to the international campaign as well with photos and reports of the protests distributed throughout campaign networks.

Due to the tenuous control by the Myanmar military in many of these regions the dams would perform a valuable function in exerting pressure on insurgent groups as well as providing much sought after foreign exchange. If the dams were built, the reservoirs behind the dams would flood large areas that provide either shelter or transit zones for insurgent groups. Around the Tasang Dam the Shan State Army South engaged in sporadic battles with the Tatmadaw, while the Dar Gwin, Wei Gyi and Hat Gyi sites provided security for the KNU and were also the busiest routes for Kayin refugees fleeing Myanmar into Thailand. The Wei Gyi Dam would also flood most of Kayah State's two river valleys that lie upstream where the KNPP is active (Kusnetz 2008). One experienced Northern activist argued that as a result of these projects ethnic IDPs in these areas were effectively

---

42   Win Myo Thu, interview with author, Yangon, Myanmar, 10 May 2011.

43   Myint Thein, interview with author, Mae Sot, Thailand, 19 January 2004.

44   Chana Maung, email to author, Chiang Mai, Thailand, 25 March 2008.

45   M. Bergoffen, interview with author, Chiang Mai, Thailand, 8 January 2009.

46   Sai Sai, interview with author, Chiang Mai, Thailand, 9 January 2009.

'held hostage' by the Myanmar military in the negotiations between themselves, ethnic insurgents and the Thai state.[47]

The Hat Gyi Dam was slated as the first Salween Dam to be constructed in Kayin State, being the only one out of the lower three dams to be entirely within Myanmar's borders and therefore likely to receive less external scrutiny (Noam 2008; Pianporn Deetes 2007; Tunya Sukpanich 2007). While security was generally more difficult for activists deeper in Myanmar, there were also Tatmadaw garrisons on the Myanmar side of the river near the Dar Gwin and Wei Gyi Dam sites (KRW 2004), and in February 2008 ten Tatmadaw soldiers were injured after being shot by Kayin insurgents while crossing the river at the border near the Wei Gyi Dam site (Saw Yan Naing 2008a).[48] Less than a month later the villagers gathered nearby for their protest.

Kayin villagers in Myanmar experienced a sense of powerlessness. This marginalization was also felt by their ethnic brethren on the Thai side of the border although their activism was less likely to be life threatening: 'if the government wants to build [the dams] it will as we are only poor people but, still, we and other villages send ... representatives to meetings in Mae Hong Son, Chiang Mai and Bangkok'.[49]

On the Myanmar side of the border the protests at secure sites were the only outlet for Myanmar villagers to voice their concerns over the dams. As with the Myanmar side of the Yadana Pipeline there was no formal public participation in any of the Salween Dam projects. Under the Thaksin government in September 2006, a senior official from EGAT announced that it would not be undertaking EIAs for the projects at all (Markar 2006; Piyaporn Wongruang 2006). Three months later, after Thaksin had been ousted by the coup and under intense public pressure, EGAT commissioned Chulalongkorn University's Environmental Research Institute to conduct an EIA for the Hat Gyi Dam. It was not a transparent procedure, however, as public participation was not part of the process and the report remained confidential with only EGAT having disclosure rights (Pianporn Deetes 2007; Tunya Sukpanich 2007).

Security in the Hat Gyi area for EGAT workers was still tenuous; in May 2006 an EGAT geologist lost his leg to a mine while surveying the area and, according to reports, died later from his wounds (KRW 2006; KRW and SEARIN 2006; Kultida Samabuddhi 2006; Tunya Sukpanich 2007). In response to this and other ongoing concerns, Thailand's Human Rights Commission recommended that the Hat Gyi Dam be abandoned (*Watershed* 2007). Nevertheless, in July 2007 30 EGAT engineers and other workers attempted work on a three-month feasibility study.

---

47   S. Green (pseudonym), interview with author, Thai–Burmese border (location withheld), 6 April 2009.

48   Pipob Udomittipong, interview with author, ERI office, Chiang Mai, Thailand, 14 January 2004.

49   Sanchai, interview with author, Ban Ta Tar Fung, Salween River, Thailand, 6 January 2009.

According to one activist the immediate area surrounding the dam was a 'brown area',[50] largely controlled by the Democratic Karen Buddhist Army (DKBA), which at that time was allied to the Tatmadaw.[51] The KNU controlled most of the area to the east towards the border, however, and, as a result of opposition amongst villagers, initially refused the EGAT team passage to the reservoir site.

After talks with EGAT in Mae Sot, the KNU relented and gave the team permission to conduct a two-day survey (Saw Yan Naing 2007). In September 2007 another EGAT employee surveying the Hat Gyi site died from an artillery shell and the remaining 42 EGAT staff were subsequently evacuated to Thailand (AP 2007b). No one took responsibility for the attack, but it highlighted that these were highly insecure sites for major projects. As Naing Htoo from ERI pointed out, 'even if [the Tatmadaw] crush the KNU [in the Wei Gyi and Hat Gyi area] they will melt into the forest and continue their fight'.[52] Likewise Ka Hsaw Wa noted that it would always be 'easy for the insurgents to take a [hand-held RPG] into the dam area'.[53] It also highlighted that in these dam projects, where official channels of public participation were effectively closed, local villagers and activists often relied on militant insurgent groups to influence the outcome of the project.

While security considerations for ethnic Kayin communities in the region were part of their precarious daily existence, these experiences are largely foreign to Northern activists who engage in activism in their own countries (Rootes 2008). While intelligence agencies in the North have been known to target environmental groups, activists have civil liberty protections unheard of for Kayin communities in Myanmar. As a result the focus of the protests in Myanmar was on survival for the communities. The content of these protests also emphasised local culture and ritual. Multi-religious prayers and ceremonies for the protection of the river by villagers and activists were aimed at fostering local solidarity among the villagers and publicising their plight (KRW 2007a). Most Kayin in this region adhered to a synthesis of animism and Christianity and undertook rituals emphasising the connection between humans and their environment, such as an 'animist ceremony [in the Wei Gyi Dam area] calling on the local spirit to protect their lands and water' (KRW 2004: 60).[54] Katie Redford of ERI related a story told by a Kayin woman to fellow Kayin Ka Hsaw Wa:

---

50   The Tatmadaw divides Myanmar into 'white' areas, which are under complete Tatmadaw control, 'brown' areas, which are essentially under Tatmadaw control but where resistance forces can occasionally penetrate, and 'black' areas, where there is regular armed resistance activity.

51   Alex Shwe, interview with author, Chiang Mai, Thailand, 8 January 2009.

52   Naing Htoo, interview with author, Chiang Mai, Thailand, 9 January 2009.

53   Ka Hsaw Wa, interview with author, Chiang Mai, Thailand, 9 January 2009.

54   A 1983 census indicated that 84 per cent of Kayin State was Buddhist with less than 10 per cent Christian and 0.2 per cent animist but this breakdown may reflect biases in reporting as a result of the severe marginalisation experienced by non-Buddhists, and particularly animists, in Myanmar (Tin Maung Maung Than 2005a: 69).

Her child had been ill ever since he was forcibly removed from their village. The woman argued 'He is away from his god'. When babies are born the placenta is buried under a tree. That tree spirit [god] looks over you but if you leave, the spirit can no longer look after you.[55]

Through both deforestation and the internal displacement of people, severed local connections can cause illnesses in animist communities (Hares 2006: 108; McGready 2003). Religion therefore played a significant role in Kayin areas but, as the Buddhist Sai Sai noted, Salween activists came from many ethnicities and religions and all were focused on saving the Salween and its people through democratic processes.[56] Aung Ngyeh of KDRG also contended that, despite the diverse ethnicities and the authoritarian policies of the military, when 'his people' fully understood the dams and their consequences 'the public action [will] come'.[57]

Shwe Gas Pipeline Project

For the same reasons as the limited local activism against the Salween Dams, for much of the campaign against the Shwe Gas Pipeline Project there were no significant local protests in Rakhine (Arakan) State or the rest of Myanmar although the proximity to India and Bangladesh in the west allowed transnational linkages with activists across the border.[58] Its distance from the activist locus of Thailand initially left local activists more isolated although linkages grew throughout the campaign. In 2004, in the campaign's early stages, the then president of the exiled All Arakan Students and Youth Congress (AASYC), who lived in Bangladesh, indicated that the organisation had learnt from the Yadana Project where the promised benefits for locals failed to materialise, but he suggested that at that stage knowledge regarding the project in Rakhine State was even lower than it was at a comparable stage in the areas surrounding the Yadana Pipeline due to restrictions on their activity.[59]

The Shwe Gas Project was based primarily around Kyauk Phyu (Kyaukphu or Kyauk Pru) on Ramree Island, the second largest town in Rakhine State[60] near the offshore Shwe gas deposit (A1 and A3 Blocks), where the gas terminal was to be built with a gas pipeline running to Yunnan Province (see Figure 4.3). Nearby Maday Island was the site for an oil terminal where predominantly Middle Eastern oil was to be offloaded and transported in a parallel pipeline to avoid the Straits of Malacca. In addition to the oil and gas projects a Special Economic

---

55   K. Redford, interview with author, Chiang Mai, Thailand, 15 January 2004.

56   Sai Sai, interview with author, Chiang Mai, Thailand, 9 January 2009.

57   Aung Ngyeh, email to author, Mae Hong Song, Thailand, 13 December 2007.

58   Soe Myint, interview with author, Delhi, India, 24 December 2004.

59   Kyaw Han, interview with author, Delhi, India, 24 December 2004.

60   The last census in 1983 estimated the population at 20,000.

Zone and deep port for container shipping were also planned as well as a railway line to Kunming.[61] These developments around Kyauk Phyu would likely bring heavy industry to a predominantly rural region currently surrounded by sensitive mangrove forests.

The projects increased militarisation in the area as well as establishing a large prostitution industry to cater for the largely Chinese workers brought in for the project.[62] Aye Tha Aung, a prominent Buddhist Rakhine politician from the NLD-aligned Arakan League for Democracy (ALD),[63] argued that the Shwe Project exacerbated existing environmental and human rights pressures in Rakhine State, where extensive land confiscation and forced labour were linked to infrastructure projects and military shrimp farms.[64] An exiled Rakhine activist recounted how two people from each house in his home town were taken as forced labour to build three helicopter pads for the military and later he himself was arrested and tortured.[65] The link between the militarisation of these projects and human rights abuses was aggravated by the rapid expansion of the Myanmar military after 1988 in its attempts to spread its influence to the more remote border regions of the country (Selth 2001). While local protests that emerged under the new government demonstrated the depth of community antipathy towards the project Tatmadaw soldiers at Kyauk Phyu during military rule expressed support for the Chinese investment, although they were unlikely to express any personal opposition to a foreigner.[66] The project and its associated militarisation threatened to increase environmental insecurity in the region but offered few improvements to local energy access.

As with the Yadana and Salween Dam projects there was no public participation in the development process and, although an EIA was undertaken, the absence of legislated requirements limited the efficacy of the process. As CNPC, the pipeline operator, was Chinese there were also fewer opportunities to bring pressure to bear than there were on the Yadana Pipeline, when Thai and Western TNCs were involved. Despite the Chinese state demonstrating an increased recognition of the need for sustainability in projects within its own borders, there was little indication of similar concerns for projects in Myanmar. As a result, with two largely authoritarian regimes playing dominant roles in the project, for much of the campaign there was little visible domestic public activism.

---

61   Winston Set Aung, interview with author, Nay Pyi Taw, Myanmar, 29 April 2013.

62   Phyo Phyo (pseudonym), interview with author, Yangon, Myanmar, 7 January 2011.

63   In mid-2013 the ALD and the Rakhine Nationalities Development Party (RNDP) announced that they would merge and that they had applied to be registered as the Rakhine National Party (RNP).

64   Aye Tha Aung, interview with author (translated by Zaw Myat Lin), Yangon, Myanmar, 27 December 2010.

65   Wong Aung, interview with author, Chiang Mai, Thailand, 6 April 2009.

66   Tatmadaw soldier, interview with author, Kyauk Phyu, Rakhine State, Myanmar, 30 December 2010.

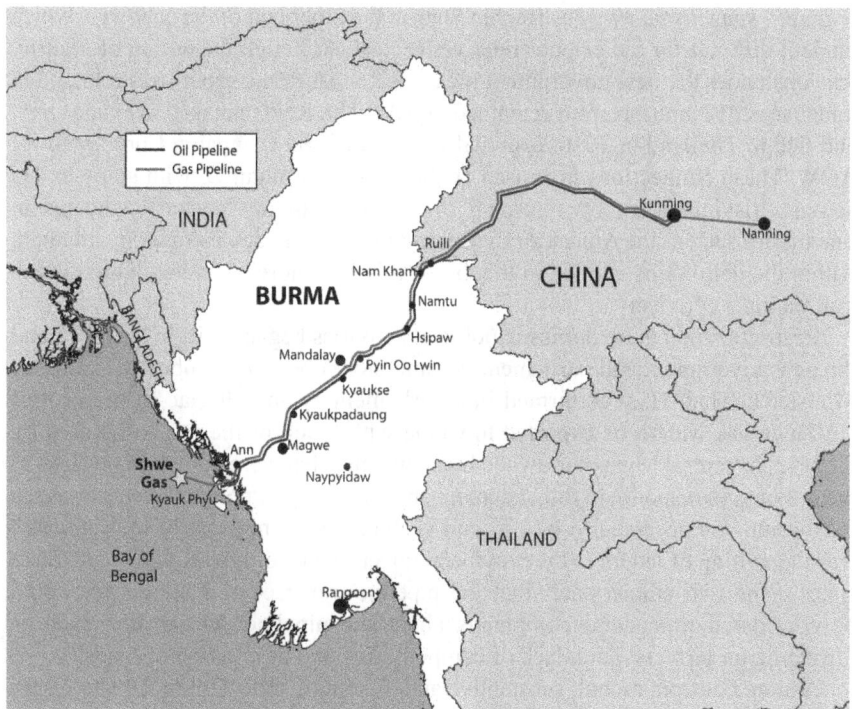

Figure 4.3   The Shwe Gas Pipeline from Kyauk Phyu to Nanning, Yunnan Province, China

Source: Shwe Gas Movement

As under most authoritarian regimes, however, political activism in Myanmar took unconventional forms so, while there were no domestic NGOs established to challenge the Shwe Project directly, concerned youth channelled their energy through alternative organisational structures. In 2004 Thu Rein, a student at Sittwe University in the state capital, set up the association of University Stipends and Social Affairs Rakhine, a student body nominally established to represent student interests. As the details of the Shwe Project gradually emerged the Association became a focal point for related debate and discussion. As well as holding covert gatherings on the topic they contacted exiled organisations, such as Arakan Oil Watch (AOW) and Shwe Gas Movement (SGM) in Thailand, to gather more information. Their activities were eventually discovered by military intelligence and in 2009 eleven association student members were arrested, receiving sentences of up to six years in prison with three years each for two illegal activities: 'contacting unlawful [e.g. exiled] associations' and 'illegal border crossings', into Thailand or India.[67] These young students were taken to prisons in other states and

---

67   Thu Rein, interview with author, Chiang Mai, Thailand, 11 February 2012.

regions – some as far away as Kachin State at the other end of the country – which made it difficult for old or poor relatives to visit. Although the easing of security tensions under the new government meant that most of the group was released in a January 2012 amnesty, four remained in jail.[68] Thu Rein managed to evade arrest and fled to Thailand in 2010, eventually becoming the Campaign Coordinator of AOW. These connections increased as the campaign matured, with some exiled activists risking arrest by returning to provide training for local students in sanctuaries such as the American Centre and the British Club in Yangon, although, within the definitions applied in this book, this was more accurately classified as transnational activism.[69]

Restrictions on these domestic political activities began to lift in late 2011 and the new government and parliament became more open. In October 2011 Energy Minister U Than Htay confirmed in a parliamentary speech that 80 per cent of the Shwe gas was to be exported to China with much of the rest to be used by factories in central Myanmar owned by companies that were known to be close to the military. In response to this statement and the easing of political restrictions the local campaign became more overt and visible. One component of the campaign was the writing of letters to the provincial governments of both Rakhine and Shan states – the two ethnic states that the pipeline traversed – making what could be considered procedural complaints about the pipeline construction such as infringing on farmers' land, lack of compensation and destruction of roads.[70]

A more confrontational, substantive and dissenting argument in the campaign, which was informed by concepts of natural resource rights, was the demand that the gas be used to provide 24-hour electricity to Rakhine State before any gas was exported. Rakhine State was one of Myanmar's poorest, with the lowest per capita electricity usage in the country according to the state-run media (see Table 3.1). The emerging permissive political atmosphere convinced local activists that it was possible to make this sort of demand openly for the first time. Across the state t-shirts, posters, stickers and calendars appeared bearing the message '24 Hours of Electricity Now. We Have the Right to Use Our Gas' (*Mizzima* 2012). Despite the easing of security concerns restrictions remained, with some activists forcibly required by police to remove these t-shirts when engaged in a *Rahta-Swe-Bwe*, or traditional tug-of-war, in Sittwe in early February 2012. Other 'protests' in Sittwe and other towns were muted, with no slogans shouted, but the visibility of the campaign had certainly increased. Even in the new political climate, however, protesting in Rakhine State, far from the international media, was much riskier than campaigning in Yangon. Indeed, although several exiled dissidents made heavily publicised visits to Yangon in early 2012 Rakhine exiles argued at that

---

68   Phyo Phyo (pseudonym), interview with author, Yangon, Myanmar, 18 February 2012.

69   Phyo Phyo (pseudonym), interview with author, Yangon, Myanmar, 7 January 2011.

70   Patrick (pseudonym), interview with author, Chiang Mai, Thailand, 13 February 2012.

time that while it may have been safe for 'prominent Burman (Bamar) activists to return it [was] still too early for ethnic activists'.[71]

The demand for 24-hour electricity resonated with the more formalised ethnic political parties in Rakhine State, the Rakhine Nationalities Development Party (RNDP), which participated in the 2010 elections and had parliamentarians in the national parliament, and the ALD, which boycotted the 2010 elections but agreed to register as a party in late 2011 along with the NLD.[72] Both parties adopted a less confrontational approach than activists, however, and saw potential developmental benefits from the pipeline projects, with the Secretary of the RNDP, Oo Hla Saw, noting that they were much less socially and environmentally damaging than the Myitsone Dam in Kachin State.[73] Indeed he 'welcomed' foreign direct investment in Rakhine State although preferring US/EU investment rather than Chinese investment due to 'greater transparency and respect for international norms'. While he argued they were unable to stop the project the RNDP argued in parliament that the 20 per cent of gas allocated to Myanmar should be used for development in Rakhine State, setting up a '24-hour Electricity Committee' to pursue this objective. The committee, although initiated by the RNDP, included other organisations, including those along the pipeline. It welcomed the project, however, and its less confrontational stance was demonstrated when it precluded a group from joining because the group opposed the project and had links to exiled groups.[74]

The ALD Secretary, Aye Tha Aung, voiced similar opinions to the RNDP from outside the parliamentary system, but argued that more public participation was necessary in the decision-making process and that Rakhine State 'received nothing' under the existing contracts. All technical labour was brought in from China and many of the manual labourers were brought in from central Myanmar.[75] Local campaigners were now very openly arguing for overt politicisation of the gas in favour of Rakhine State, with some activists arguing for it all to be kept in the state rather than being sold to China in international energy markets.

For all the growth of free market discourse, national energy security concerns are seen as too important to the smooth running of capitalist economies to be left simply to market forces. The gas and oil pipelines were central to China's search for energy security for both Yunnan Province and the rest of south-west China. Although the decision to suspend the Myitsone Dam illustrated the Myanmar government's increased responsiveness to activists' campaigns, it also

---

71   Jockai Khaing, interview with author, Chiang Mai, Thailand, 11 February 2012.

72   Soon after the ALD decided to participate in the elections the plainclothes security personnel who had been posted outside the secretary's house for over two decades were removed. Aye Tha Aung, interview with author (translated from Arakanese by Zaw Myat Lin), Yangon, Myanmar, 20 January 2012.

73   Oo Hla Saw, interview with author, Yangon, Myanmar, 20 January 2012.

74   Aung Aung Naing, interview with author, Yangon, Myanmar, 12 June 2013.

75   Aye Tha Aung, interview with author (translated from Arakanese by Zaw Myat Lin), Yangon, Myanmar, 20 January 2012.

paradoxically increased the importance of the pipelines to China's energy security and therefore greatly reduced the likelihood that the contracts for the pipelines would be amended. Indeed, a Northern activist who worked on the campaign inside Myanmar had argued nine months before the Myitsone Dam decision that, while 'the Myitsone and other dams might be negotiable, the pipelines are of such importance to China's national security that if they were threatened we are likely to see [Chinese] PLA troops in Myanmar to secure them'.[76] Thus, even with emerging local activism joining the transnational campaign the forces ranged against the activists were daunting.

Conclusion

Analysis of the four local campaigns against the energy projects suggests that the ability to provide domestic activist environmental governance for large-scale energy projects in the South is very much dependent on domestic political openings. In these cases the ability to undertake public participation, and in particular environmental protest, was determined by the local political regime with the opportunities for genuine and effective public dissent under authoritarianism extremely limited. Both public protest and participation in Myanmar were severely constrained while the public expression of dissent under a more competitive regime in Thailand was somewhat easier. Despite nascent political openings emerging in Myanmar following an end to direct military rule most of the local activism discussed in this chapter occurred in Thailand and was directed towards change within the Thai state and Thai TNCs.

Over the 1990s the avenues available for the airing of public grievances in Thailand increased, particularly under the 1997 constitution, although these outlets were often subverted. Public participation in the development process and the pursuit of an ecological rationality was regularly undermined by powerful political forces within business, government and the military, that took a tokenistic attitude to public participation and used it as 'legitimising cover for business as usual' (Dryzek 2005a: 82). Processes such as EIA in Thailand can therefore be seen as a strategy of accommodation by the state; one that defuses public opposition without surrendering the underlying philosophy of large-scale industrial development. For activists, participation in formalised state activities could be disempowering. Nevertheless, the experience of projects in Myanmar supported the proposition that, while EIAs with public consultation processes may be imperfect in practice, they were certainly better than nothing at all (Howes 2005: 108).

The campaign case studies largely followed the model of most other environmental campaigns in Asia (Kalland and Persoon 1997: 7) and were predominantly over environmental security concerns tied to specific conflicts

---

76   Founder of Myanmar-based environmental NGO (details withheld on request), interview with author, Yangon, Myanmar, 6 January 2011.

due to resource use or environmental degradation, but there were also links with broader social movements and philosophies. Some of these movements resonated with local cultural or religious influences and represented a form of identity politics. Buddhism heavily influenced many activists and was manifest in the activist philosophies and practices of Engaged Buddhism, including nonviolence. Buddhism was used both strategically as a cultural symbol and tactically during street protests, although often the state competed for a hegemonic Buddhist legitimacy. In the campaign against the Yadana Project this struggle for legitimacy also led activists to use the symbol of the monarchy, itself a competitor for Buddhist legitimacy, to validate their protests. While Buddhism was frequently employed, Muslim villagers in the south of Thailand fighting the TTM Pipeline increasingly framed their opposition through an Islamic identity, with the project depicted as undermining the community's Islamic integrity through the confiscation of *wakaf* land. In the campaigns against the Salween Dams the animist beliefs of the Kayin were highlighted. In these cases activists and communities, while strategically employing the symbols of their culture and religion, also saw the radical essence of their beliefs as a bulwark against the authoritarian tendencies of governments and complicit business interests. Despite the global reach and essential transnationalism of Islam and Buddhism, communities used these religions throughout the campaigns as local symbols of identity and differentiation.

The local activism also identified both social and environmental goals. Achieving justice for communities, primarily for ethnic minorities in Myanmar and Muslim fisherfolk in southern Thailand, was intimately linked to issues of ecological health. On the Thai side of the Yadana campaign, issues of forest ecology were prominent but, equally importantly, activists linked business dealings with the Myanmar military to the suppression of human rights in Myanmar and the delay of democratic reform. Initially, relatively affluent local activists from Kanchanaburi focused particularly on preservation of the forest but through their activism their awareness of linkages to wider social and political issues tended to radicalise their approach. A similar but different transformative process also characterised the campaign against the TTM Pipeline. Local Muslim villagers were on the whole less affluent than the activists from Kanchanaburi and, as fisherfolk, they were concerned to preserve their way of life, but as the campaign progressed they began to view the project both as an attack on their religion and as part of a wider pattern of globalised capitalist industrialisation.

As well as deforestation and impacts on livelihoods one of the major concerns in the limited local activism against the Salween Dams in Myanmar was the impact on internally displaced peoples (IDPs), not just to clear the reservoir zone itself but through the military relocating villages as part of their ongoing civil conflict with insurgents. The Shwe Gas Pipeline campaign also had limited local activity under military rule but what did exist focused on the energy insecurity that afflicted local communities, which was unlikely to be ameliorated following completion of the project. The level of this activism increased as political openings emerged under the new government with more visible challenges to the project developing.

While in the 1980s the successful campaign against the Nam Choan Dam demonstrated that, at times, the Thai environment movement appeared powerful enough to challenge the construction of large dams in Thailand itself the Yadana and TTM campaigns demonstrated that it was still extremely difficult for public and community opposition to impose constraints on other large transnational energy projects. Even this influence may have been exaggerated as most of the high quality dam sites in Thailand had already been used and EGAT found it simple enough to start sourcing its energy from Thailand's more authoritarian neighbours (Simpson 2007: 540–41; Sovacool 2009: 472). The dynamic movement within Thailand achieved some notable goals, such as entrenching EIA processes, while there was little evidence of any impact on public policy in Myanmar during military rule. This analysis suggests, therefore, that political regimes and their degree of openness had significant impacts on the level and impact of local environmental activism. Local cultural and ethnic factors were shown to influence campaign tactics and philosophies at the local level but more precarious living conditions also induced a greater focus on social or postcolonial issues rather than post-materialist environmental concerns. The impacts of these various influences on the transnational campaigns are now explored in the following two chapters.

Chapter 5

Bridging North and South: EarthRights International

Introduction

This chapter is the first of two in which I deal explicitly with the transnational aspects of activist environmental governance but here I focus specifically on the transnational NGO EarthRights International (ERI). In the previous chapter I examined local informal groups, such as the Kanchanaburi Conservation Group (KCG), and in the next chapter I investigate other transnational actors, such as the Shwe Gas Movement (SGM) and coalitions against the Salween Dams, to complete the multilevel, multiscalar analysis. I chose ERI as the central case study of this book, however, because as a transnational social and environmental movement organisation (Caniglia 2002; Rucht 1999: 207), straddling North and South, it provided a compelling case study of an emancipatory governance group (EGG) engaging in activism against environmental insecurity in the South. Having been co-founded by a Kayin exile, it was a key player in the formation of Myanmar's activist diaspora and had a significant influence on the wider activist exile community.

The underlying philosophies of ERI can be strongly identified with the four core green pillars of participatory democracy, ecological sustainability, social justice and nonviolence. These values were manifest in both its organisational structure and its aims and activities. ERI offered a particularly valuable case study for this book because it operated primarily from Thailand but it was a contributor to all of the campaigns involving Myanmar and it engaged in covert activism inside Myanmar itself. It also provided an exemplar of an EGG forming due to personal networks between Northern and Southern activists. Its focus was on human rights and the environment but it could also be seen as part of a global justice movement. It focused on the 'nexus' of human rights and the environment so, unlike some organisations within the environmental governance state (EGS), 'pure conservation' for its own sake was not its concern.[1] Its focus on both human rights and environmental protection from its inception in the 1990s accorded with research that demonstrated a large increase in multi-issue transnational social movement organisations, when compared with single-issue organisations, between the 1970s and 2000 (Bandy and Smith 2005: 6).

---

1   T. Giannini, interview with author, Chiang Mai, Thailand, 21 January 2004; K. Redford, interview with author, Chiang Mai, Thailand, 9 January 2009.

ERI's activities derived from the philosophies that underpinned the four green pillars and were divided into three strategic areas – legal programs, training and campaigns – which were based on the integrated advocacy strategy of its founding Echoing Green grant proposal (Giannini and Redford 1994). In tandem with its fieldwork in the civil conflict zones of Myanmar's mountainous borderlands ERI also undertook precedent-setting litigation against TNCs in US courts. Although unable to completely halt the projects its relative success in other areas demonstrated how environment movements could influence the behaviour of TNCs and large business interests. ERI also addressed many of the cultural, philosophical and political dilemmas that face organisations which traverse North and South and it therefore warranted detailed analysis within discussion of the broader transnational campaigns.

As demonstrated through the example of Friends of the Earth International (FoEI) in Chapter 2, EGGs that cohabit North and South must be conscious of the potential for acute differences in foci between activists from the affluent and less affluent worlds. Unlike FoEI, ERI was not a federation of pre-existing groups but rather a single organisation founded by activists from the North and South with staff from both. This history also set it apart from most influential human rights NGOs, which, even if they have global operations, are usually based in the North (Wong 2012: 61). Its office in the US only opened several years after the founding of its office in Thailand and it ran two activist schools in Southeast Asia, while also having small operations in South America. This resulted in multi-ethnic, multilingual activities that sometimes involved dozens of languages.[2] Previous research on international solidarity work has found that centralised organisations with greater resources tend to reinforce power imbalances among organisational participants (Smith and Bandy 2005: 11), but from its inception the founders and staff of ERI were extremely conscious of the potential for North–South difficulties, including the perceptions of Northern 'imperialism' that are sometimes elicited by an exclusive focus on the environment (Doyle 2005; Dyer 2011). The sensitivities to North–South power relationships were reflected in both the organisation's structure and activities. ERI therefore had introspective elements to its organisation despite its formal structure. In contrast to FoEI, which was founded in the North and grew to include groups in the South, ERI operated initially in the South and was founded by both Northern and Southern activists on issues primarily affecting the South. This Southern focus permeated all aspects of the organisation with the management of the organisation's activities in the South eventually transferred to Southern activists. It was, therefore, well placed to negotiate any North–South tensions that arose.

ERI did not, however, adopt the characteristics of ultra-radical introspective groups such as Earth First! (Doherty 2002) and it would be considered conservative by such self-consciously anarchic groups. It consciously adopted the structure

---

2   K. Redford, interview with author, Chiang Mai, Thailand, 9 January 2009.

of an NGO, rather than an informal group, and was incorporated in the US.[3] Nevertheless, it engaged in a wide variety of activities in very diverse geographical, cultural and political settings. It is unlikely that an ultra-radical introspective group could have undertaken the scope of these activities. Doyle argues that non-introspective groups 'are often not conscious of their political form, their structure or their ideology' (2000: 34) but it is clear that ERI was conscious of all these elements. The founders decided to formalise the structure, which Doyle sees as an attribute of a non-introspective group, but ERI activists – both in their organisation and their activities – demonstrated that they were well aware of societal power imbalances and particularly potential North–South tensions.

These acknowledgements were indicative of ERI's broader campaigns for justice in the South in coalitions with local groups and networks. ERI was firmly entrenched, therefore, as an EGG within the green public sphere, rather than a compromise governance group (CGG) or a perpetuator of the status quo as part of the EGS. Doyle and Doherty argue that in their model emancipatory groups 'construct themselves ... often in rugged opposition to what they perceive to be a global neoliberal project [and] through grassroots networking, develop shared techniques, strategies and repertoires of action alongside more localised networks and groups' (Doyle and Doherty 2006: 883). As I will demonstrate, ERI fulfilled these criteria, but it also went further, adhering to the four green pillars in both organisation and activities and providing diverse contributions to activist environmental governance in its legal programs, training and campaigns through participation in various local, national and transnational fora. An analysis of ERI's organisational philosophy and practices and its progress towards its justice goals also provides some indication of links between the practice of equity and justice within an organisation and the achievement of justice in the wider world. These characteristics contributed to ERI's successes in improving environmental governance and security around transnational energy projects for marginalised communities in Myanmar.

Origins and Aims

ERI was founded in 1995 by a Kayin exile from Myanmar, Ka Hsaw Wa, and two American lawyers, Katharine (Katie) Redford and Tyler Gianinni. Ka Hsaw Wa defined their mission as follows:

> EarthRights International (ERI) combines the power of law and the power of people in defence of human rights and the environment. We focus our work at the intersection of human rights and the environment, which we define as earth rights. We specialise in fact-finding, legal actions against perpetrators of earth rights abuses, training for grassroots and community leaders and advocacy

---

3   K. Redford, interview with author, Chiang Mai, Thailand, 15 January 2004.

campaigns. Through these strategies, ERI seeks to end earth rights abuses, and to promote and protect earth rights.[4]

The abuse of these 'earth rights' leads to human and environmental insecurity for marginalised groups and in Myanmar this burden has fallen particularly on ethnic minorities. Ka Hsaw Wa had been a Kayin student involved with the protests in Yangon in 1988, where a fellow student died in his arms after being shot by the Myanmar military. He himself was arrested and tortured for three days following the demonstrations. His immediate reaction to the massacre was anger and a desire for revenge through armed rebellion: 'I wanted to shoot the military [men], kill them'.[5] After seeing the mutilated body of a woman, however, he developed a more nonviolent philosophy and adopted nonviolent methods to achieve justice for the Kayin and other ethnic groups of Myanmar. In 1992 he started working with a Canadian human rights group, gathering evidence of human rights abuses from interviews both inside Myanmar and in the border regions, adopting various names, including Ka Hsaw Wa, that allowed him to use his passport to slip in and out of Myanmar incognito. According to Redford he took 'enormous personal risks' over ten years to document the facts and bring them into the international media and courtrooms.[6] A common thread in the horrific stories he collected was the destruction of village environments by the Myanmar military, precipitating an emergent theory of the linkages between human rights and environmental protection.[7]

This background contrasted with that of Giannini and Redford, who met each other at the University of Virginia School of Law in the early 1990s, but they displayed a similar interest in the symbiotic nature of human rights and the environment. Redford came from a human rights background, having spent two years teaching English in a Kayin refugee camp near Mae Sot on the Thai–Myanmar border, where nearby shelling from the Myanmar military and the Tatmadaw-aligned Democratic Karen Buddhist Army (DKBA) was not uncommon, and which the DKBA twice burnt down.[8] During her law degree she returned to Thailand in 1993 on a fact-finding mission for Human Rights Watch Asia for which she travelled to jungle villages in the conflict zones of Myanmar to conduct interviews with victims of abuses perpetrated by the Tatmadaw. In

---

4   Ka Hsaw Wa, interview with author, Chiang Mai, Thailand, 14 January 2004.

5   Ka Hsaw Wa, interview with author, Chiang Mai, Thailand, 9 January 2009.

6   K. Redford, interview with author, Washington, DC, USA, 21 March 2011.

7   In 1999, as recognition for his efforts in highlighting the plight of ethnic minorities and their environments within Myanmar, Ka Hsaw Wa was awarded the Goldman Environmental Prize, the Reebok Human Rights Award and the Conde Nast Environmental Prize. Having won both environmental and human rights awards, demonstrating the linkages between the two areas, he was then awarded the Sting and Trudie Styler Award for Human Rights and the Environment in 2004. In 2009 he won the Ramon Magsaysay Award for Emergent Leadership.

8   K. Redford, interview with author, Chiang Mai, Thailand, 15 January 2004.

the grant submission to Echoing Green in 1994 for funds to establish ERI she described how these experiences motivated her to pursue justice for these people:

> After my experiences in Thailand ... I knew that I would focus my law school career on international human rights. Indeed, I think it is impossible to live among victims of systematic human rights abuses, as I did with Karen refugees, and not want to do everything possible for them. (Giannini and Redford 1994: 8)

These experiences sharpened Redford's awareness of the plight of ethnic minorities inside Myanmar, but her dedication to resolve the problems nonviolently derived directly from a commitment to human rights. In 1994 Redford wrote her final-year law school paper on suing oil companies using the *Alien Torts Claim Act* (ATCA), long before any court cases of this nature had been undertaken, and this litigation became a central goal in the formation of ERI in 1995.[9] In 2006 Redford's role in initiating this legal action was recognised by her election as an Ashoka Fellow for 'introducing a simple and powerful idea into the human rights movement: that corporations can be brought to court for their role in overseas abuse' (Ashoka 2006).

Giannini had inherited a strong commitment to the green pillar of nonviolence from his father but during his time in Thailand this philosophy was also influenced by Thai Buddhist activists in the International Network of Engaged Buddhists tradition.[10] Engaged Buddhist activists such as Pipob Udomittipong later worked for ERI and a 'cross-fertilisation' (della Porta and Mosca 2007) of philosophies and tactics took place between Northern and Southern activists within the organisation. While Giannini had worked with Haitian refugees he had also focused on environmental issues during his law studies. During an externship with the Sierra Club Legal Defense Fund (renamed Earthjustice in 1997) between September and November 1994 he examined the strategies that environmental NGOs used to influence the World Bank to change its environmental protection policies. Although NGOs such as Earthjustice focused particularly on environmental issues, they also promoted a strong participatory element for local groups and used disputes in the South such as the Narmada Dam as case studies linking environmental problems to a lack of participation, transparency and accountability (Giannini 1994). Giannini's paper from this externship formed the basis for ERI's integrated advocacy campaign (Giannini and Redford 1994).[11]

Redford met Ka Hsaw Wa in Thailand during her work for Human Rights Watch Asia in early 1993 when he acted as a Kayin–English translator. They became a couple and later started a family. When Redford and Giannini finished their law degrees the three activists incorporated EarthRights International in February

---

9   Ibid.

10   T. Giannini, interview with author, Chiang Mai, Thailand, 21 January 2004.

11   T. Giannini, interview with author, Bangkok, Thailand, 22 January 2000.

1995, starting operations in July of that year.[12] For the rest of the 1990s the three co-founders ran the organisation largely from Thailand, although a Washington, DC office was also established. In 2004 Giannini left ERI to become a Clinical Advocacy Fellow in the Human Rights Program at Harvard Law School where he later became a Clinical Professor of Law and also Director of the International Human Rights Clinic. The connection between human rights and the environment was once again evident with his early experience at an environmental organisation feeding into his work in the human rights program, resulting in new courses being offered that combined both perspectives.

The initial funding to set up ERI was provided by an Echoing Green fellowship of US$35,000 per annum for Giannini and Redford between 1995 and 1999.[13] Echoing Green awarded fellowships 'to individuals with innovative ideas for creating new models for tackling seemingly unsolvable social challenges' (Echoing Green 2007). In their original Echoing Green grant proposal the activists argued that the ERI would be formed to accomplish two goals:

> to ensure that indigenous voices will not be silenced, and to simultaneously fight for the vital objective of environmental protection [to be achieved by conducting an] integrated advocacy campaign to empower the local unrepresented people and guard their rights … utilising education media strategies, political advocacy, cooperative efforts, litigation and coalition building to effect change [by joining an] existing alliance of local human rights and environment groups. (Giannini and Redford 1994: i–ii)

From the outset, therefore, ERI worked towards the emancipatory goals set out by Doyle and Doherty of 'building regional and global networks in a manner which increases the power resources of the poor and environmentally degraded' (2006: 883).

While some components of ERI's strategy involved radicalism, other elements of its platform could be considered largely reformist. Half the organisation's motto expressed confidence in the 'power of law' suggesting a reformist approach through legislation and litigation even though in countries such as Myanmar there was little semblance of an independent judiciary to adjudicate fairly on cases. ERI demonstrated an interest in restricting corporate power in general but not in overthrowing the capitalist system. In Redford's words, 'I don't think ERI is trying to break down the capitalist system, just make it more just and fair'.[14] It is common within the global justice movement to challenge neoliberal globalisation, US imperialism or global capitalism (Epstein 2001) but there are rarely attempts to do away with market mechanisms altogether. Despite ERI's apparently reformist elements, in the particular milieu in which it operated and in its day-to-day

---

12   Ka Hsaw Wa, interview with author, Chiang Mai, Thailand, 14 January 2004.
13   K. Redford, interview with author, Chiang Mai, Thailand, 15 January 2004.
14   K. Redford, interview with author, Chiang Mai, Thailand, 9 January 2009.

activities and organisation ERI could be considered a radical organisation, being both introspective and emancipatory.

Like the founders, other ERI activists understood that the earth rights concept always linked environmental issues to some fundamental breaches of human rights. Naing Htoo, a Kayin exile who had been driven out of Myanmar by the Tatmadaw in 1997 and became ERI's Myanmar Project Coordinator in 2004, put it in the context of his personal experiences: 'I learnt about the philosophy of earth rights, the connection between human rights and the environment, in 1998. It confirmed and crystallised what I had seen and understood from the Burmese military actions in Burma. In Burma most people live off the land'.[15]

Marco Simons, who became ERI's Legal Director, wrote a paper on the 1988 protests in Myanmar while at school and later combined the two perspectives with degrees in environmental science and human rights law.[16] Another staff member, an Australian activist, became ERI's Assistant Team Leader in the Southeast Asian Office in 2004. She was frustrated with the Australian environment groups that divorced human rights from the environment, including the Australian Conservation Foundation which had actively lobbied the federal government to implement a zero immigration policy: 'Humans should be integrated into the environment, part of the environment. Also indigenous peoples are often the most trampled over, even though they tread the lightest on the earth'.[17] Working with ERI allowed her to combine what she considered was necessary in the pursuit of justice, human rights and environmental protection. This underlying commitment to social justice, a core green pillar, permeated the values of the staff and founders of ERI.

While some introspective groups would consider ERI mainstream, organisation members certainly saw themselves as radical, particularly during Thaksin's premiership; as another Thai activist noted, 'under Thaksin all NGOs in Thailand [were] targeted, particularly those who work on Burma and particularly those who receive foreign funding'.[18] ERI satisfied all these criteria with one activist arguing that for ERI in Thailand 'just existing' was civil disobedience, although it consciously avoided Thai politics to preserve its operational centre in Thailand.[19] ERI had been raided by police seeking access to their computers in the mid-1990s but during Thaksin's period in office it avoided the attention given to other NGOs.[20] Nevertheless, ERI's office remained a secluded and anonymous house in suburban Chiang Mai where only trusted individuals were taken. It was

---

15 Naing Htoo, interview with author, Chiang Mai, Thailand, 10 January 2005.

16 M. Simons, interview with author, Washington, DC, USA, 21 March 2011.

17 ERI Assistant Team Leader (name withheld), interview with author, Chiang Mai, Thailand, 10 January 2005.

18 Wandee Suntivutimetee, interview with author, Chiang Mai, Thailand, 11 January 2005.

19 ERI Assistant Team Leader (name withheld), interview with author, Chiang Mai, Thailand, 10 January 2005.

20 T. Giannini, interview with author, Bangkok, Thailand, 22 January 2000.

moved in 2008 because, as Redford observed, 'after twelve years we were getting too many visits from the Thai authorities, just to let us know they were there'.[21] This point illustrated a wider paradox that Redford highlighted between the costs and benefits of media coverage and visibility in the North and South: 'In the US media coverage of NGOs is considered great and a key measure for funders but in Thailand it's actually problematic as it attracts attention to our work and therefore attention from the authorities'.[22]

As an organisation ERI's budget recorded significant growth from an initial US$35,000 from Echoing Green in 1995 to approximately US$1.5 million a decade later (ERI 2006a: 12–13; 2007b: 11–12). According to Rucht (1999: 218) such growth may reflect or induce self-interest in organisational growth and maintenance and may cause co-option and deradicalisation. There are also instances where North–South transnationalism has resulted in NGOs forgoing principled structures and practices in favour of increased funding opportunities (Duffy 2006; Fagan 2006; Kerényi and Szabó 2006); NGOs undertake 'rent seeking behaviour, [when they] are structured to meet the expectations of Western funders' (Doherty and Doyle 2006: 699). In contrast ERI's structure and activities continued to appear wholly consistent with the justice-oriented values of its founders and staff. In a later email Redford defended ERI's growth and the fundraising efforts to pay staff salaries:

> When I think of what corporate CEOs are getting for messing up the world, and then what we're getting for cleaning up their messes ... you get the point. The service that we provide is worth it – and I think ERI does a great job, and is an effective organisation, and so there is nothing that I'm uncomfortable about in any of our fundraising activities.[23]

Giannini explained that flat pay scales at ERI reflected an egalitarian justice-oriented philosophy, making the understatement – from a future Harvard Law School Professor – that 'we don't work at ERI for the money'.[24]

Nevertheless, as a non-profit organisation ERI required continuous fundraising just to cover modest staff remuneration. While Redford suggested that it was 'easy' to fundraise due to believing in ERI's 'product' (her inverted commas), she also admitted: 'I didn't always feel this way – it took a long time for me to get over feeling embarrassed asking for money'.[25] Despite ERI's unremitting fundraising efforts there was little evidence of it debasing the original aims of the organisation. In the rest of this chapter I will demonstrate that these emancipatory

---

21   K. Redford, interview with author, Chiang Mai, Thailand, 9 January 2009.
22   Ibid.
23   K. Redford, email to author, 21 January 2008.
24   T. Giannini, interview with author, Chiang Mai, Thailand, 21 January 2004.
25   K. Redford, email to author, 21 January 2008.

values, reflecting a commitment to the four green pillars, were evident in both ERI's organisational structure and activism, qualifying it as an effective EGG.

Organisational Structure

Adapting Gole's (2000) analysis of Western modernity to environmental movements, Northern NGOs should aim to 'decentralise' the North and adopt a view from 'modernity's edge'. ERI's organisational structure – eventually with full operations in both the South and the North but initially only the South – was designed from the beginning to ensure that the North was 'decentred'. While EGGs usually maintain less institutional forms of organisation, ERI's attempts to address structural power imbalances through its internal politics, despite its formalisation, qualified it as an introspective and emancipatory group. Giannini argued that the ERI co-founders recognised the inherent disparity in power relations between Northern and Southern activists in areas such as formal education and expected remuneration and they therefore took measures to counter this imbalance.[26] The requirement of a formal tertiary education for Northern employees was often waived for Southern activists in lieu of experience, and all ERI staff in Thailand or Washington were paid 'local rates', whether they were from the North or the South. In some situations assistance was provided to Northern employees who had regular payments to make on student loans – issues that Thais and Myanmar exiles generally did not have – but the emphasis was on equity and providing a living wage. As Redford clarified,

> the lawyer or US-trained PhD does not get a higher salary or bigger title than the Burmese field staff who speaks four languages and can get to the regions that we are working on. We have a complex salary structure that values relevant life experience and educational/job experience equally.[27]

Despite ERI's formal structure, which required directors and a nominal hierarchy of responsibility, ERI staff indicated that at a practical level there was extensive consensus decision making with significant autonomy provided to each project team.[28] As the organisation grew it became a challenge to keep the communication channels open between the two main offices – with a 12-hour difference between them – but cooperation was easier due to the 'shared values' that Keck and Sikkink (1998) identify.[29] One staff member found when she joined the organisation that all employees were considered equal 'no matter what their position' with all staff providing input into long-term decision making

---

26   T. Giannini, interview with author, Chiang Mai, Thailand, 21 January 2004.

27   K. Redford, email to author, 21 January 2008.

28   Naing Htoo, interview with author, Chiang Mai, Thailand, 10 January 2005.

29   M. Simons, interview with author, Washington, DC, USA, 21 March 2011.

in the organisation's strategic planning process.[30] This approach reflected the core green belief in participatory democracy within the organisation where, as Dryzek put it, 'hierarchy ... is recognised and condemned' (2005b: 216). There were legal difficulties, however, that prevented an organisation incorporated in the US with a budget of over $1.5 million from removing hierarchy completely but the significant attempts to minimise it in practice demonstrated a commitment to participatory democracy and equality. As the then Assistant Team Leader of ERI's Thai office noted, 'if there are positions of seniority, ERI has ensured that these positions have gradually been taken up by [ethnic minority] activists'.[31] This organisational structure, with high-level North and South representation, both geographically and in terms of personnel, is quite unusual within NGOs; studies demonstrate a tendency for organisations to be isolated either within the North or the South (Bandy and Smith 2005: 6). ERI was extremely conscious of this characteristic, clearly stating in an employment advertisement that the 'staff is ethnically diverse (evenly divided between people from the Global North and South)' (ERI 2007d). The Program Coordinator also indicated that throughout the selection process and her subsequent employment at the organisation 'cross-cultural understanding is critical ... [I needed] awareness of the limitations of the English language in a multi-cultural setting where power-sharing is important ... Tyler [Giannini] emphasised he wants culturally sensitive staff'.[32]

In 2004 the management of the Southeast Asia office shifted from Northern to Southern hands with the appointment of Chana Maung, a Kayin exile, as Director of the office.[33] In the same year the management of the EarthRights Schools was also transferred into Southern hands when Da Do Wa, another Kayin exile, took on the leadership role. These shifts reflected the organisation's commitment to empowering members of marginalised ethnic minorities of Myanmar, but it also reflected an acknowledgement that the specific linguistic, cultural and personal skills that these activists accrued through their personal experiences and histories were characteristics that Northern activists were unlikely ever to replicate fully. As a result, ERI considered its diversity in ethnicity, language, culture and gender to be a major source of strength. Ka Hsaw Wa's pivotal position as a Kayin man and co-founder of the organisation provided powerful leadership from an ethnic grouping otherwise marginalised within Myanmar. Likewise women comprised a significant proportion of ERI's employees, with Redford as a co-founder providing a strong role model:

---

30   ERI Program Coordinator (name withheld), interview with author, Chiang Mai, Thailand, 10 January 2005.

31   ERI Assistant Team Leader (name withheld), interview with author, Chiang Mai, Thailand, 10 January 2005.

32   ERI Program Coordinator (name withheld), interview with author, Chiang Mai, Thailand, 10 January 2005.

33   Chana Maung, interview with author, Chiang Mai, Thailand, 10 December 2010.

Funders ... love you to fill out 'diversity charts', but they don't necessarily ask or care WHY you practice diversity. For us, diversity is about building power – diversity brings different values, approaches, ideas, ways of doing things, and while it takes longer to really maximise this, in the end it makes groups like ERI more powerful – and this kind of diversity of experience and expertise is what groups like ERI, which is made up of such vastly diverse people ... more powerful than corporations. In our view, this is how North-South collaboration ... will enable local communities to rise up and resist the corporate powers, because we have what they don't – diversity and [the] power that comes from that.[34]

This diversity brought its own challenges, however. Redford noted:

At any given time, we have a minimum of 5 languages that our organisation is working in, and if you add that together with the students at our schools, it's dozens of languages ... When we started, there were three of us in one room in Thailand and so, we could speak directly to each other, and take the time to make sure we were all on the same page ... Now, with over 30 staff, 2 offices ... plus two schools, there is the potential for miscommunication, because of the lack of face time, and also the heavy reliance on email.[35]

Southern activists generally had advantages in the field such as fluency in local languages and immersion in local cultures, but most Northern staff in the Thai office were also multi-lingual. Nevertheless, as the then Assistant Team Leader noted, 'most meetings are held in English and in this context ERI is aware that sometimes "consensus" does not necessarily mean "equality" because of power imbalances due to access to university education and ability to think "on the spot" in English'.[36]

As a result, the Myanmar-speaking staff were often absent from English-speaking meetings and other meetings were held in the Myanmar language with English summaries. Regardless of attempts to mitigate language barriers within the offices the reliance on email was an additional avenue for miscommunication despite the 'crucial' role the internet played in its campaigns.[37] Ka Hsaw Wa acknowledged as much with a postscript at the end of each email: 'If I said something in this email that seems insulting, I might not mean it. Please give me another chance. Because English is not my native Language'.[38]

English is still the dominant language for communication via email or the internet (Lewis 2006: 115; Tadros 2005: 186). Many studies have examined the

---

34   K. Redford, email to author, 21 January 2008.

35   Ibid.

36   ERI Assistant Team Leader (name withheld), interview with author, Chiang Mai, Thailand, 10 January 2005.

37   Ibid.

38   Ka Hsaw Wa, email to author, 1 May 2007.

role of the internet and email in transnational environmental activism (Castells 2003: 187; O'Neill and VanDeveer 2005: 208), yet their cultural impact on North–South transnational organisations has not received much attention. Email exacerbates the dominance of English as the activist *lingua franca* but Redford was very aware of the power imbalances this introduced:

> Email is a very comfortable form of communication for 1) native English speakers 2) good typists and 3) people who are culturally used to this kind of communication (i.e. the American, Canadian, Australian staff at ERI!) For example, KSW [Ka Hsaw Wa] would NEVER respond to this email from you because it would take him 2 days, whereas it's going to take me 2 hours. So, my voice gets heard, his doesn't.[39]

To improve Southern activists' transnational networking opportunities and promote their empowerment, Myanmar exiles were often chosen to represent ERI at conferences and international meetings. These activists had the added advantage of being more sensitive to security concerns than Northern activists.[40] ERI's Assistant Team Leader acknowledged the vast cultural and political chasm that existed between Northern and Southern activists when asked whether nonviolence could be interpreted flexibly: 'Living as a Westerner with a "cushy" background, it's impossible to answer – I can't comment on armed struggle. But nonviolent social change produces or creates "mass movements" which can effect change. But ERI does not deal with armed groups'.[41]

While Ka Hsaw Wa only briefly flirted with the idea of armed struggle and did not act on it others in ERI and the broader activist diaspora had taken up arms with insurgent armies before they adopted the nonviolent path. Chana Maung, for instance, who was born in Kayin State two days' walk from the Wei Gyi Dam on the Salween River, left Myanmar with his family when he was 7 or 8 only to return as an adult to fight with the Karen National Liberation Army (KNLA). He did not regret his activities and many years later still saw the armed groups as necessary under direct military rule to mitigate the repression meted out by the Tatmadaw, despite the insecurity caused for villagers by the insurgents' minefields (South 2012).[42]

In addition to the social aspects of ERI's organisational structure, ERI's founders also ensured that the organisation's activities were as ecologically sustainable and sensitive as possible. In 2004 staff joined Kayin student refugees on a meditative forest walk up Doi Inthanon, the highest mountain in Thailand one hour south of Chiang Mai, which combined support for exiled ethnic minorities

---

39   K. Redford, email to author, 21 January 2008.

40   ERI Assistant Team Leader (name withheld), interview with author, Chiang Mai, Thailand, 10 January 2005.

41   Ibid.

42   Chana Maung, interview with author, Chiang Mai, Thailand, 10 December 2010.

and an almost ecocentric or spiritual approach to nature (Eckersley 1992).[43] Later ERI launched a capital campaign to build their own ecologically sensitive office in Chiang Mai, which would use natural airflow for cooling, although there were delays due to the difficulty in procuring the necessary ecologically sound materials in Thailand. In all their activities they endeavoured to buy products from local suppliers, conscious of both minimising the energy consumed and supporting local communities.[44]

ERI's founders and staff, therefore, demonstrated a sensitivity to its internal operations that reflected a central concern with promoting emancipatory values. In this sense ERI can be considered a radical emancipatory group because it is in these groups, as with radical networks, 'that the internal power relations ... have been addressed most self-consciously' (Doherty and Doyle 2006: 699). ERI's organisational structure clearly created an environment of participatory democracy, nonviolence, sustainability and social justice for its entire staff, whether they originated from the North or the South. The next section demonstrates that ERI also applied these four green pillars to its three activities: legal programs, training and campaigns.

ERI's Activities

Legal Programs

ERI's legal programs, particularly its precedent-setting court cases in the US, were the most globally visible of the three strategic program areas drawn from its original integrated advocacy strategy (Giannini and Redford 1994). While ERI had a broad array of legal programs, the successful litigation against Unocal (now Chevron) in US courtrooms over its involvement with the Yadana Project stimulated international media coverage and a shareholder backlash, and was considered by Redford to be one of ERI's two most important achievements.[45] The most significant aspect of the court proceedings for the broader global justice movement was the use of the *Alien Torts Claim Act* 1789 (ATCA) to sue a US corporation involved in 'egregious human rights violations' outside of US territory (Fahn 2003: 198). Indeed central to the formation of ERI, according to the Echoing Green proposal, was the aim of exploring 'litigation possibilities based on emerging principles of human rights and international law' (Giannini and Redford 1994: 5).

The ATCA grants original jurisdiction to any civil action claimed by an alien for a tort committed in violation of 'the laws of nations or a treaty ratified by

---

43 ERI Assistant Team Leader (name withheld), interview with author, Chiang Mai, Thailand, 10 January 2005.

44 Naing Htoo, interview with author, Chiang Mai, Thailand, 13 February 2012.

45 K. Redford, interview with author, Chiang Mai, Thailand, 9 January 2009.

the US' (Christmann 2000: 209). The most widely accepted international law norms, also referred to as *jus cogens*, include the prohibition of genocide, torture, systematic racial discrimination and slavery. The ACTA had lain virtually dormant for almost two centuries until *Filitarga v. Pena-Irala* (1980) when a Paraguayan plaintiff successfully sued for civil damages against a Paraguayan police officer living in New York who had earlier tortured his son to death in their home country (Haas 2008: 240–41). The Filitarga case provided Redford with the inspiration for suing Unocal, a corporate body rather than an individual. The plaintiffs in the Unocal case were fifteen Myanmar villagers who claimed that they were subjected to forced labour, rape and torture during the construction of the Yadana Pipeline. Soldiers from the Tatmadaw allegedly committed these abuses while providing security and other services for the pipeline project, making Unocal vicariously liable (Mariner 2003). In October 1996 ERI, along with other human rights NGOs, filed a class action lawsuit in a Los Angeles federal district court on behalf of the villagers. In 1997 the court set a new precedent by agreeing to hear the case and concluded that corporations and their executive officers could be held legally responsible for violations of human rights norms in foreign countries and that US courts had the authority to adjudicate such claims (Christmann 2000: 209–10). As the International Commission of Jurists noted in their later report on transnational corporate complicity, the court found that the evidence suggested that 'Unocal knew that forced labour was being utilised and that the [corporation] benefited from the practice' (International Commission of Jurists 2008: 37).

The implications of the case for US TNCs, particularly energy companies that often do business with authoritarian regimes in the South, were significant. The flood of litigation that followed in the wake of the Unocal case caused corporations to examine more closely the human rights record of the governments with whom they do business (*The Economist* 2003; Markels 2003). The Bush administration, due to its close relationship with big business in general and the energy sector in particular, attempted to stifle both the Unocal case and the ATCA in the form of a 'friend of the court' brief from then Attorney General John Ashcroft (Chomsky 2004: 154–5; Mariner 2003). According to Human Rights Watch, it was a 'craven attempt [by the administration] to protect human rights abusers at the expense of victims' (Human Rights Watch 2003). As a measure of the concern over the ATCA in business circles, the National Foreign Trade Council paid for a full-page advertisement in *The New York Times* during the deliberations by the Supreme Court on another ATCA case arguing that the court should 'reign [sic] in' use of the ATCA and that corporations should not be held liable for the human rights abuses of foreign governments because, among other reasons, 'it discourages foreign investment' (ERI 2004b: 56).[46] In June 2004 the Supreme Court Justices ruled

---

46   ERI and the other prosecuting lawyers in the Unocal case were asked by the Court of Appeals to brief them on the impact of this Supreme Court case. T. Giannini, email to author, 17 September 2004.

that the ATCA permitted foreigners to sue in US courts for violations of certain international laws, meaning that the Unocal case could proceed (Girion 2004).

While the administration attempted to eviscerate the federal ATCA law, a parallel tort case had been proceeding under California state law in which Myanmar plaintiffs again alleged Unocal was liable for their injury in relation to construction of the Yadana Pipeline. In September 2004 the judge agreed that a jury should hear the case (*New York Times* 2004). Giannini was confident that the accumulation of evidence and harrowing personal testimonies, including a woman whose baby was thrown into a fire by the Myanmar military, would be sufficient to reach a conviction.[47]

Facing two ongoing court battles that appeared to be favouring the Myanmar villagers, Unocal made the stunning announcement that they would settle both state and federal cases out of court. The payout was finally agreed on 21 March 2005 for an undisclosed sum, although *The Irrawaddy* estimated it to be in the region of $30 million (ERI 2005; Parker 2005). The ERI co-founders refused to divulge the Unocal settlement or their fees due to a confidentiality clause but the settlement is likely to have caused a significant expansion of ERI's finances,[48] raising the question of whether any NGO could look at this potential financial boon and still make the best decisions for their clients. After all, the prosecution had settled out of court rather than pursue Unocal all the way to a conviction. In response to questions regarding the appropriateness of the settlement the ERI co-founders made several points. Redford argued that

> first of all, many of the legal precedents were already set ... The one outstanding legal question that didn't get decided was ... the appropriate standard for corporate complicity in human rights abuses. Because we settled, that decision never came down – however, the decision that the 9th circuit originally made in our case (that the standard should be aiding and abetting liability, similar to the Nuremberg standards following WWII) is the one that courts have since adopted in other cases.[49]

Giannini suggested a more complex relationship between the final agreement and legal precedent but raised a more compelling argument for the settlement: 'It was not EarthRights or the lawyer's decision but the plaintiffs'. Ethically speaking, it was easy to weigh the plaintiffs' interests against the movement's interest of

---

47 T. Giannini, interview with author, Chiang Mai, Thailand, 21 January 2004.

48 As a potential indication of ERI's fees, in 2005 ERI received over US$2.9 million in temporarily restricted income and net assets increased over the year from just under $0.5 million to almost $2.8 million. This compares with temporarily restricted income in 2006 of $245,000 and net assets almost unchanged. It appears possible, therefore, that much of this $2.9 million related to the Unocal settlement (ERI 2006a: 12–13; 2007b: 11–12).

49 K. Redford, email to author, 10 February 2007.

having the legal precedent. The plaintiffs' interests trump the latter. Having said that, it was still not easy'.[50]

Redford took this point further:

> It was always the plaintiffs' case and it was their decision to settle. I think people forget that these folks had been living in hiding for over 10 years, not knowing whether they would have to run the next day, not knowing where their next meal was coming from, not knowing whether their kids would be safe. Had they decided to go to trial (and it was a tough decision for them), even if we had won, Unocal would have appealed and we would have been in litigation for the next 5–7 years – that's 5–7 years of continued poverty, fear, inability to move on with their lives. So, it was easy for me (for example) to be like 'let's nail them in court' when I had a home, safety, security. Not so for our clients. People need to understand the conditions that they were living in to understand their decision.[51]

These were persuasive arguments and there was no evidence that ERI's emancipatory principles were in any way compromised by the decision. Indeed, the above arguments were particularly consistent with ERI's philosophy, internal politics and organisational structure.

The result of the settlement with Unocal was that ERI achieved, to some extent, the main aim of their integrated advocacy campaign, which was 'to empower the local unrepresented people and guard their rights' (Giannini and Redford 1994: i). Despite the Yadana Project continuing to operate the financial resources the settlement made available to the marginalised communities via the plaintiffs had the potential to improve their environmental security, although Redford did not want to identify the organisation in Thailand that most of the funds were channelled though.[52] The enormous publicity that this case generated, according to the ILO Liaison Officer in Yangon, also improved the performance of Total, the French pipeline operator, in the pipeline region in succeeding years.[53] More significantly, for the South in general, it also increased the likelihood of more rigorous human rights assessments by US TNCs in weighing up investments in large-scale energy projects in the South.

ERI's success can be attributed, at least in part, to aspects of its organisational structure and *modus operandi*. Although there were several NGOs and legal firms involved, the prosecution case relied largely upon ERI's field research. Ka Hsaw Wa's groundbreaking research throughout the 1990s was supported by other ethnic Kayin ERI activists who undertook extensive fieldwork in the ethnic Kayin conflict zones in the pipeline region. Much of this field research may not have been possible had exiled activists and their project teams not been

---

50　T. Giannini, email to author, 14 February 2007.
51　K. Redford, email to author, 10 February 2007.
52　K. Redford, interview with author, Washington DC, 10 March 2011.
53　S. Marshall, interview with author, Yangon, Myanmar, 7 January 2011.

given the empowerment and operational autonomy they were afforded in ERI.[54] Along with an effective lack of hierarchy, these characteristics provided activists with greater licence to carry out their work. ERI's collaboration with many ethnic minority activist groups and NGOs also increased its legitimacy in the eyes of the communities in Myanmar's borderlands.

Setting legal precedents in the Unocal case also vindicated the legal tactics ERI had adopted. Giannini argued that in the other areas of ERI's work – training and campaigns – they emphasised the links between the environment and human rights. Yet, when it came to legal proceedings under the ATCA in the US, they focused primarily on human rights:

> It was hard enough us [litigating against] a corporation like Unocal with business and the Bush administration trying to bring down the ATCA so we really needed to focus on an established area of rights. If we'd focused on environmental issues it would have provided more ammunition for opponents to say that this was not a valid area for US courts to be looking at ... By focusing on human rights there were plenty of precedents and we knew that if we made it to a jury trial with Burmese villagers in the witness box talking about the human rights abuses undertaken by the military we'd have a good chance [of victory].[55]

Having achieved what ERI described as 'a historic victory for human rights and for the corporate accountability movement' (ERI 2005), Redford argued that it would significantly change business practices, at least forcing corporations to take the potential financial costs of human rights into consideration:

> The message that was sent from the settlement to corporations was incredibly powerful. Before the settlement, corporate executives told their shareholders, the media, policy makers that these lawsuits were crazy – 'nobody has ever paid a dime for these suits, it's just a bunch of crazy activists, don't worry ...'. They can no longer say that ... Companies now have to figure liability into their bottom lines, and hopefully they'll start to realise that it's cheaper to just not commit abuses, rather than commit them and pay for them after the fact.[56]

Later in 2005 in a separate court case in France that had been running for three years, Total announced that it had reached a US$6 million settlement with eight plaintiffs from Myanmar who also alleged human rights abuses related to the Yadana Project. Almost $5 million was put aside for Myanmar refugees in Thailand while over $1 million was allocated for 'the people who could claim and justify that they were subjected to forced labour' (Aung Lwin Oo 2005).

---

54 Naing Htoo, interview with author, Chiang Mai, Thailand, 13 February 2012.
55 T. Giannini, interview with author, Chiang Mai, Thailand, 21 January 2004.
56 K. Redford, email to author, 10 February 2007.

Although these rulings had the potential to impact on all TNCs it was likely that corporations in the extractive industries, including oil, gas and mining, would be most affected since in the South the host governments' armed forces are often used to protect company operations. Unocal later sued both its insurers and re-insurers, guaranteeing that insurance companies would also start to scrutinise more closely clients who operated in countries with repressive or authoritarian regimes; as Daphne Eviator wrote in *The Nation*, 'the costs of genocide and slavery insurance could be pretty high' (2005: 2). These impacts should eventually improve the environmental security of marginalised communities throughout the South by limiting the number of adverse projects undertaken by US TNCs.

In 2005 Unocal's non-operating interest in the Yadana Project was transferred to Chevron. The project thus challenged implementation of the 'Chevron Way': 'We conduct our business in a socially responsible and ethical manner. We respect the law, support universal human rights, protect the environment, and benefit the communities where we work' (Chevron 2005). More specifically, the corporation argued that '50,000 people along the Yadana Pipeline now have free and improved healthcare' (Chevron 2007). There was little independent evidence to support such claims although there was little doubt that conditions along the pipeline had improved from the egregious human rights abuses that afflicted ethnic minorities in the 1990s during the clearing and construction phases of the project. In order to support the claims of benefits for local communities Total engaged the organisation CDA Collaborative Learning Projects to assess the human rights situation in Total's area of activity around the pipeline. In a report entitled *Getting it Wrong* ERI was scathing about CDA's methodology and findings, describing them as 'deeply flawed' (Naing Htoo et al. 2009). A more optimistic analysis of CDA's work and influence on Total was provided by Cerletti (2013). This difficulty in genuinely independently verifying corporate social responsibility (CSR) claims supports Blowfield's assertion that, despite significant information on CSR as a business tool, it 'tells us little about the real outcomes ... in terms of the impact on its stated beneficiaries' (2007: 685).

Chevron was also the subject of several other environmental and human rights controversies that were still subject to court proceedings.[57] ERI submitted an *amicus curiae* brief in support of the plaintiffs in a case involving land and water contamination in the Ecuadorian Amazon. With Myanmar added to its list of operations in 2005 Chevron was listed as one of Global Exchange's 'Most Wanted' Corporate Human Rights Violators for that year (Global Exchange 2005).

In addition to ERI's flagship Unocal case and the Chevron litigation, it also acted as co-counsel with the Center for Constitutional Rights for plaintiffs in the South in several other major court cases against TNCs. Its most significant success in these actions came in June 2009 when Royal Dutch Shell settled three ATCA lawsuits alleging complicity in the torture, killing and other abuses of Ogoni leader

---

57   Despite the settlement Chevron could also still be sued over its ongoing activities in Myanmar (ERI 2008c: 53–55).

Ken Saro-Wiwa and other nonviolent Nigerian activists in the mid-1990s in the Ogoni region of the Niger Delta (CCR and ERI 2009).

While ERI engaged in litigation, largely in the US, it also attempted to empower activists in the South by providing the tools for activists to pursue their own legal avenues. As part of this process it published a litigation manual for non-lawyers that emphasised strategies for bringing cases on transnational environmental degradation and human rights abuses to US courts (Simons 2006). In 2007 ERI expanded its work in South America to include legal training for judges and lawyers in Iquitos, Peru: 'The idea for the program was born from the recognition that many of the students at our EarthRights Schools were lawyers whose traditional legal training in their home countries had not included human rights, environmental or indigenous law' (ERI 2007c).

By 2011, with Daniel King appointed to a new position of Asia Legal Director, ERI had established a Mekong Legal Network of public interest lawyers in six countries that aimed to play a communication role 'from the local to international and the international to local'.[58] In April 2012 the first training for Myanmar lawyers, similar to that held in Iquitos, was held in Chiang Mai and with the easing of restrictions in Myanmar future training courses were planned for inside the country. Despite the often conservative nature of relying on legal redress, ERI's legal activities epitomised the emancipatory values that characterised the entire organisation. Notwithstanding the ad-hoc legal training sessions, however, it was the EarthRights Schools that provide the core component of training for activists in the South.

Training

The most comprehensive training schemes ERI operated were within its EarthRights Schools for environmental and human rights activists in Southeast Asia and South America. Redford considered the success of these schools, along with the Unocal case, to be ERI's two most important achievements that reflected its aims and goals: 'If you ... look at our legal program as "the power of law" piece (harnessing the power of international law as a tool for people, movements, affected communities) the EarthRights Schools are about enhancing "the power of people"'.[59]

The first EarthRights School opened in Thailand in 1998 with fourteen students of different ethnicities from Myanmar. The second school opened in Ecuador in 2001 and a third opened in the Mekong Region in July 2006. The schools worked to 'create local human rights and environmental activists in Southeast Asia and the Amazon [and] teach that earth rights promotion and protection are the cornerstones of democracy, which requires an engaged civil society to secure and defend these rights' (ERI 2008d).

---

58   D. King, interview with author, Chiang Mai, Thailand, 13 February 2012.
59   K. Redford, email to author, 21 January 2008.

In 2004 Redford wanted to create schools that worked specifically on five petroleum-producing countries.[60] This narrow focus broadened over the years, and the Mekong School dealt with more diverse development issues connected to the Mekong River region, particularly hydro-electric development, other water issues and the Asian Development Bank. Each intake generally took two students from each of the countries of the Greater Mekong Subregion: Yunnan Province and Guangxi Zhuang Autonomous Region (China), Myanmar, Laos, Thailand, Cambodia and Vietnam (ERI 2008d). The EarthRights Myanmar School[61] generally took sixteen students from a variety of ethnic minorities across Myanmar; the 2009 cohort included Kachin, Chin, Kayin, Kayah, Shan, Mon and Rakhine students.[62] To promote the empowerment of women within the ethnic communities there was an equal intake of women and men. Khin Nanda, the Program Coordinator at the Myanmar School, noted that although it was initially difficult to fill the eight women's positions the demand had gradually increased so that by 2009 applications outstripped the positions available.[63] As a graduate herself she argued that this sort of empowerment had challenged established norms within ethnic communities, which were often reluctant to promote women into prominent or influential roles. Mainstream civil society organisations across Myanmar, such as the KNU, were dominated by elites that replicated the unequal power relations that existed in broader society, which often included gender imbalances in participation and decision making (South 2009: 174–99).

Although there were practical courses at the schools that developed skills in information technology and dealing with the media much of the schools' curriculum employed a rights-based approach to activism and development that provided a legal basis and normative framework for campaigns. The schools' training, as with all ERI's work, connected these rights-based approaches with what Tsikata describes as a 'normative stance on the side of the oppressed and excluded' (2007: 215).

Due to security concerns in the early years the staff were cautious about discussing their roles within the organisation and the schools were absent from the ERI website.[64] These concerns were valid, as several of the schools' students and graduates had been held by Thai police.[65] Security problems increased as improved border surveillance technologies, such as passport microchips, made it more difficult for students as well

---

60   K. Redford, interview with author, Chiang Mai, Thailand, 15 January 2004.

61   This school was known as the EarthRights School Burma until early 2012 when ERI started to change its terminology and use the name Myanmar in its operations to reflect its increasing operations inside the country with local groups, who already used this terminology. P. Donowitz, email to author, 6 September 2013.

62   I was invited to teach a class on International Relations to this cohort during a fieldtrip in April 2009.

63   Khin Nanda, interview with author, Chiang Mai, Thailand, 4 April 2009.

64   R. Wolsak, interview with author, Chiang Mai, Thailand, January 2005.

65   K. Redford, interview with author, Chiang Mai, Thailand, 9 January 2009.

as activists to travel unobtrusively between Thailand and Myanmar. In earlier years simple tactics such as changing photos on passports allowed border crossings but post-9/11 border controls made this activity more difficult.[66] These security concerns therefore informed both the teaching content and processes in the schools and each student had a comprehensive 'security plan' associated with their research project, particularly when undertaking research in Myanmar.[67]

In 2005 Holly Melanson, a Northern activist, was a Teacher and Conflict Transformation Coordinator at the Myanmar School who helped develop ERI's 'respected insiders' model of conflict resolution.[68] This model once again demonstrated ERI's concern for developing culturally sensitive and culturally specific forms of training set out in one of ERI's publications:

> The approaches to conflict resolution that interviewees described to us set it distinctly apart from Western methods and the techniques taught in academic and conflict studies settings. Instead, we have found that respected insiders who are normally elders or those in higher positions are the primary third parties for resolving serious conflict in Burma. By contrast, impartial outsiders – the traditional Western conflict 'resolver' – are much less likely to play central roles.
> (Leone and Giannini 2005: 1)

This form of conflict mediation was taught at the school within the context of participatory management processes. The report found that, despite the authoritarian environment in Myanmar, 'at least in some instances, elders and other community leaders do conduct inclusive decision making process; such practices may serve as models for community-based natural resource management over the long term to ensure earth rights protection' (Leone and Giannini 2005: 2).

This culturally specific approach was emphasised by ERI activist Matthew Smith in a later book chapter, where he argued that '[e]ffective local participation in environmental governance in Myanmar will necessarily involve a unique tradition-based paradigm developed by local Burmese themselves' (Smith 2007: 239). Another Northern activist, Rebecca Wolsak, managed the Alumni Program of the school.[69] Redford believed this program played a central role in ERI's long-term strategy in the region: '[our] long term, experiential learning approach, and the ongoing support we provide to our alumni through a formal alumni program, is a demonstration of our commitment [to] long-term solutions to deep rooted problems'.[70] These alumni also created a network of activists across Asia which became increasingly useful in ERI's campaigns, although Ka Hsaw Wa denied that

---

66 Ka Hsaw Wa, interview with author, Chiang Mai, Thailand, 9 January 2009.
67 Khin Nanda, interview with author, Chiang Mai, Thailand, 4 April 2009.
68 H. Melanson, interview with author, Chiang Mai, Thailand, January 2005.
69 R. Wolsak, interview with author, Chiang Mai, Thailand, January 2005.
70 K. Redford, email to author, 21 January 2008.

the schools were set up for this reason.[71] Nevertheless, as the number of graduates increased, so too did the pool of individuals available to work on ERI's projects. Naing Htoo, a Kayin graduate of the Myanmar School in 2001, left his village near the Yadana Pipeline route in 1997 when it was attacked by the Tatmadaw. His family fled to a refugee camp in Thailand but he stayed in the jungle for six months and then joined ERI in 1998. His personal experiences and fluency in four languages symbolised ERI's grassroots strength and he eventually became Myanmar Project Coordinator.[72] He later worked on publications on forced labour (Naing Htoo et al. 2002) and the Shwe Pipeline (Smith and Naing Htoo 2005; 2006), providing a notable exemplar of the aims of the activist schools. As with the management of the Thailand office the management of the schools was largely transferred into local hands when Khin Nanda, a Shan graduate, took on a lead role as Program Coordinator. She heard about the school from a graduate of the first Myanmar School program while in a Thai refugee camp in 2000.[73] She was successful in her application for the second intake, and after graduation she and Sai Sai, another Shan exile, set up Sapawa, the first Shan environmental group.[74]

By handing over management to various ethnic minorities, ERI demonstrated increased confidence in the administrative abilities of activists from Myanmar. While training at the schools provided some tools in this regard, Redford emphatically denied that the training increased the activists' 'capacity':

> It is NOT 'capacity building' which is what most NGOs like to call training of this sort – no, the students in our Burma and Mekong Schools have capacity already – and they have capacity that we, as Northern, educated, whatever, people don't have. But they don't have the tools, experience, information and skills that they need to deal with the new phenomenon of globalisation, of corporations coming into their lands, and violating their rights and harming their environments. So, it's like they need to learn a new language, and learn new strategies to resist or deal with these new threats and abuses. The model of the ERS program is a real emblem of what ERI is all about.[75]

As Fagan demonstrates in his study of the environment movement in Bosnia-Herzegovina, some environmental NGOs are dismissed by donor organisations as lacking 'capacity' largely because of 'an inability to complete project grant application forms' (2006: 794). The reticence to classify the training as 'capacity building' indicated, once again, ERI's concern to respect the diversity of skills held by activists from the South.

---

71   Ka Hsaw Wa, interview with author, Chiang Mai, Thailand, 9 January 2009.
72   Naing Htoo, interview with author, Chiang Mai, Thailand, 10 January 2005.
73   Khin Nanda, interview with author, Chiang Mai, Thailand, 4 April 2009.
74   Sai Sai, interview with author, Chiang Mai, Thailand, 9 January 2009.
75   K. Redford, email to author, 21 January 2008.

Although many ethnic minority activists from these schools already challenged hegemonic forces in their own home environments the schools provided training and networking and they learnt about wider issues such as the forces of economic globalisation and the impacts of transnational processes on local economies and societies. The opportunity to engage in activism over recognition and resource redistribution also helped radicalise ethnic minorities in the same way that activists have been in other contexts (Maddison and Scalmer 2006: 73).

Campaigns

ERI's first banner campaign promoted human rights and environmental security in Myanmar by opposing construction of the Yadana Gas Pipeline from Myanmar to Thailand. Over time the organisation maintained a core interest in Myanmar but it also broadened its geographic focus with campaigns in other parts of the South including the Mekong region, South and Central America, India and Nigeria (ERI 2008a). Its litigation program and training through the EarthRights Schools were major components of its campaigns but it also collaborated with local groups in preparing reports and organising protest actions. While its more formalised activities, such as litigation and report writing, received a high profile its involvement with protest was more discreet. Emancipatory social movements engage in forms of protest to challenge existing social structures and ERI's co-founders had a long history of protest.

The most prominent example was the involvement of Ka Hsaw Wa in the student protests of 1988 in Myanmar, during which he was tortured by the military.[76] Giannini also engaged in a nonviolent democracy protest in Myanmar although with the protection of US citizenship. On the tenth anniversary of the 1988 uprising he travelled to Myanmar and was one of eighteen people who were detained and later sentenced to five years' hard labour. Half the group had the protection of Western passports and the other half were from Southeast Asian countries.[77] All their sentences were later commuted, however, and they were deported after six days (Fink 2009: 263). For the people of Myanmar engaging in nonviolent street protests was an extreme form of manufactured vulnerability but there were also some risks attached to these activities for Northern activists under this type of authoritarian regimes.

While ERI staff largely refrained from street protests themselves they provided organisational support for the transnational protests against the various energy projects in this book. According to Matthew Smith, at the time a Senior Consultant with ERI, part of the organisation's long-term strategy was to retreat from its leadership position and play a more supportive role allowing Southern

---

76   Ka Hsaw Wa, interview with author, Chiang Mai, Thailand, 14 January 2004.
77   T. Giannini, interview with author, Bangkok, Thailand, 22 January 2000.

environmental activists, in this case mostly ethnic minority groups, to be the most vocal and visible participants in protest actions.[78]

ERI's cooperation with Southern groups in preparing reports on environmental and human rights abuses dated back to its first major publication. In 1996 ERI and the Southeast Asian Information Network published the first edition of *Total Denial* (ERI and SAIN 1996), which, Ka Hsaw Wa argued, 'comprehensively documented atrocities along the Yadana Gas Pipeline for the first time'.[79] As part of a broader campaign promoting human rights and justice in Myanmar ERI also wrote or co-authored several major reports during this period on the extensive use of forced labour in Myanmar, primarily in the eastern region of Myanmar near the Yadana Pipeline and Salween Dams (Giannini and Friedman 2005; Mahn Nay Myo et al. 2003a; 2003b; Naing Htoo et al. 2002). It also published a report entitled *Energy Insecurity*, which argued that 'relatively little of the gas or the revenue [the Yadana Pipeline] generates is used to benefit the people of Burma or the country's own energy security' (ERI 2010: 7). This publication of reports was a significant element in ERI's campaigning. While many of these reports focused specifically on issues in Myanmar others addressed environmental security issues in the South more generally, including an *Alternative Annual Report* for Chevron which examined its 'egregious corporate behaviour' in countries such as Colombia, Ecuador, Kazakhstan, Nigeria and the Philippines as well as Myanmar (ERI 2011).

Although due to security concerns ERI had tried to minimise its exposure in the news media in Thailand, media attention in the US and internationally was fundamental to its campaign strategy. In 2000, when Thailand was required to pay for Yadana gas it was not receiving, Pipob Udomittipong was interviewed as a representative of ERI by *The New York Times*, arguing that '[i]t's altogether a failure ... a mismanagement of the energy policy in this country' (Arnold 2000). The progressive US magazine *The Nation* also picked up the campaign, writing prominent articles that supported ERI's legal action and critiqued Unocal's investment in Myanmar (Eviatar 2003; 2005). The later campaign against the Salween Dams also received attention, with Richard Parry (2006) of *The Times* in the UK posting a copy of the confidential MoU between Thailand and Myanmar online. This media coverage in the North increased public pressure on politicians and corporations but also provided evidence for ERI's 'funders' that the campaigns were effective.[80] ERI's activities also gained attention beyond the news media. When Caroline Kennedy and the Robert F. Kennedy Center in New York put on a play 'Speak Truth to Power', actor Woody Harrelson played the role of Ka Hsaw Wa.

Although the campaign against the Salween Dams gained international attention after the publication of a front-page article in the *Bangkok Post* in 2003, Redford and Giannini had already conducted research on the human rights and environmental implications of dams on the Salween River a decade earlier. At this

---

78   M. Smith, interview with author, Chiang Mai, Thailand, 6 April 2009.

79   Ka Hsaw Wa, interview with author, Chiang Mai, Thailand, 14 January 2004.

80   K. Redford, interview with author, Chiang Mai, Thailand, 9 January 2009.

time the World Bank was supporting Salween Dam projects and they intended to file claims with the bank's Inspection Panel before the bank withdrew its support under pressure from NGOs and some governments (Giannini and Redford 1994: 4). Myanmar was in arrears to the bank and this provided the 'official reason' to deny Myanmar further funding.[81] In reality, however, it provided welcome political cover for the World Bank to avoid being targeted by NGOs for their engagement.

Although ERI was engaged in the Salween Dams campaigning and also contributed to the major report *The Salween Under Threat* (Salween Watch and SEARIN 2004), it appeared to have played a much greater role in providing training and experience for Southern activists who later established the anti-dam campaign networks. Sai Sai graduated from the Myanmar School in 2001 then worked for ERI between 2001 and 2003 before founding the Shan Sapawa Environmental Organisation (Sapawa) and taking up a key position in Salween Watch.[82] Between 2002 and 2004 Pipob Udomittipong worked at ERI campaigning against the Salween Dams until he left and joined the Advisory Committee for Salween Watch. ERI, therefore, consciously stayed in the background to promote local activists despite playing a central training and coordinating role in the campaign.[83]

In the campaign against the Shwe Pipeline in western Myanmar ERI activists conducted research in Rakhine State but, as with the Salween campaign, ERI endeavoured to promote local, ethnic minority or exiled Myanmar groups as the drivers and visible face of the campaign. Although it was one of the five original core members of the Shwe Gas Movement (SGM) in 2009 ERI relinquished its membership to play a more supportive background role.[84] This retreat from frontline activities increased as domestic political openings emerged under the new government and domestic groups were better able to publicise their views, although throughout 2013 some staff of ERI itself began to be based inside Myanmar.[85] While previously ERI had undertaken much of its own research by interviewing farmers or local villagers its strategy shifted to 'amplifying' the demands or needs of local communities to an international level by, for example, linking to the reports published by local environmental groups in Myanmar itself.[86]

In 2006 the then Assistant Team Leader of ERI's Southeast Asia office claimed that the distinction between ERI and some other NGOs was that ERI's work was 'solidarity work' rather than 'issues based'.[87] Matthew Smith contrasted the relatively pragmatic approach ERI took on the Shwe Pipeline to that of other NGOs such as Burma Campaign UK. ERI were endeavouring to ensure that forced

81   World Bank Myanmar Country Program Coordinator, interview with author, Washington, DC, USA, 22 March 2011. See also Simpson and Park (2013).

82   Sai Sai, email to author, 17 December 2007.

83   K. Redford, interview with author, Chiang Mai, Thailand, 9 January 2009.

84   M. Smith, interview with author, Chiang Mai, Thailand, 6 April 2009

85   R. Ryrie, interview with author, Yangon, Myanmar, 26 April 2013.

86   P. Donowitz, interview with author, Chiang Mai, Thailand, 18 April 2013.

87   ERI Assistant Team Leader (name withheld), email to author, 19 January 2006.

labour monitoring regimes were in place along the whole Shwe Pipeline while Burma Campaign UK considered such activity tacit acceptance of the project. Smith also noted that, while both NGOs were trying to stop the Shwe Pipeline, ERI no longer adopted a disinvestment policy regarding Total and the Yadana Pipeline as Total's role would have been taken over by Chinese TNCs resulting in less effective oversight.[88] NGOs in Washington also tended to adopt a more hard-line policy than ERI's pragmatic approach.[89] The predominance of ethnic minorities within ERI and its proximity to the Myanmar border regions rather than the Northern political centres of London and Washington contributed to a more nuanced approach to their campaigns.

Although ERI did not focus uniquely on women's interests it did have a Myanmar Women's Rights Project that recognised the particular vulnerability of women in Myanmar. It also acknowledged through all its Myanmar campaigns that the security and well-being of ethnic women in Myanmar were doubly at risk:

> It means that not only are they refugees, but they are refugees in charge of providing homes for their families ... They are subjected to all the dangers the military poses to men, such as forced labor, torture, and murder, and then additional horrors based on their gender, including rape, forced marriage, and forced pregnancy. (ERI 2006c)

In one of the earliest reports for ERI Betsy Apple, the Women's Rights Project Director, wrote the report *School for Rape: The Burmese Military and Sexual Violence* (Apple 1998), which served as a launching pad for numerous reports on sexual violence in the country.

Apart from ERI's core work on Myanmar, which in 2005 made up 80 per cent of its campaign efforts,[90] it was also a member of the International Network for Economic, Social and Cultural Rights (ESCR-Net), which wrote reports for NGOs, the UN and other agencies, often advocating rights for indigenous peoples or ethnic minorities (ESCR-Net 2007). It was also a central player in global campaigns such as International/Community Right to Know (Faber 2005: 54) and Publish What You Pay (PWYP), which experienced significant success in the US with section 1504 (the Cardin-Lugar Amendment) of the Dodd-Frank Wall Street Reform and Consumer Protection Act 2010. This section required US TNCs in the extractive industries to disclose payments made to governments of the countries in which they operate. In stark contrast to the opacity of domestic business activities within Myanmar itself this information will now be made available to the public online.

---

88   M. Smith, interview with author, Chiang Mai, Thailand, 11 April 2010.

89   J. Quigley, interview with author, Washington, DC, USA, 22 March 2011.

90   ERI Assistant Team Leader (name withheld), interview with author, Chiang Mai, Thailand, 10 January 2005.

Conclusion

Transnational social and environmental organisations that traverse the North and South often experience stark differences in foci between their Northern and Southern activists. The precarious living conditions and authoritarian governance that frequently accompany existence in the South results in concerns that are closely aligned to postcolonial societal critiques (Doherty and Doyle 2006: 707; Torgerson 2006: 717). This materialist critique is contested by many activists in the North, although it is being rediscovered within the context of identity politics (Routledge 2003: 335). Even the most progressive and introspective NGOs can, however, experience tension between the strategies and activities of their Northern and Southern nodes. ERI appeared to have negotiated this North–South divide better than most. Its co-founders from Myanmar and the US established the NGO to address their common concerns relating to 'earth rights' – the nexus between human rights and environmental protection and security – primarily in Myanmar. While it expanded its activities significantly from its modest origins in 1995 ERI maintained an emancipatory introspection regarding its organisational structure and activities that attempted to ameliorate structural power imbalances between Northern and Southern activists (Elliott 2009: 404). By recognising the importance of Southern management in Southern issues and the pursuit of consensus decision making, it avoided the major conceptual conflicts that afflicted other NGOs.

Some of ERI's activities could be considered conservative or mainstream from a more radical anarchist perspective but in other important respects, including the approach to its activities and organisation within the cultural and political milieu in which it operated, it retained a radical edge. In its awareness of cultural sensitivities and focus on lack of hierarchy within the organisation it demonstrated introspective and emancipatory characteristics. In addition to promoting the central role of co-founder Ka Hsaw Wa as a Kayin exile ERI made significant efforts to achieve equity between its Northern and Southern activists to redress the inherent disparity in power relations in areas such as formal education, English proficiency and remuneration. It also made a conscious attempt to achieve gender equity within its organisational structure.

The increased management of ERI's Thai activities by exiled ethnic minorities was matched by its collaboration with many ethnic minority activist groups and NGOs throughout the region, increasing its legitimacy in the eyes of the marginalised communities of Myanmar's borderlands. It also ensured that Southern environment groups took the lead in campaigns against the Salween Dams and Shwe Pipeline, concentrating its own efforts on training and its broader strategies rather than on the 'banner campaigns' as it did during the early years. The philosophy of nonviolence underlined all these activities and ERI's organisational structure, derived from an amalgam of human rights and Engaged Buddhist ethics.

ERI's establishment of the EarthRights Schools and the success in the Unocal case empowered marginalised communities in the South in three important ways. First, the schools offered information and education on the effects of globalisation in the South and provided crucial training and tools for managing or challenging the associated processes. Second, the court settlement provided significant resources to improve the health, well-being and security of ethnic minority communities in the Thai–Myanmar border region. Third, the precedents set in the court case were likely to encourage other TNCs to reconsider their engagement with authoritarian regimes or conflict zones in the South or at least oblige them to consider the potential for significant related costs.

This ERI case study has demonstrated that formalised NGOs that engage in legal practice as part of their strategy – normally a very conservative area – can also claim status as introspective and emancipatory groups if their philosophy and organisational structure reflect the core green values of social justice, nonviolence, ecology and democracy. With a focus on these core values both in the structure of the organisation and in its activities ERI provides a key exemplar of a North–South EGG engaged in emancipatory activist environmental governance.

ERI's quest for equity between its Northern and Southern activists appeared intimately related to its success in both achieving its organisational aims and improving environmental security for the marginalised communities of Myanmar. This linkage, which could be employed by other transnational NGOs with a focus on human rights and the environment, suggests that promoting justice within an organisation can make a substantial contribution to achieving broader organisational goals and ameliorating the effects of marginalisation on communities in the South. With these findings in mind, the next chapter continues the analysis of transnational activism associated with the four energy projects but broadens the investigation to include other transnational actors.

Chapter 6
Transnational Campaigns

Introduction

In this chapter I continue my analysis of transnational activist environmental governance but I shift the focus from a single organisation to the remaining transnational organisations, coalitions and networks that comprised the case study campaigns. While much of the activism dealt with in this chapter was, as with the local activism, based in Thailand, it was of a very different nature. It was activism with an overt transnational dimension: much of it arose in response to projects across the border in Myanmar and it was aimed at international governments, TNCs and international media. The activism over these projects often focused on issues that affected the environmental security of marginalised ethnic minorities in Myanmar's border regions. As in other parts of the South, the postcolonial milieu of poverty and marginalisation provided the context for the development of movements for human rights and environmental justice (Torgerson 2006: 717; Williams and Mawdsley 2006: 662). The campaigns therefore focused on issues such as the insecurity of IDPs and ethnic minority women in the face of systematic sexual assault by the Myanmar military. Ecological issues also came into play, but were considered just one component of a more integrated justice-oriented approach.

The nature of these campaigns attracted many groups that qualified as EGGs. Activists within these groups had diverse backgrounds, including as soldiers in ethnic armed groups, but through a process of education and radicalisation based on green values they adopted nonviolent methods that also emphasised democracy, justice and sustainability. Although most groups in these campaigns were EGGs, Northern conservation groups within the EGS sometimes engaged with the Myanmar regime to promote ecological ends, instigating conflict with the more emancipatory campaigns. Although less common within the campaigns than EGGs, there also existed CGGs with conservative and hierarchical organisational structures. These structures contrasted with their more emancipatory aims and activities with the contradiction providing tensions both within the organisations and in their relations with other activist groups; reducing the effectiveness of their activities as a result.

Examination of these transnational activities helps elucidate differences between local and transnational activism in the deployment of cultural symbols. Religious and royal symbols were central to the campaigns at a local level but were less significant within the transnational campaigns. Symbols were still essentialised, but appeals to the international community were often couched in more universal concepts such as human rights and democracy. Local cultural influences were often therefore de-emphasised, unless they provided greater returns under a rights-based approach. Using more universal concepts has been successfully applied in other Southern campaigns, such as the human rights approach of the women's movement for *maquila* workers' rights in Central America (Mendez 2002). Once rights are legislated, however, both feminist and environmental movements still face difficulties over the 'gulf that exists between elegant laws and the indignities of ... everyday realities' (Cornwall and Molyneux 2006: 1183). Despite this gap, the advantage for activists of these rights-based approaches is that they have a legal basis and a normative framework that supports 'the oppressed and excluded' (Tsikata 2007: 215). As a result transnational campaigns often take these rights-based approaches to global governance institutions, particularly the UN, in addition to national governments. Despite the rationalist approach adopted in much of the transnational campaigns described in this chapter activists from marginalised ethnic minorities retained strong links to their cultural heritage, which provided an obstacle to what Christoff argues are the 'dangers of policy professionalisation' for activists (2005: 301).

The organisations in this chapter often formed transnational coalitions, in which they pooled their resources and expertise to provide more focused and effective campaigning. ERI is also discussed here but only within the context of its relationships with various other 'reference groups' (Rucht 1999: 221) in coalitions or networks. The impact of transnational activist coalitions has received some attention, particularly in relation to the global justice movement (Faber 2005; Tarrow 2005; Yanacopulos 2005b), but there are fewer studies focused particularly on environmentalism in the South. The transnational coalitions discussed here were expedited by the growing availability of inexpensive communications technologies and they also signified both the growing sophistication of the campaigns and, for the Myanmar-centred campaigns, the impact of an activist diaspora. These coalitions were largely comprised of Southern activists and organisations although they found support in Northern environment activists with strong connections in the South. While these Northern activists generally demonstrated a cooperative and sensitive relationship with Southern groups, more conservative Northern environmental organisations within the EGS persisted in prioritising conservation over human rights. While ERI played a positive supporting role some Northern conservation groups, such as the New York–based Wildlife Conservation Society (WCS), were at odds with Myanmar's activist diaspora over collaboration with the Myanmar military regime.

Justice-oriented networks or coalitions in the South may be more enduring than those that are focused exclusively on ecological grounds. Pieck's (2006) study of the transnational networks between indigenous Amazonians and Northern activists suggested that support for indigenous activists was directly linked to the political importance associated with the Amazon's ecological significance in the North, which declined over time. She also suggested, however, that 'Northern environmentalists often tend to interpret Southern movements as ecological ones' (2006: 310). While this may have been the view of Northern conservation groups such as the WCS, the Northern activists who were intimately involved with the transnational energy campaigns exhibited a keen awareness of the links between human rights and environmental protection in the South.

The analysis in this chapter also suggests that TNCs can play a central role in perpetuating authoritarian governance and resultant insecurity for marginalised populations, even if the TNC's home country is democratic. While the roles of Unocal and Total, of the US and France respectively, are expanded upon, I also demonstrate that the economic engagement of South Korea's Daewoo International with the Myanmar military transcended its business interests in the Shwe Project to include illegal arms shipments. This provided more evidence for South Korean activists that the TNC was unlikely to make investment decisions based on the welfare of Myanmar's citizens and bolstered their calls both for the prohibition of foreign investment in Myanmar under military rule and for greater transparency in the activities of TNCs. These transnational campaigns therefore provided useful case studies for analysing the relationships and factors that facilitated or inhibited transnational activist environmental governance of cross-border energy projects in the South.

Yadana Gas Pipeline Project

The Yadana Project provoked one of the most extensive environmental campaigns in the region, which eventually extended all the way into US and European courtrooms (Dale 2011). Fahn labelled it 'a global campaign for the 90s' (2003: 197). In Chapter 4 I demonstrated the intense local activity against the project in Thailand and in Chapter 5 I analysed the involvement of ERI and its court action against Unocal in particular. ERI was the central transnational NGO in the campaign but many other groups, mostly ethnically oriented exiled groups in Thailand and its borderlands, also participated. As with most actions involving Myanmar a central focus was the egregious human rights abuses committed by the Myanmar military. The transnational campaign highlighted both the connection between these abuses and environmental degradation and the complicity of TNCs in facilitating the gas revenues that entrenched the authoritarian military regime. As an editorial in the exile news website, *The Irrawaddy*, put it, '[a]s dreadful as the forced labor, torture, rape and murder committed along the pipeline route was in human terms, the greater part of the Yadana and Yetagun consortiums' responsibilities with relation to Myanmar are that the gas revenues help perpetuate and enrich a brutal, incompetent government' (*The Irrawaddy* 2004).

Since Myanmar's independence ethnic minorities in Myanmar's mountainous border regions have been the particular targets of repression. The world's longest running civil war was fought in the Thai–Myanmar borderlands that surrounded the Yadana Pipeline Project (Lintner 1999; Smith 1999), although precarious ceasefire agreements were negotiated by the new government in 2012. As Soe Myint, founder and editor of exiled news agency *Mizzima News* argued, with activists unable to campaign openly in Myanmar activism in international fora became the only route for civil society activism.[1] When Ka Hsaw Wa left Yangon in 1988 he was one of the few activists with the access and language skills to undertake field research for the Yadana campaign and communicate it effectively outside the country.

The activists within the Burma Lawyers' Council (BLC) underwent a similar experience. The organisation was established in the 'liberated area' of Kayin State in 1994 but retreated to Mae Sot in western Thailand in the face of increased Tatmadaw offensives (Piya Pangsapa and Smith 2008: 496). From there they provided quarterly human rights and security training in the Thai–Myanmar borderlands for Yadana activists and others promoting democracy and the rule of law in Myanmar, setting up a law school in 2005 for both ethnic minorities and Bamar (Burman) students.[2] The BLC worked closely with ERI during the Yadana campaign, with the secretary considering Giannini and Redford 'good friends' of the BLC.[3] Despite a commitment to nonviolence senior members of the BLC nevertheless maintained links with ethnic insurgent groups such as the KNU as they were seen as providing security for minorities inside Myanmar against the Tatmadaw.[4]

The Tatmadaw offensives around both the Yadana Pipeline and the Salween Dams prompted groups to speak out about the systematic rape and sexual assault that afflicted women in ethnic minority communities. Due to the increasing affordability of desktop publishing various transnational NGOs, drawn mostly from exiled ethnic communities, produced reports on this issue that were published in Thailand and aimed at transnational audiences. In 2002 the Shan Human Rights Foundation and the Shan Women's Action Network published a report entitled *Licence to Rape*, documenting 173 incidents of rape and other forms of sexual violence by the military within Shan State near the northern Salween Dams sites (SHRF and SWAN 2002). According to a long-term activist in the region this publication caused 'a storm' and other ethnic communities who had experienced similar repression began to compile similar documentation of their experience.[5]

In April 2003 ERI's Betsy Apple, who was serving as a consultant to Refugees International, co-authored another report, *No Safe Place: Burma's Army and the Rape of Ethnic Women*, documenting 43 rapes of ethnic women by Tatmadaw soldiers with anecdotal evidence of hundreds of further cases also gathered (Apple and Martin 2003: 57). In 2007 the Karen Women's Organisation (KWO) published a further report, *State of Terror*, which documented hundreds of cases of ongoing forced labour and sexual assault, often linked to the Yadana Pipeline:

---

1   Soe Myint, interview with author, Delhi, India, 24 December 2004.

2   Thein Oo, interview with author, Mae Sot, Thailand, 3 December 2010.

3   Aung Htoo, interview with author, Mae Sot, Thailand, 19 January 2004.

4   Myint Thein, interview with author, Mae Sot, Thailand, 19 January 2004.

5   S. Green (pseudonym), interview with author, Chiang Mai, Thailand, 11 January 2005.

> This forced labour ... continues to include the construction of roads and bridges, the clearing of landmines, often [costing] their lives, carrying of military supplies, ammunition and rations, and the guarding of military installations and equipment and the gas pipeline. Women and girls are at particular risk of being forcibly recruited since men and boys often flee from the villages and hide in the jungle in order to avoid arrest, torture or killing by the SPDC soldiers ... in many cases women taken as porters are also raped. (KWO 2007: 13)

Women were often taken into bondage for weeks at a time and the Tatmadaw were, at best, careless with their lives (KWO 2007: 53). The evidence in these reports demonstrated that sexual violence was not simply a by-product of Tatmadaw operations, such as clearing the corridor for the Yadana Pipeline, but was rather a systematic political strategy to marginalise and terrorise its political opponents.[6] Insecurity along the pipeline corridor therefore stimulated activism particularly concerned with the impacts on women as vulnerable IDPs.

The contrast between these emancipatory Southern organisations and more conservative Northern NGOs was made clear in the attempts to establish the Myinmoletkat Nature or Biosphere Reserve in the area surrounding the pipeline in Tanintharyi (Tenasserim) Region. Conservation projects were used by the military to pursue their own narrow interests and several Northern conservation groups became complicit in their activities by promoting particular projects. The million-acre Myinmoletkat Reserve was touted by Myanmar's leaders in the mid-1990s as a major conservation project and was supported by the New York–based WCS and the Smithsonian Institution in Washington. Emancipatory NGOs had refused to engage with the military regime due to its human rights abuses and scepticism over its objectives but WCS announced on its website that in 1993 it had become 'the first International NGO to initiate a program in the Union of Myanmar' (WCS 2003). At this time the organisation, led by Alan Rabinowitz, promoted 'people-less' post-materialist approaches to conservation (Martinez-Alier 2002) that were closer to concepts of eco-imperialism than emancipation (Dyer 2011).

By 1996 the two American NGOs were working closely with the Myanmar military on the Myinmoletkat Reserve, which was expanded to include the pipeline route. In reality the reserve was being used as an excuse to clear out Kayin villages to secure the pipeline route and provide access to forests for logging by the military and its business interests (Brunner, Talbott and Elkin 1998: 8; Mason 1999: 7). A team from *The Observer* newspaper visited Myanmar in 1997 and gathered first-hand accounts of the rape, execution and enslavement of Kayin villagers in the vicinity of the pipeline, while representatives of the two conservation NGOs effectively placed the importance of biodiversity conservation above that of human rights in the region (Levy, Scott-Clark and Harrison 1997). Once the pipeline route had been cleared the area became the front line of military engagements between the military and insurgents and the proposed reserve failed to materialise.

---

6   K'nyaw Paw, email to author, 9 March 2007.

A similar pattern followed the establishment of the Hukaung (or Hukawng) Valley Tiger Reserve in Kachin State in which WCS again cooperated with the Myanmar military. It praised the efforts of the Ministry of Forestry in promoting conservation, despite acknowledging that the reserve resulted in increased militarisation in the area (Lamb 2007; Rabinowitz 2007; WCS 2006). Forced relocations of villagers out of the reserve cleared the area to allow deforestation, gold mining and mono-crop agriculture by companies with close relations to the military such as Yuzana (KDNG 2007; 2010). WCS refused to criticise the regime and clearly failed to satisfy the criteria of an EGG or CGG, as its aims and activities in Myanmar tended to undermine the four green pillars. While it ignored the social aspects of the pillars, its support for the regime had little chance of contributing to an overall ecological sustainability either. It was, therefore, clearly part of the EGS.

Although the vast majority of organisations involved with the Yadana campaign, and indeed all the campaigns in this book, qualified as EGGs this was not always the case. As an example, one transnational NGO based in Bangkok was founded at the time of the Yadana campaign and undertook what could be considered emancipatory work through this and other pro-democracy and environmental security activities on Myanmar, but the nature of its internal organisation was less wedded to green pillars such as participatory democracy and justice.[7] Employees argued that the organisation was highly hierarchical with a vertical authority relationship between the founder and the rest of the staff.[8] There was little attempt to institute democratic or consensual decision making in the organisation and a single individual made all important decisions. This arrangement was particularly conservative for an organisation that promoted democracy as the best governance arrangement in Myanmar. One Northern former employee of the NGO suggested that the work environment was oppressive and autocratic, resulting in a high turnover of staff although many of the ex-employees went on to hold long-term and successful NGO positions elsewhere.[9] Not only did the organisation's work environment result in it losing valuable staff but its reputation meant that, according to another long-term activist, 'it doesn't receive the best information from the border [and therefore it] isn't that effective'.[10] These responses contrasted strongly with past and present staff of EGGs such as ERI who emphasised the importance of consultation and participation in decision making and the recognition, and

---

7   This organisation remains anonymous so that the past and present employees who were interviewed are not identifiable and the work of the organisation is not prejudiced. Some former employees refused to be interviewed on this topic, even anonymously, as they still worked in Thailand and were concerned that they could be identified. It remains a useful case study for this analysis, however, with all information held on file.

8   CGG employees (details withheld), interview with author, Thailand, March 2010.

9   CGG, former employee (details withheld), interview with author, Thailand, December 2010.

10   CGG, external activist (details withheld), interview with author, Thailand, December 2010.

attempted amelioration, of power imbalances within the organisation. Due to its lack of adherence to the green pillars in its organisational structure this NGO failed to satisfy the criteria of an EGG but, due to its generally emancipatory aims and activities, it did qualify as a CGG. It was clear from the responses of those closely associated with the organisation, however, that the contradiction between the values it espoused and the values it implemented reduced the effectiveness of its activist environmental governance and, despite being well funded, it failed to reach its emancipatory potential.

Despite international interventions that were occasionally counter-productive or of limited benefit, transnational activism that extended beyond the exiled Myanmar community was nevertheless essential in gaining international political support. While much of the international interest in the campaign was due to the Unocal case unfolding in US federal courts, a valuable campaign addition appeared in March 2006 when *Total Denial*, a film by Milena Kaneva (2006) that documented the Unocal case, won the Vaclav Havel Award at the One World International Human Rights Documentary Film Festival in Prague (ERI 2006b; One World 2006). Borrowing the title from ERI's original report on the pipeline (ERI and SAIN 1996), the film was shot over 2000–2005 and included footage taken in the jungles of Myanmar. Further footage from Thailand, Europe and the courtrooms of America demonstrated the global nature of the campaign. After the settlement with Unocal in 2005 media attention waned but the film renewed global interest in the now-Chevron-owned project and its impacts, particularly in relation to the September 2007 Myanmar protests (Goodman 2007; Sanger and Myers 2007).

Although the campaign against Unocal in the US was well underway by the mid-1990s the transnational campaign against Total, the French operating partner of the pipeline, only rose to prominence after the pipeline's completion in 1999. It took until February 2005 before the Total Oil Coalition, a group of 53 organisations based in 18 countries, was formed (USCB 2005). That year Burma Campaign UK released *Totalitarian Oil*, a report that documented the abuses surrounding the project and Total's associated complicity (BCUK 2005b). Protests in the same year were undertaken in the US at the French embassy and five consulates and at Total's AGM in Paris (AFP 2005a; USCB 2005). According to John Jackson, Director of the NGO,

> The board of TOTAL are out of touch with the modern world. They need to realise the Cold War attitude of supporting dictators as long as they are 'our' dictators is long gone. Customers and shareholders don't want to be involved with companies helping to prop up military dictatorships. (BCUK 2005a)

The majority of shareholders appeared opposed to these considerations, however, as a shareholder was widely booed in response to a question about whether Total was considering pulling out of Myanmar (AFP 2005a). Despite some successes in shareholder activism it therefore remained a difficult route for engendering concern for human rights and environmental protection within oil and gas companies.

In 2005 Total also settled a court case that had been brought in 2002 by eight plaintiffs in the Nanterre District Court in France, who claimed that they were used as forced labourers on the Yadana Project by the Myanmar military. In November 2005, Total agreed to pay €10,000 to each of the claimants as part of a €5.2 million 'solidarity fund' to finance humanitarian projects near the Yadana Pipeline and at the Thai border (AFP 2005b; Lassalle 2005). Despite the settlement, Total upheld its 'categorical denial of any involvement in forced labour and all accusations of [that] nature' (Total 2005). Nevertheless, it accepted that forced labour had occurred because €1.12 million of the settlement fund was set aside for, as Lassalle put it, 'people who can justify that they've been subject to forced labour at the time of the work in pipeline areas' (2005).

The campaign against Total continued in 2006 with an International Day of Protest on 3 February. At least fifteen countries took part and the Burma Centre Netherlands held demonstrations at ten different Total petrol stations throughout their country asking drivers not to fill their tanks at a Total station 'as a sign of solidarity with the repressed peoples of Burma' (BCN 2006). In a press release the National Coalition Government of the Union of Burma (NCGUB), the 'government in exile', expressed its solidarity 'with all pro-democracy and support groups for Burma' who were participating in the day of protest against 'investments and business deals which only encourage the generals to continue their iron-fisted rule' (NCGUB 2006). In 2008 the campaign continued at the AGM in Paris where activists again queried Total management about the role of the Myanmar military in securing the pipeline and the links to ongoing human rights violations.

Despite the long-term efforts of Unocal and Total to embellish the living conditions around the pipeline the figures from the *New Light of Myanmar*, the military regime's own mouthpiece, demonstrated that energy security for local communities had not increased. The per capita use of electricity in Tanintharyi Region was still the second lowest in Myanmar, less than 1 per cent of that in Yangon, six years after the pipeline's completion (see Table 3.1).

In 2011 Nicolas Terraz, who had been the General Manager of Total in Myanmar for three years, argued that the working environment in Myanmar for corporations had been 'much more difficult' in the 1990s during the pipeline construction than it was fifteen years later.[11] He implicitly acknowledged that there had been problems, arguing that 'Total's practices have changed'. As some activists accepted, conditions in the area had improved and dealing with a European corporation was preferential to one from China.[12] Total's practices in the pipeline area had improved since the 1990s and increased some aspects of environmental security for local villagers. This was, however, only after severe insecurities in the early years of the project and it was only the transnational activist environmental governance – in the form of international pressure generated by EGGs such as ERI – that precipitated these improvements.

---

11   N. Terraz, interview with author, Yangon, Myanmar, 9 May 2011.
12   M. Smith, interview with author, Chiang Mai, Thailand, 11 April 2010.

Trans Thailand-Malaysia (TTM) Gas Pipeline Project

In contrast to the extensive global campaign associated with the Yadana Pipeline, the transnational campaign against the Trans Thai–Malaysian (TTM) Pipeline was somewhat limited. There were no international days of action although in the UK and Europe reports were published criticising the role of UK-based Barclays Bank in partially financing the project. Part of the reason for the lack of transnational activism from the North was the relative absence of Northern corporate interests overall. The two main TNCs involved in the project were PTT of Thailand and Petronas of Malaysia. There was, therefore, no significant potential for court cases in the North similar to those against Unocal and Total. A more important contributing factor was that most of the project was undertaken in Thailand, which had a more competitive political environment and established environment movements allowing more opportunities for domestic participation and dissent. That the project did not involve Myanmar also minimised the interest of international NGOs who might focus on issues related to authoritarian regimes. Nevertheless, Thai activists still endeavoured to transnationalise the campaign by forming transnational networks and appealing to universal concepts such as human rights. A distinctive feature of this campaign, however, was the linking of the local marginalisation of Muslim villagers to the wider victimisation of Muslims that took place under the 'War on Terror'. Local cultural symbols were therefore transnationalised and exploited to highlight the injustice and insecurity wrought upon the village communities.

Despite opposition to the TTM Pipeline from local communities from 1998 activism in Europe only gathered momentum after June 2004 when Barclays agreed to be the lead arranger of almost half the $524 million required for the project. In that same month Ida Aroonwong, an activist with the Alternative Energy Project for Sustainability who was living with the villagers in Chana District, helped villagers draft a letter in English to the president of Barclays Capital in London.[13] The letter urged the bank to withdraw its support for the project based on: impacts on the ecology and local livelihoods; a flawed EIA process; failure of the government to listen to opponents of the project; mock 'public hearings'; violence against local communities; and the take-over of religiously significant public lands (Lohmann 2007: 24–28).

Barclays was one of the ten original signatories to the Equator Principles, a voluntary industry code launched in 2003 that aimed to ensure projects were 'socially responsible and reflect sound environmental management practices' (Equator Principles Secretariat 2008). In response to a later letter from the Alternative Energy Project for Sustainability questioning their adherence to the principles, Barclays argued that the project had begun before the implementation of the principles and it was therefore not possible to apply each principle 'retroactively' but that changes made relating to the principles had 'resulted in

---

13   Ida Aroonwong, interview with author, Bangkok, Thailand, 25 November 2008.

substantive improvements to its social and environmental profile' (Birtwell 2005). As FoE UK demonstrated, however, the activities surrounding the management of the project discussed in Chapter 4 breached not only several of the social and environmental principles but also Barclays' own human rights policy (FoE 2005). In a report on CSR in the finance industry in Europe the Corporate Responsibility Coalition cited Barclays' involvement with the pipeline as evidence of the ineffectiveness of voluntary finance-sector CSR initiatives such as the Equator Principles (CORE 2005: 24). As in other campaigns activists took the CSR statements made by Barclays and used them against it.[14] Oilwatch Southeast Asia argued that participation in the project would therefore harm Barclays' reputation (Oilwatch SEA 2004).

English-speaking Thai activists translated some messages from local activists for international audiences. This led in 2007 to a report published by the Corner House in the UK which had documented the campaign over the previous five years (Lohmann 2007).[15] While it covered the European campaign against Barclays much of the report dwelt on the issues surrounding the loss of the community's *wakaf* land and the Islamic communities' campaign to oppose construction of the gas separation plant (GSP). Although few of the Islamic villagers could write in English with the help of other activists in Thailand and abroad they were able to communicate their plight through these international publications. They appealed for their human rights to be respected but they also identified their oppression with that of the global campaign being waged against Muslims (Funston 2006: 87–88), by 'talking as Muslims rather than as poor people'.[16] The reports of protests that were sent via activist networks all over the world had titles such as 'Pipeline opponents insist "State has no right to force Muslims to commit a sin"' and 'Stop destroying Islam, pipeline opponents demand' (Lohmann 2007: 59; Oilwatch SEA 2005a; 2005b).

Thai activists furthered their use of rights-based universalist language when they took the campaign to a transnational audience via UN bodies. When the UN Special Envoy for Human Rights, Hina Jilani, visited Thailand activists ensured that she was made aware of the TTM campaign. This communication proved successful as she noted, in her final report, both the violent response to the Hat Yai protest and that those opposing the TTM had reported that they were afraid to highlight human rights violations for fear of retaliation from local authorities (Jilani 2004: 18). Thailand received other unwanted and, for the country, unusual attention at the 59th Session of the UN Commission on Human Rights early in 2003 when Pax Romana, a Geneva-based Catholic human rights organisation, raised similar issues during an oral intervention. Pax Romana accused the government of having failed to take effective measures to ensure the right to life of its people while at the same time creating a 'culture of impunity' (Pax Romana 2003). In 2005 activists also produced a detailed shadow report in response to the Thai government's submission to the UN

---

14   Patrick (pseudonym), interview with author, 30 November 2010.

15   L. Lohmann, email to author, 27 August 2007.

16   Varaporn Chamsanit, interview with author, Canberra, Australia, 3 February 2006.

Human Rights Committee under the International Covenant on Civil and Political Rights the previous year (Royal Thai Government 2004). The shadow report was written by six experienced activists and highlighted the shortcomings of the Thai government report by including details of the crackdown on protesters in Hat Yai (Pibhop Dhongchai et al. 2005: 44).[17]

Despite the rapid increase in linkages and networks among activists in the South during this time there was very little coordinated activism in Malaysia, which was involved in the project on several levels. Penang-based Third World Network founder and director Martin Khor had emphasised that monitoring TNCs and their projects in Southeast Asia was vital for coordinating activism in the region,[18] but well after the campaign had begun the NGO was not involved with the TTM campaign at all.[19]

The other prominent environmental organisation based in Penang, Friends of the Earth Malaysia (Sahabat Alam Malaysia), provided moral support but was not active in the campaign in any substantial way.[20] Despite the role of Malaysian TNC Petronas as a co-partner in the project there were no other Malaysian organisations involved with the campaign. This lack of activity south of the border was partially a result of the main construction work being undertaken in Thailand, as the pipeline slotted into Malaysia's pre-existing network, but it also emphasised the difficulty in transnationalising a campaign that lacked a ready-made activist diaspora, such as that against the Salween Dams.

Salween Dam Projects

The decision by Myanmar's President Thein Sein, in September 2011, to suspend the Chinese-backed Myitsone Dam on the Ayeyarwady (Irrawaddy) River in Kachin State provided a fillip for the transnational campaign against the dams on the Salween River, although the inability of local villagers to return to their homes long after the decision had been made suggested that the victory was somewhat incomplete. Nevertheless it hinted that the new government could be responsive to effective activist environmental governance, in stark contrast to the situation under military rule.

Although already underway by the early 2000s, the transnational campaign against the Salween Dams was given a significant boost by the prominent front page article in the *Bangkok Post* discussed in Chapter 4, which highlighted

---

17   The authors included several activists involved with a variety of campaigns examined here including Sulak Sivaraksa, Penchom Tang of Campaign for Alternative Industry Network and Phinan Chotirosseranee (Bhinand Jotiroseranee) of Kanchanaburi Conservation Group (KCG).

18   M. Khor, interview with author, Devon, UK, 8 February 1998.

19   Beng Tuan, email to author, 29 January 2005.

20   Nizam Mahshar, email to author, 2 February 2005.

plans for 13 dams on the Nu River in China upstream from the Salween (Kultida Samabuddhi and Yuthana Praiwan 2003). This kind of media coverage was considered essential for transnationalising campaigns in the region. The Thai print media was considered the region's most open (McCargo 2000), with an NGO activist in Phnom Penh labelling the *Bangkok Post* the 'activists' paper'.[21] This kind of publicity was certainly not possible in Myanmar's restricted media environment under military rule (Lewis 2006: 51–52).

The exodus of activists from Myanmar as they fled oppression, particularly after the 1988 protests, assisted in transnationalising the campaign against the Salween Dams. These ready-made activist networks were initially ethnically segregated but over time cooperation across ethnic boundaries increased, creating a more cohesive and potent activist diaspora. The Salween Dams campaign provided a boost to this cross-ethnic cooperation as the river traversed several states and numerous ethnic regions. A key element of this activity was the formation of transnational coalitions that pooled resources and formed strong organisational ties. The campaign emphasised the universal rights, in the form of human rights, of the affected ethnic minority communities. Despite cross-ethnicity cooperation, however, the campaign also promoted culturally specific identities. This cultural particularism extended beyond rights debates into the ecological realm, highlighting the importance of indigenous knowledge of biodiversity (KESAN 2008).

The campaign demonstrated the depth and vitality of the Myanmar activist diaspora, particularly within Thailand. Salween Watch, Karen Rivers Watch (KRW) and Burma Rivers Network (BRN) were all formed during the decade after 2000 as coalitions of smaller environmental groups that opposed large dams in Myanmar. They were staffed primarily by expatriate ethnic minority activists from Myanmar and operated mainly from Chiang Mai and the Thai–Myanmar borderlands. Despite the name BRN was actually a coalition rather than a network. Its website noted that '[t]he new coalition ... is comprised of [ten] civil society groups representing communities from different regions of Myanmar being impacted by at least 20 large dams planned by the military regime' (Burma Rivers Network 2007).

There is a relative paucity of studies on the nature of coalitions in the literature on environmental activism, although some studies have demonstrated their growing importance in transnational campaigns (Bandy and Smith 2005; Carter 2007: 162). Yanacopulos (2005a: 259) has argued, however, that coalitions afford economies of scale and the anti-dam, or pro-river, coalitions of Myanmar worked effectively by pooling their minimal resources and exploiting the growing availability of inexpensive communications technologies. They all worked closely with ERI, particularly the Kayin Team Leader of its Southeast Asia Office, Chana Maung, although ERI maintained a low profile in the campaign, leaving ethnic organisations to drive the operation.[22]

---

21   K. Lazarus, interview with author, Phnom Penh, Cambodia, 30 January 2005.
22   K. Redford, interview with author, Chiang Mai, Thailand, 9 January 2009.

KRW, a coalition of Kayin organisations based in Thailand including the Karen Office of Relief and Development (KORD), was formed in June 2003 and organised protest actions in Kayin State along the Salween River where it formed the border with Thailand (Saw Karen 2007). KORD had formed a decade earlier and, based in Mae Sariang, it brought expertise to KRW in both emergency relief and community development. KORD's director, Nay Tha Blay, argued that it took this two-pronged approach in both its fieldwork in Kayin State – 'we give them fish but we also teach them to fish' – and also in the development of both local and international networks.[23] BRN was formed later, in May 2007, and brought together organisations of various ethnicities across Myanmar. Its mission was to 'protect the health of river ecosystems and sustain biodiversity, and to protect the rights and livelihoods of communities affected and potentially affected by destructive large-scale river development' (Burma Rivers Network 2007).

The pooled expertise from the various component organisations was particularly useful in the launch of the comprehensive BRN website in January 2009 which examined dam issues related to six rivers (Burma Rivers Network 2012). Despite Yanacopulos's assertion that coalitions 'have broader strategic aims than single-issue thematically focused networks' (2005b: 95), the aims of these coalitions were relatively specific, although with different geographic foci. The secretary and coordinator of BRN, Aung Ngyeh, outlined the rationale for the formation of the BRN when other, more localised, coalitions such as Salween Watch already existed:

> The plans of dam construction are not only found on Salween River but the plans are also found on the other rivers in other ethnic lands such as Kachin, Arakan, Chin. Therefore, it's very important to have network group of Burmese civil society organizations which are working on environmental issues. As a result, initiated by Salween Watch, Burma Rivers Network (BRN) is formed to carry our advocacy campaigns against dams construction inside Myanmar.[24]

Although the general aim of these coalitions was to protect particular rivers this aim was embedded in the context of promoting security for ethnic minorities against the militarisation of dam areas. While these coalitions were important further transnational networks were also formed with other activists and NGOs such as International Rivers in the US and TERRA in Thailand. The coalitions made particular attempts to create networks with activists and NGOs in China such as Green Watershed, which campaigned against dams on the Nu River (Xie 2009: 97–105). One NGO in Yunnan Province, Three Rivers Guardians, was set up by Lao Zhang, an alumnus of ERI's Mekong School for

---

23 Nay Tha Blay, interview with author, Mae Sariang, Thailand, 7 January 2009.
24 Aung Ngyeh (pseudonym), email to author, 13 December 2007.

activists, demonstrating the importance of the schools not only in training but in establishing networks in the region.[25]

The emergence of affordable desktop publishing allowed many ethnic minority groups to publish professional reports on the dams and their potential impacts to disseminate information to a wider audience. Most reports were published in English in Thailand and aimed at transnational audiences but others were also published in various Myanmar languages to be covertly imported into Myanmar and distributed within ethnic communities. Under military rule these efforts remained covert, as Sai Sai the other Co-Coordinator of BRN, emphasised: 'Members of our coalitions Salween Watch and Burma Rivers Network are creating awareness with local communities inside Burma, but the action must be underground'.[26]

The first of these professional reports appeared in 2004 in which KRW and Salween Watch provided the first detailed analyses of the plans for the Salween: KRW published *Damming at Gunpoint* (KRW 2004) while Salween Watch co-authored *Salween Under Threat* with the Southeast Asia Rivers Information Network (Salween Watch and SEARIN 2004).[27] These publications argued that large-scale environmental damage and human rights abuses would flow from construction of the dams. In *Damming at Gunpoint* KRW highlighted that the Yadana Pipeline caused oppression of Kayin communities while providing huge revenues for the Myanmar military. It regarded the Salween Dams as a continuation of this process: 'It can be seen that the regime's plans to exploit the water resources in the Salween River, by building dams and selling hydropower to Thailand, fit into its ongoing strategy of subjugating the ethnic areas and exploiting the natural resources there' (KRW 2004: 10).

Salween Under Threat also cited the Yadana Project as a precedent to be avoided making the connection between the resultant gas revenues and the purchase of fighter jets by the Myanmar military (Salween Watch and SEARIN 2004: 42). It also provided a central rationale for the transnationalisation of the campaign. Not only were TNCs and governments from various countries involved with the projects but activists from outside Myanmar had much more chance of voicing their opposition without violent retribution and were better able to influence their governments:

> There is an urgent need for people to speak out, as local potentially-affected people face dangers if they choose to protest ... because dissidence is met with fierce and often fatal retaliation. Those who are able to express concerns, including indigenous communities and international NGOs working outside of Myanmar, should therefore challenge these projects. (Salween Watch and SEARIN 2004: 12–13)

---

25   R. Morris, email to author, 28 May 2008.

26   Sai Sai, interview with author, Chiang Mai, Thailand, 9 January 2009.

27   Most NGO reports mentioned in this section can be accessed from the BRN website (Burma Rivers Network 2012).

The vulnerable political and social environment in which these communities lived ensured that the focus of these reports was on issues of human rights and social justice rather than simply ecological issues, while retaining an emphasis on nonviolent solutions.[28]

When dealing with ecological issues, however, the organisations also focused on the importance of indigenous knowledge. KESAN focused on 'the relationship between social and environmental issues, [doing] so in a way that reflects Karen priorities' (MacLean 2003: 1). In its report on the biodiversity of Khoe Kay on the Salween River near the Wei Gyi Dam site, the primary author makes connections between his Western university scientific knowledge and his indigenous heritage:

> Because many plants are toxic to humans, the local people need to know the species well before using them … Since I was young my parents have taught me how to identify plants and animals so I can survive in the forest. They taught me to make a fire when there is no lighter by using bamboo chits or stones, and how to extract water from plants … This knowledge is important and useful … when travelling deep in the forest. (KESAN 2008: 5)

While written by a Kayin team and extolling the virtues of local indigenous knowledge, the report's centrality to the transnational campaign was emphasised by the fact that it was not available in the Kayin language and it was therefore inaccessible to most Kayin communities inside Myanmar. Forsyth and Walker write extensively on the environmental narratives developed by various actors in land use disputes and they argue that an emphasis on 'local' knowledge may not empower ethnic minorities and may actually 'reinforce preexisting power structures' (2008: 15). The term 'local' may indeed be a simplification of complex social and ecological histories but in this case the narratives associated with such shorthand terms are useful in providing guidance on aspects of historical relationships that would be left hidden if the narrative was left to the Myanmar state, which was largely the alternative here.

BRN emphasised the vulnerability of these communities to ecological crises in its press release in response to Cyclone Nargis in which it argued that the military regime's energy policies, together with its forest and mangrove destruction, would only exacerbate climate change and its impacts: 'people in Burma are now starting to die as a result of climate change in unprecedented numbers' (Burma Rivers Network 2008). The Karen Human Rights Group (KHRG) similarly highlighted that the reservoirs of the three proposed dams in Kayin State would remove a major escape route to Thailand for Kayin refugees but would also cut the principal supply line from Thailand for aid to Kayin IDPs displaced by conflict and dams (KHRG 2007: 38–39). In another report ERI and KESAN contended that ethnic communities forced from their homes by the military were obliged to engage in 'environmentally

---

28   Eddie Mee Reh, interview with author, Ban Nai Soi, Mae Hong Son, Thailand, 7 December 2010.

destructive practices ... instead of their traditionally more sustainable rotational techniques' to provide food for their families (MacLean 2003: 61).

The two Salween publications in 2004 provided the stimulus for various other, mostly ethnic-centred, groups to publish reports on their particular areas of interest. The Karenni Development Research Group (KDRG) was, itself, a coalition of nine Kayah (Karenni) civil society organisations, including the environmental group Karenni Evergreen (KEG), and was a member of both Salween Watch and BRN. As with ERI and other groups, members of KDRG and its member organisations had diverse backgrounds and religious influences that affected their approach to activism over the dams. Nge Reh, from KEG, was an artist who joined the Karenni Army, the military wing of the Karenni National Progressive Party (KNPP), at the age of 16 for 5 years after the 1988 protests.[29] After leaving the militia he lived in refugee camps on the Thai side of the border where Steve 'Green', a Northern environmentalist from the Images Asia E-Desk, visited and taught him about environmental issues and community forestry. His views evolved until he saw the protection of the environment through nonviolent activism as a path consistent with his Buddhist faith and he joined KEG as its artist and, later, as coordinator of its 'inside projects', which required negotiation with the Karenni Army. This greening of his views was accompanied by a paradigm shift in authority structures from the hierarchical top-down organisation of the militia to a participatory democratic system in the KEG, whose six members made decisions by consensus.

In contrast Eddie Mee Reh, another Kayah activist, had been a Christian Pastor in northeast Kayah State before undertaking a covert 9-day trek with the Karenni Army across the state to Thailand after the village was attacked by the Tatmadaw for providing the militia with food.[30] Under military rule monks and pastors were allowed greater leeway in discussing community issues so he had distributed CDs and MP3s from KDRG that contained messages and information about the dams 'so they could talk not only about the Bible'. Despite an underlying animism in Kayah communities, therefore, the adopted religions of Buddhism and Christianity became nonviolent emancipatory vehicles for ethnic Kayah activists in their struggles for justice and environmental security. The first KDRG report based on this research and activism, *Dammed by Burma's Generals*, appeared in 2006 and drew parallels between the proposed Salween Dams and the earlier experience of Kayah communities affected by the Mobye Dam and the Lawpita Hydropower Project in the west of their state (KDRG 2006).[31]

The Salween Delta at the southern end of the river spills into the Gulf of Mottama (Martaban) in Mon State. The exiled Mon Youth Progressive

---

29   Nge Reh, interview with author, Ban Nai Soi, Mae Hong Son, Thailand, 8 December 2010.

30   Eddie Mee Reh, interview with author, Ban Nai Soi, Mae Hong Son, Thailand, 7 December 2010.

31   Graduates of ERI's Earth Rights School have also published a report on this project (EarthRights School of Burma 2008: 81–97).

Organization had its office in Sangkhalaburi in Kanchanaburi Province, where Mon State abutted Thailand. The police raided the office in 2002, when Thaksin's government was putting many NGOs under pressure, but it continued operations and in 2007 published a report entitled *In the Balance* that called on all parties to stop their investments in the Salween Dams (MYPO 2007). Numerous reports and updates from these various ethnic groups followed this initial flurry as the campaign continued with each publication demonstrating greater professionalism as activist skills increased and desktop publication software improved. Although these individual reports were important, more critical for the campaigns, according to Aung Ngyeh, were the networks that grew out of cooperation with parallel campaigns on the Ayeyarwady (Irrawaddy) River in Kachin State and elsewhere within Myanmar.[32] These networks were often facilitated by well-resourced North–South organisations such as ERI, the Soros Open Society Institute or the Burma Relief Centre.[33]

As well as the published reports, there were a number of websites dedicated to the campaign against the Salween Dams and listserves were regularly used to keep activists around the world informed of recent events (Burma Rivers Network 2012; Salween Watch 2012). As Reitan (2007: 80) notes in her study of Jubilee 2000, these forms of electronic activism can create multiple forms of information diffusion that stimulate growth of transnational networks. Most groups in the Salween campaign therefore ensured a prominent online presence but, as della Porta and Diani (2006: 133) found in other cases, many who signed up for listserves were already either part of the campaign or had other links to activists in the campaigns. The international protests against the involvement of TNCs in the Salween Dams were organised through this global network of online activists.

These results are consistent with evidence drawn from broader environment and justice movements that the internet has revolutionised the movements' development and tactics (Curran 2006: 75; Doherty 2002: 172; Eschle 2005: 21; Klein 2001). While these technologies can be used for activism and to seek out alternative media perspectives, high internet penetration in a society does not necessarily reflect greater activism or social awareness. As Lewis notes in his examination of the internet in Southeast Asia, Thais are much less comfortable with the English language than their neighbours in Malaysia and Singapore and, partially as a result, much of the internet use in Thailand is for game playing rather than engaging in activism or searching out alternative media (Lewis 2006: 115). With limited English literacy in Thailand the campaigns faced difficulties in the Thai language press which, although relatively open compared with some neighbouring countries, rarely covered the transnational campaigns in Myanmar that featured so prominently in the English language press. As the editor of the Salween News Network noted 'Thais depend on vernacular dailies for the news

---

32   Aung Ngyeh (pseudonym), email to author, 13 December 2007.

33   Eddie Mee Reh, interview with author, Ban Nai Soi, Mae Hong Son, Thailand, 7 December 2010.

about Myanmar. But neither the high-circulation papers nor the specialist dailies print much Myanmar news. When they do, however, it is usually negative [and] not supported by concrete evidence' (Wandee Suntivutimetee 2003). Despite these obstacles, Thai environment groups such as TERRA played key roles in supporting the Salween campaign (TERRA 2011).[34]

While globalising technologies were employed, there were great benefits in these transnational contacts remaining as loose networks rather becoming a single organisation with a central authority. These sorts of networks, often 'greatly facilitated by the internet, can ... enable relationships to develop that are more flexible than traditional hierarchies' (Routledge 2003: 335). Keck and Sikkink (1998: 30) argue that the motivation to form these transnational advocacy networks is primarily shared principled ideas or values which, in this case, were based on justice and the protection of human rights and the environment even if the campaign vehicle focused on the Salween River. International organisations who engaged in the Salween campaign, such as International Rivers (formerly International Rivers Network), were focused on the health of rivers but their support was also couched in terms of justice for riverine inhabitants (International Rivers 2007).

The ability to tap into global communications and networks – which was strictly limited in Myanmar – was an important determinant in the development of transnational activism for exiles. In the Thai–Myanmar border region this often resulted in movement out of the jungles, villages and refugee camps to towns and cities where this access could be expedited. The environmental activism of these exiled communities can be generalised from O'Kane's study of women in this area: 'Intersections between globalisation processes and women's activism occur in border locations via INGOs, communication technologies and resources attracted to the borderlands for economic, political, military and humanitarian reasons' (O'Kane 2005: 20).

As a result, Myanmar exiles inhabited many of the Thai towns along the border roads, and there was a vast congregation of activists in Chiang Mai. In addition to the online environment the congregation of Myanmar exiles in these borderlands also provided a conduit for transnational actors from outside the region to become involved. Some of these actors were Northern activists who were transformed or radicalised through witnessing the precarious existence in the margins of the South.[35] Unfortunately for NGOs in the region, however, the conduit also operated in the reverse direction. Many Myanmar exiles on the Thai side of the border were awaiting resettlement in third countries and this acted as a 'brain drain' for NGOs. Throughout 2007–8 this exodus was particularly acute with NGOs losing up to half their staff as highly trained activists were resettled in Northern countries such as the US, Canada and Australia. Although this could have presented an ideal opportunity to extend the transnational activism into the North the difficulty in refugees finding work, gaining fluency in English and adapting to their new lives

---

34   Premrudee Daoroung, interview with author, Bangkok, Thailand, 13 May 2011.
35   L. Acaroglu, interview with author, Melbourne, Australia, 27 September 2007.

in the North often meant that they lost touch with the campaigns.[36] Other activists argued that the subtle nuances of working with ethnic communities in Myanmar could only be fully appreciated by maintaining a presence in the region as oral communication took precedence over written forms.[37]

As discussed in Chapter 4, KRW did organise some protest activities within Myanmar's borders in Kayin State (Cho 2008; KRW 2007a; Paw Wah 2008), but these only occurred in the politically grey fuzzy zones usually associated with the KNU-controlled liberated area as it was 'impossible to protest under the ruling [military] regime'.[38] These events near the Wei Gyi Dam site and the area upstream of the Hat Gyi Dam site were, therefore, somewhat transnational as they occurred in a territory that Giddens would argue was 'ill-defined' (1987: 18–19). These areas therefore became part of a borderlands region that, in effect, crossed three areas of jurisdiction: Thai, Myanmar and Kayin.

The protests in Kayin State occurred on 14 March every year as part of the International Day of Action Against Dams, which was very much linked into a global campaign (Cho 2008; International Rivers 2008; Saw Karen 2007).[39] The events were primarily instigated by KRW activists, with organisational assistance from ERI, and were aimed at an international audience, as evidenced from press releases and websites. As the campaign progressed these borderland protests were supplemented by collaboration with emerging domestic activist groups such as ECODEV.

While organising the global protests against the dams, activists also directly addressed the formal institutions involved in the dam decision making. To coincide with an International Day of Protest against the Salween Dams in February 2007, an open letter and petition against the Salween Dams was delivered to the Thai prime minister.[40] The request for the withdrawal of cooperation with the Myanmar military regime over the dams was signed by 124 Thai organisations, 56 Myanmar organisations, over 1,400 individuals and 52 other organisations (NGO-COD-North and Salween Watch Coalition 2007). In March 2007 KRW met with the ASEAN Inter-Parliamentary Myanmar Caucus (AIPMC) during its visit to the Thai–Myanmar border to argue that all available channels should be used to oppose the dams until Myanmar democratised (KRW 2007b).

Another major tactic of the regional campaign was to put pressure on the Asian Development Bank (ADB), although some groups adopted a more cooperative approach than others.[41] As Myanmar was in arrears, and it was the subject of

---

36   Alex Shwe, interview with author, Chiang Mai, Thailand, 8 January 2009.

37   Paul Sein Twa, interview with author, Thai–Myanmar border (location withheld), 6 April 2009.

38   Sai Sai, interview with author, Chiang Mai, Thailand, 9 January 2009.

39   Paw Wah, email to author, 18 March 2008.

40   This International Day of Protest included events in Thailand, India, the Philippines, Australia, the US and Japan (Tunya Sukpanich 2007).

41   M. Aung-Thwin, interview with author, New York, USA, 25 March 2011.

sanctions, it did not receive direct assistance but it did receive indirect technical support through the Greater Mekong Subregion Program which formed the basis for a regional development agenda drawing largely on transnational capital (Hirsch 2002: 150; Oehlers 2006: 465; Simpson 2014c; Simpson and Park 2013). In Myanmar itself this technical assistance was limited and only received by non-state actors,[42] but it also funded infrastructure projects, including highways and electricity transmission lines between Myanmar and Thailand, which increased the viability of the Salween Dam projects. In May 2007 at the ADB's 40th Annual Meeting of the Board of Governors in Kyoto, Japan, representatives from groups working on Myanmar issues, including ERI, called on the ADB to cease providing this technical support and to address activists' concerns regarding the ADB's involvement with the Myanmar military (ERI 2007a). This activism put pressure on the institutions, governments and TNCs that continued to engage with Myanmar on dam-related activities. It demonstrated that, in the absence of entrenched democratic reforms in Myanmar, transnational activism would remain the most effective scale of activist environmental governance promoting environmental security in the Salween River region.

Shwe Gas Pipeline Project

Similarly, the potential impacts on human and environmental security, and the issue of the ownership and control of Rakhine (Arakan) State's gas, were addressed most visibly through the transnational campaign against the Shwe Gas Pipeline Project. This campaign drew particularly on the experience of activists and groups previously involved in the Yadana and Salween campaigns. The campaign was initiated through the formation of a transnational coordinating organisation, which identified itself as the Shwe Gas Movement (SGM). According to the organisation's website, it comprised 'individuals and groups of people from western Burma who are affected by the plans to extract natural gas from Arakan State as well as regional and international friends who share our concerns' (Shwe Gas Movement 2012b).

In essence the SGM was more of a coalition than an entire movement, albeit one that included networks of individuals and one that, unlike the definition of coalition provided by Yanacopulos (2005b: 95), was particularly focused on a single issue. The SGM had a relatively formal membership with the original core in Thailand comprising Arakan Oil Watch (AOW), the All Arakan Students and Youth Congress (AASYC) and ERI. It was also what Tarrow defines as a 'campaign coalition', rather than an instrumental, event or federated coalition, due to the combination of 'high intensity of involvement with long term cooperation' (2005: 168). As with most organisations in the Myanmar activist diaspora engaged in activist environmental governance the organisation and its members adhered

---

42   ADB Myanmar Country Coordinator, interview with author, Bangkok, Thailand, 12 May 2011.

to the four green pillars in organisation and activities and can be classified as EGGs. Decision making within the organisation emphasised not only democratic principles, with a lack of hierarchy and communal decision making, but also a focus on justice. ERI was central to the initial development of the coalition but it later diluted its role, and eventually left the coalition in favour of the mostly Rakhine ethnic groups to acknowledge their particular interest in, and ownership of, the campaign.[43]

Bringing these various organisations together within the SGM provided the economies of scale that coalitions afford (Yanacopulos 2005a: 259) and, by drawing upon diverse sources, allowed the publication of regular reports. In addition to its group membership the SGM also had its own staff based in Chiang Mai: Wong Aung, the SGM Global Coordinator, who focused on international advocacy; and Phyo Phyo, who joined the SGM in 2010.[44] The other core members of SGM outside of Thailand were SGM-Bangladesh and SGM-India, mostly comprised of individuals from various exiled Rakhine groups. In the early days of the campaign these core members were particularly active as the pipeline was initially intended to traverse either Bangladesh or India.

The movement against the Shwe Project was similar to others based in the South in that it facilitated a local–global connection and linked into broader justice and human rights campaigns (Chatterjee and Finger 1994: 76). The linking and networking with other campaigns such as those against the Yadana Pipeline and Salween Dam Projects, both in eastern Myanmar, resulted in a cross-fertilisation of ideas and more effective activities (della Porta and Mosca 2007). The campaign illustrated the extensive cross-border linkages with both Northern and Southern activists cooperating across the region, although Northern activists consciously adopted a supporting role. One Northern activist who had been based in Thailand for almost two decades noted that for the SGM 'there are a number of very close support groups, including myself and my organisation that participate in most activities but [we] stay away from formal membership as we also facilitate financial support'.[45] Other formal members of the SGM who communicated regularly across the region included the Korean House for International Solidarity (KHIS), the Korean Federation for Environmental Movements and other NGOs in India and Bangladesh.

While these network members highlighted their concerns over ecological damage associated with the Shwe project, the overriding concerns of the network related far more closely to the inability of the people of Myanmar, and particularly ethnic minorities in Rakhine and Shan States, the two ethnic minority regions traversed by the pipeline, to participate in any decision-making processes regarding the project, with their involvement likely to be 'limited to forced labour

---

43   M. Smith, interview with author, Chiang Mai, Thailand, 6 April 2009.

44   Wong Aung, interview with author, Chiang Mai, Thailand, 6 April 2009.

45   Patrick (pseudonym), interview with author, Chiang Mai, Thailand, 30 January 2007.

and land confiscation'.[46] As many Rakhine exiles had personally experienced forced labour and torture under the military regime and the gas was to be extracted from Rakhine State the project was considered by Northern activists as primarily an 'Arakan national issue', inseparable from broader justice concerns.[47] Groups like the Burma Lawyers' Council (BLC) therefore provided human rights training for Shwe activists in Chin State to enable more effective reporting of human rights abuses.[48] Concerns over human rights were manifest in any ecological issues with the campaign goals being to

> postpone the extraction of the Shwe natural gas deposit until a time when the affected people in Western Myanmar can participate in decisions about the use of their local resources and related infrastructure development without fearing persecution [and to] withdraw or freeze all current business with the military regime, and [for TNCs and governments to] refrain from further investment until dialogue can be held with a democratically elected government. (Shwe Gas Movement 2012b)

The All Arakan Students and Youth Congress (AASYC), one of the SGM founders, was an exiled Buddhist Rakhine group with offices in Bangladesh and Chiang Mai and Mae Sot in Thailand. It was largely these Buddhist Rakhine from Rakhine State who were at the forefront of the Shwe campaign although, as the pipeline crossed the whole of Myanmar, activists saw it as a potential unifier for different ethnicities: 'the Shwe Pipeline is now an opportunity to bring together Arakanese [Rakhine], Burmese [Bamar] and Shan activists'.[49] As an indicator of this emerging multi-ethnic cooperation Sai Sai, the Shan coordinator of Salween Watch, attended meetings with the SGM in Chiang Mai in 2008 and 2009, which also indicated increased cross-campaign cooperation.[50] Other exiled groups from Shan State including the Ta-aung Student and Youth Organisation[51] and Palaung Women's Organisation,[52] both based in Mae Sot in Thailand, also provided expertise on the impacts of the pipeline project in their own regions.

With the AASYC at the vanguard of the campaign from the beginning San Ray Kyaw, an AASYC Central Executive Committee member, contrasted the development of the Shwe campaign with the Yadana campaign, which was slow to initiate transnational links: 'We learnt from the Yadana campaign. There is [in 2005] cooperation between activists in Bangladesh, India, Myanmar, Thailand and

---

46   Patrick (pseudonym), interview with author, Chiang Mai, Thailand, 2 January 2009.

47   M. Smith, interview with author, Chiang Mai, Thailand, 6 April 2009.

48   Myint Thein, interview with author, Mae Sot, Thailand, 19 January 2004.

49   Patrick (pseudonym), interview with author, Chiang Mai, Thailand, 2 January 2009.

50   Jockai Khaing, interview with author, Chiang Mai, Thailand, 2 January 2009; Sai Sai, interview with author, Chiang Mai, Thailand, 9 January 2009.

51   Mai Aung Ko, interview with author, Mae Sot, Thailand, 3 December 2010.

52   Lwin Lwin Nao, interview with author, Mae Sot, Thailand, 3 December 2010.

Korea before [the Shwe project even] gets underway'.[53] Campaign committees were set up in all these countries and also the US with ERI participating but providing space for the ethnic groups to drive the process. Nevertheless, some ERI activists were regularly involved in SGM meetings in Thailand while others helped set up the committee in South Korea. ERI founder Ka Hsaw Wa himself travelled to Yunnan in China to help establish the campaign there with alumni of ERI's EarthRights Mekong School.[54]

Soe Myint, editor of *Mizzima News,* argued that there was no shortage of volunteers for the campaign committee in India as many of the NGOs in India's northeast opposed the pipeline, which was then destined for India, as they saw little benefit for local communities in either Myanmar or their own region.[55] Kyaw Han, then president of AASYC, also emphasised the lessons learnt from the Yadana campaign, specifically that promises to the communities around the Yadana Pipeline were broken, that the communities reaped few benefits and that promises of electricity to local communities remained unfulfilled.[56] Rakhine State's per capita electricity consumption was the lowest in Myanmar so there was a desperate need for greater electricity provision but the government had no plans for local consumption in Rakhine State. As a result of the lessons learnt and shared by activists in the Yadana, Salween and Shwe campaigns – some of whom had been involved in all three – the strategy in the Shwe campaign focused on 'approaching companies and governments asking [them] not to do business in Myanmar [... The activists] don't approach the Burmese regime as they have made it clear they will proceed whatever the concerns'.[57] San Ray Kyaw likewise emphasised that they attempted to 'stop [the project] through other means than contact with [Myanmar's military government]'.[58] The campaign therefore originally targeted the Indian TNCs, GAIL and ONGC, and Daewoo in Korea.[59] When it became clear throughout 2006–07 that the gas would be sold to China rather than India and that a pipeline would be built to Kunming the activists' focus shifted eastwards and Chinese TNCs were targeted, including CNPC and its subsidiary PetroChina.

The other core Rakhine organisation of the SGM was Arakan Oil Watch (AOW), a small group with only three formal staff, which, from the early stages of the Shwe campaign in March 2005, produced regular Shwe Gas Bulletins on a bimonthly basis (Shwe Gas Movement 2012b). These publications were an essential conduit for disseminating information throughout the region via email

---

53   San Ray Kyaw, interview with author, Chiang Mai, Thailand, 11 January 2005.
54   Ka Hsaw Wa, interview with author, Chiang Mai, Thailand, 9 January 2009.
55   Soe Myint, interview with author, Delhi, India, 24 December 2004.
56   Kyaw Han, interview with author, Delhi, India, 24 December 2004.
57   Soe Myint, interview with author, Delhi, India, 24 December 2004.
58   San Ray Kyaw, interview with author, Chiang Mai, Thailand, 11 January 2005.
59   Soe Myint, interview with author, Delhi, India, 24 December 2004.

and hard copy.[60] In 2009 AOW began leading covert workshops in Rakhine State. These workshops comprised only 6–7 participants to avoid attracting unwanted attention from the military and promoted the concept of natural resource rights for local communities. They catered to two types of participants, either 'affected communities' or 'intellectual activists', such as writers, academics or journalists, indicating the importance of communicating their concerns with a broad range of societal actors.[61]

Phyo Phyo – one of 'many names' of a SGM activist – also ran workshops inside Myanmar to provide education and training for Rakhine youth on land confiscation and community development in relation to the project but these were mainly held in the American Centre and British Club in Yangon,[62] which could more easily be understood as foreign islands within the country rather than open domestic political spaces. In 2010 she travelled undercover in Rakhine State as a teacher in the Kyauk Phyu Education Program to undertake research and educate local villagers on the project. A close relative was a senior civil servant in Sittwe and provided her with useful government information that she could employ in her activism, indicating the covert forms of dissent that were undertaken under military rule, even within the state itself.[63] Although AOW cooperated with SGM and SGM coordinated AOW's international campaigns, their activists worked separately from each other on the ground in Myanmar to maintain security. Even if they were undertaking research or dissemination activities in Rakhine State at the same time the groups did not communicate about where they were active and their activists feigned unfamiliarity if they happened to meet.[64] This fear of arrest was well placed. Thu Rein, an AOW activist and the founder of a student association in Sittwe, only avoided arrest by absconding to Thailand in 2010.[65] As the political situation eased under the new government the links between local and transnational activists started to increase although security was still more precarious in Rakhine State and other remote ethnic regions compared with the urban centre of Yangon.[66]

Apart from the regular Shwe Gas Bulletins, one of AOW's main publications was *Blocking Freedom* (Arakan Oil Watch 2008), which examined Chinese investments in oil and gas in Myanmar. The development and enhancement of ICT skills by activists over the duration of the campaigns was a key driver in

---

60   Patrick (pseudonym), interview with author, Chiang Mai, Thailand, 20 January 2007.

61   Jockai Khaing, interview with author, Chiang Mai, Thailand, 11 February 2012.

62   Phyo Phyo (pseudonym), interview with author, Yangon, Myanmar, 7 January 2011.

63   Further details are withheld here to avoid identification and retribution by the Myanmar authorities.

64   Phyo Phyo (pseudonym), interview with author, Yangon, Myanmar, 18 February 2012.

65   Thu Rein, interview with author, Chiang Mai, Thailand, 11 February 2012.

66   Jockai Khaing, interview with author, Chiang Mai, Thailand, 11 February 2012.

the increasing sophistication of the information dissemination. As a Northern activist noted: 'doing the Shwe Gas Bulletins really helped with our ability to put together *Blocking Freedom*'.[67] The other main reports published by the wider SGM, including *Supply and Command* (2006), *Corridor of Power* (2009), *Sold Out* (2011) and *Drawing the Line* (2013), once again demonstrated the benefits to environmental movements of inexpensive desktop publishing. These reports listed ecological concerns such as the destruction of mangroves, coral reefs and rainforests due to construction of the onshore gas terminal, port facilities and pipelines and the potential for future devastation due to tanker spills once the pipelines were in operation. As was seen throughout these campaigns, however, much of the reports focused on the human rights abuses in the region that resulted from increased militarisation around the project including forced labour, land seizure and physical and sexual abuse. Land confiscation was exacerbated for fishing communities when fishing grounds around Ramree and Maday Islands were classified as restricted access zones resulting in reduced access for fisherfolk and increased on-the-spot 'taxation' from navy vessels in the surrounding waters. Mangroves had been disappearing at an accelerating rate in the region since the mid-1990s when the military started confiscating coastal land to create shrimp farms for the local battalions and, although the mountainous terrain of the Rakhine coast afforded greater protection from the elements than the flat Ayeyarwady (Irrawaddy) Delta, the ASEAN secretary-general argued that mangrove destruction was largely responsible for the enormous death toll from Cyclone Nargis in 2008 (Kinver 2008). Similarly, a United Nations Environment Programme report on the 2004 Indian Ocean Tsunami noted that coral reefs could absorb and dissipate up to 90 per cent of the energy from waves, and regions without intact reefs were more likely to be devastated (UNEP 2005: 26). These intimate connections between the protection of ecosystems and the environmental security of coastal or low-lying communities was particularly pertinent for vulnerable and marginalised communities, both in Rakhine State and the South in general.

While these reports were available and distributed around the region in hard copy the main online medium that greatly facilitated the dissemination of information was the SGM website (Shwe Gas Movement 2012b). As has been the case for other social movements, mobile communication technologies and the internet were crucial in the development of the activist networks (Eschle 2005: 21; Klein 2001). Although Northern activists assisted with the setting up of the site it was predominantly local groups such as the AASYC that provided much of the information and research.[68] As in the reports the website highlighted intimate connections between human rights and environmental protection and juxtaposed threats to endangered species and rainforests with threats to the Rakhine way of life, land confiscation and forced labour. As with many other websites of exiled groups and media organisations the website was blocked in Myanmar itself until September 2011.

67　Patrick (pseudonym), interview with author, Chiang Mai, Thailand, 2 January 2009.
68　San Ray Kyaw, interview with author, Chiang Mai, Thailand, 11 January 2005.

Although the campaign was primarily driven by Myanmar exiles Friends of the Earth assisted in highlighting the campaign to a broad international audience through a report which assessed CNPC's social and environmental performance across the South. The report noted that the Chinese state-run media actively supported the company's operations; for example the *Global Times* argued that SGM's *Sold Out* report was a product of Western influence and other news outlets labelled it 'slander' against China (Matisoff 2012: 24). More favourable coverage of the campaign was received outside China when articles were published in media such as *The New York Times* (Perlez 2006) and *Al Jazeera*:

> While the gas goes abroad and profit goes to the military, locals live without electricity for all but two hours a day ... In this oil and gas-rich area, once the sun goes down and the region's resources are piped across the border to China, the locals will once again be left in the dark. (*Al Jazeera* 2008)

The SGM sought out this media coverage by regularly distributing press releases and in September 2011 hired the Foreign Correspondents Club of Thailand to launch *Sold Out* (Shwe Gas Movement 2011). Two years later, however, evidence of the dramatic shift in the domestic political environment was provided by the launch of another report, *Drawing the Line* (Shwe Gas Movement 2013), which was held at the Royal Rose Restaurant in Yangon. That formerly exiled ethnic activists now felt secure enough to hold a press conference launching a report that called for a halt to the Shwe Gas Project (despite gas deliveries having already commenced two months earlier in July) suggested domestic activities were becoming, under the new government, as important as the transnational activism. These launches also demonstrated the effectiveness and efficiency associated with forming coalitions (Yanacopulos 2005a: 259). Each organisation within the SGM had extremely limited funds so it was only able to rent out these premises with the pooled resources of all members. The attention from regional and global media organisations also provided publicity for the campaign throughout the North and South, communicating their struggle for natural resource rights far beyond the existing Shwe activist network.

From 2005 onwards central elements of the transnational campaign, and ones that grew in importance, were the International Days of Action against the Shwe Project (Doyle and Simpson 2006: 757). These actions were effective in raising consciousness in many countries, but South Korean activists became central players due to the key involvement of Daewoo International, with its 60 per cent share in the Shwe A-1 Block. Environment groups existed in South Korea in the 1970s but their activity significantly increased in the 1980s with the formation of the Korean Federation for Environmental Movements, the largest and most active environmental organisation in the country (Lee 1999: 92–93). This group and the Korean House for International Solidarity (KHIS) became partner organisations of the SGM and participated in several International Days of Action.[69] Environmental issues were highlighted by the groups while concerns over democracy, justice and related social issues were given equal focus. Mikyung Choe, the executive director of KHIS, indicated that she joined the Shwe campaign because 'human rights violations by companies, especially TNCs, are increasing [so] one of KHIS's main [activities] is monitoring Korean companies abroad'.[70]

KHIS promoted a philosophy of nonviolence and a key argument against the involvement of TNCs in Myanmar was their complicity in the Myanmar military's oppression of its own people. NGOs suspected that, in addition to its involvement with the Shwe Project, Daewoo International was sending military hardware to the Myanmar military regime in violation of South Korean law. In an open letter to the government and media outlets questioning Daewoo's military links, KHIS and 25 other NGOs argued that the government of South Korea kept 'silent about the Burmese junta's human rights violations and actions against democracy, and [lent] support to the regime under the slogan of energy security' (Korean House for International Solidarity 2006: 2). This suspicion was validated throughout 2006–7 when 14 high-ranking officials from Daewoo and associated companies, including Daewoo International President Lee Tae-Yong, were convicted on charges relating to illegally exporting production facilities and weapons technology to Myanmar while fabricating export documents (DVB 2007; Shin-who 2006; Yeni 2006). Lee Tae-Yong was fined around $50,000 – a pitifully small amount for a president of a TNC as large as Daewoo – while Lee Dukgyu, Daewoo's executive director, was given a suspended one-year jail sentence. While activists supported the conviction they contrasted the leniency of the sentence with the impacts of the offences: 'These sentences should more closely reflect the seriousness of the crime. Burmese civilians struggling for democracy are being killed by weapons sold to the Burmese army' (Shwe Gas Movement 2007). The SGM also argued that 'such lenient sentencing will only invite similar corporate crimes in the future' (AP 2007a). Mikyung Choe argued that this was a longstanding problem in South Korea: 'We Korean NGOs always condemn the Korea judicature [for being] very lenient to business and [companies]'.[71]

The court case did demonstrate, despite the relatively lenient sentencing, that the rule of law could be applied in South Korea. Conversely, the case also demonstrated the complete lack of effectiveness of voluntary CSR commitments. In contrast to its unethical and illegal activities Daewoo International's rhetoric in its 2005 annual report was one of ethics, social justice and transparency: 'With the four principles of profit-oriented management, transparency management, ethical leadership, and social commitment as the core foundations of Daewoo International, we strive to establish customer satisfaction and complete trust as the core values associated with our company' (Daewoo International 2006: 7).

---

69   Naing Htoo, email to author, 21 October 2004.
70   M. Choe, email to author, 25 October 2006.
71   M. Choe, email to author, 22 November 2007.

Due to changes in Myanmar's domestic political situation and conciliatory statements from the new government the International Day of Action on 1 March 2012 added new tactics to the day's activities. On 27 January 2012, in response to both years of transnational activism and the gathering domestic campaign, the Minister for Energy, Than Htay, made the unprecedented announcement that any new gas development projects from 2013 would be for domestic consumption and that Rakhine State would receive 24-hour electricity (Szep 2012). Companies such as Thailand's PTT continued to be awarded exploration licences at a rapid rate (*Mizzima* 2012) and both activists and opposition politicians remained sceptical but it was clear that the tone of the government's discourse had changed and that they may be open to persuasion.[72] While the previous protests had only been held outside Chinese, South Korean or Indian embassies, protests were now held outside Myanmar embassies and consulates. During previous actions the campaigning had been solely focused on foreign corporations and governments because the Myanmar government was seen as wholly unresponsive. The earlier 2006 report, *Supply and Command*, ignored the government completely and called on 'foreign governments, institutions, and civil society … to pressure businesses involved in the Shwe gas project [and for] all corporations and businesses involved in the Shwe project, either state or privately owned, to freeze all current business with the military regime' (Shwe Gas Movement 2006: 3).

The gradual opening of domestic political space and the announcements on new gas projects, the Myitsone Dam and the Dawei coal-fired power station suggested that the government was now more amenable to community pressures. Although the SGM engaged in protests outside Myanmar embassies they also addressed the president directly in a March 2012 letter signed by dozens of mostly exiled and international NGOs: 'Dear President U Thein Sein [using the honorific 'U'] … We are calling for the government to postpone this project until rights are protected and negative impacts are prevented within a sustainable framework for national development' (Shwe Gas Movement 2012a). During the following year SGM began operating more openly inside the country itself by setting up the Thazin Development Foundation, which joined with over fifteen local groups from along the pipeline route to form the Myanmar China Pipeline Watch Committee. Wong Aung and staff of Thazin Development Foundation, such as Aung Kyaw Soe, began campaigning with other newly formed groups such as the Northern Shan Farmers Committee and Myanmar Green Network (Magway and Mandalay), including for the launch of its *Drawing the Line* report.[73]

Despite the historical animosity between the government and activists an emerging *détente* suggested that the new government was open to

---

72   Aye Tha Aung, interview with author (translated from Arakanese by Zaw Myat Lin), Yangon, Myanmar, 20 January 2012; Jockai Khaing, interview with author, Chiang Mai, Thailand, 11 February 2012.

73   Aung Kyaw Soe, interview with author, Sydney, Australia, 23 May 2013.

greater engagement with civil society.[74] Evidence was emerging that activist environmental governance in the form of sustained transnational, and emerging local, environmental campaigning was, for the first time, having some impact in the decision-making processes of the Myanmar government. The new government, being entirely unfamiliar with the practical implications of concepts such as transparency and civil society consultation,[75] received encouragement in this regard via aid agencies from Australia, the UK and the US.

In December 2012 President Thein Sein had announced Myanmar's intention to join the Extractive Industries Transparency Initiative (EITI), which required significant transparency of revenue flows within the extractive industries. In response AusAID sponsored a Myanmar delegation to Australia for conferences on Mining for Development and the EITI as well as fieldtrips to Australian mines in May 2013. The delegation included the Ministers for Mines and Energy as well as civil servants and environmental activists and was the first time that Union ministers were in such close contact with activists for an extended period. The Minister for Mines, Myint Aung, was particularly interested in the views of foreign academics on the Shwe Gas Project and appeared open to new strategies.[76] He was exposed to new approaches to governance throughout the trip and although significant renegotiation of the Shwe contract was considered unlikely, given its intense importance to China's energy security, the minister announced to parliament two months after the Australian trip that the contract for the Letpadaung Copper Mine had been re-negotiated to provide more favourable terms for the government and local communities at the expense of China's Wanbao Mining Ltd (AFP 2013).

Another opportunity for the ministers to engage with civil society was scheduled for the operation of the EITI Multi-Stakeholder Group (MSG) where has equal votes for civil society, industry and government. In the membership accession process the first significant activity is the formation of an MSG, which must develop a fully-costed work plan for EITI implementation. This emergent collaboration between government and activists was unprecedented in Myanmar and the Myanmar Development Resource Institute (MDRI), which was the designated national EITI coordinator, saw the EITI as a vehicle for getting tripartite cooperation between these actors, rather than as a stand alone

---

74　In July 2013 Than Htay attended a ceremony in Yangon to mark the launch of the pipeline and the start of natural gas deliveries. The *New Light of Myanmar* (30 July 2013), reporting on this event, suggested that of the 20 per cent of gas for domestic use (100 million cubic feet (MMcf) of 500 million total), 20 MMcf would go to Kyauk Phyu, 23 MMcf to Yenangyaung in Magway Region and 57 MMcf to Taungtha in Mandalay Region.

75　Winston Set Aung, interview with author, Nay Pyi Taw, Myanmar, 29 April 2013.

76　In a meeting with the minister I recommended that the new government re-examine and re-negotiate the Shwe Gas contract and all other large resource contracts signed by the former military regime to provide legitimacy both for the government and the projects themselves. Myint Aung, interview with author, Sydney, Australia, 23 May 2013.

process.[77] As trust-building activities these processes provided opportunities to not only improve environmental governance across the country but to also facilitate the evolving peace process, which was intimately tied to the equitable sharing of natural resources.[78] These developments provided hitherto absent optimism amongst Shwe activists, who were likely to be represented in the MSG,[79] although despite pockets of collaboration the impediments to genuine participation in development decision making in Myanmar with the legislature, bureaucracy and military remained significant (Simpson 2014a).

Conclusion

This chapter examined the transnational activist environmental governance related to the campaigns against the four cross-border energy projects and found that, as with the local campaigns, there were substantial differences in the intensities at this scale. The intensity of transnational activism over the TTM dispute was considerably less than that over the other projects, which all originated from Myanmar. Most of the activism described in this chapter therefore involved actions aimed at the Myanmar-centred projects. In contrast, the extent of local activism in Myanmar discussed in Chapter 4 was strictly limited and most of the activism was undertaken in Thailand in the TTM and Yadana campaigns. The transnational activism created an activist diaspora, which was characterised by multi-ethnic cooperation not often found in the rest of the ethnic exiled communities.

The campaigns examined in this chapter were dominated by transnational coalitions that were predominantly comprised of Southern activists and organisations, although they were often assisted by justice-focused Northern activists. Although the main aims of these coalitions and their related organisations – which were often for the energy projects to be abandoned until genuine democratic processes were in place – were not wholly achieved, EGGs often focus on processes rather than outcomes. As a result, activists often considered that developing organisational structures and environments that adhered to the four green pillars was as important as the immediate outcomes, particularly as it was an aspect that was within their control. The benefits of these emancipatory characteristics appeared to extend beyond the organisations themselves, resulting in more effective emancipatory activist environmental governance than that of CGGs, with their more conservative organisational structure, and organisations within the EGS, which exhibited no emancipatory characteristics.

This chapter has demonstrated that the transnational activism, while having similarities to local activism, also used different strategies, applied different

---

77   Min Zar Ni Lin, interview with author, Yangon, Myanmar, 2 May 2013; Zaw Oo, interview with author, Sydney, Australia, 23 May 2013.

78   Aung Naing Oo, interview with author, Yangon, Myanmar, 13 June 2013.

79   Kyaw Thu Aung, interview with author, Sydney, Australia, 23 May 2013.

analytical perspectives and, using the technologies of globalisation, created more far-reaching and effective coalitions and networks. In the final chapter I draw together the various strands of this book and, based on an analysis of the theoretical and empirical discussions herein, provide findings and conclusions that may prove useful in the analysis of environmental politics elsewhere in the South.

Chapter 7
Environmental Politics in the South

Introduction

The exploration of local and transnational campaigns against cross-border energy projects in Thailand and Myanmar throughout this book suggests that the nature of the political regime, its relationship to both local and transnational business interests and the impacts on environmental security for marginalised communities in the vicinity of the projects significantly impacted on the level and type of activism undertaken at both local and transnational scales. This activism, which constituted 'activist environmental governance', encompassed both formal and informal activism and included protests on Thailand's beaches, research in Myanmar's jungles and court cases brought by NGOs in the US and France. The degree of authoritarianism in the two core countries was closely linked to living conditions for local communities, with more severe authoritarianism in Myanmar resulting in more precarious living conditions and greater insecurity.

The campaigns in the two countries were analysed over two overlapping epochs under very different political regimes. Myanmar's direct military rule until 2011 qualified it as a traditional authoritarian regime, with few sites of competition. In contrast, the governments in power in the decade from the mid-1990s to the mid-2000s in Thailand were more open and competitive, although the government of Thaksin Shinawatra qualified as competitive authoritarian due to its systematic undermining of democratic institutions and human rights.

The local activism discussed in Chapter 4 took place almost entirely in Thailand against the Thailand-based activities of the Yadana and Trans Thai–Malaysia (TTM) Projects while the transnational campaigns described in Chapters 5 and 6 took place in Thailand and around the world but were focused almost entirely on the projects based in Myanmar. Thailand-based projects were therefore characterised by high levels of local activism while transnational activism dominated the campaigns against Myanmar-based projects. The Yadana Project, which had significant operations in both Thailand and Myanmar, was therefore the only campaign to experience high levels of both local and transnational activism (see Table 7.1).

In the case of Myanmar, expatriate activists escaping authoritarianism under military rule were distributed throughout the world but were particularly active in Thailand, India, Bangladesh and the 'liberated area' of the Thai–Myanmar border region, beyond the reach of the Myanmar military. This evidence suggests that local repression under traditional authoritarianism increases the likelihood of activists emerging from within exiled communities to develop transnational networks and coalitions to undertake transnational campaigns. Authoritarianism

Table 7.1   Relative intensity of local and transnational campaigns

Project	Level of local activism	Level of transnational activism
Yadana Gas Pipeline Project (Thailand/ Myanmar)	High	High
Thai–Malaysian Gas Pipeline Project (Thailand)	High	Low
Salween Dam Projects (Myanmar)	Low	High
Shwe Gas Pipeline Project (Myanmar)	Low	High

may therefore result in an 'activist diaspora', comprised of expatriates who engage in activism transnationally from outside their home country. In Myanmar's ethnically diverse expatriate community this activist diaspora transcended ethnic divisions and provided cohesive forces that were less visible in the broader exile community. As 'divide and conquer' was one of the Myanmar military's main strategies in neutralising the opposition of ethnic minorities, the resultant activist diaspora stimulated multi-ethnic collaboration, which may contribute to more cooperative multi-ethnic governance in Myanmar in the future.

This activist diaspora, despite the difficulties it posed for activists, also provided opportunities. Training that would not have been accessible in local settings was made available while improved access to other transnational activists and media in cosmopolitan environments also created opportunities to develop strategies and tactics. The contacts and training also resulted in increased proficiency in English, the lingua franca of transnational activism and media, which facilitated the communication of campaign messages to a wider audience. Prior to formation of the new government in 2011 most activists did not petition the military regime directly, as previous experience had demonstrated they were unlikely to influence decision making at all. They therefore focused their energies primarily on transnational activities, hoping to influence international businesses, governments or publics to support their cause. These transnational networks could also form under more liberal or competitive regimes, such as Thailand's but, due to greater opportunities and outlets for dissent at home, activists generally focused on local modes of activism. In the case studies examined here widespread human rights abuses were occurring in Thailand, but the more competitive regime encouraged the vast majority of Thai activists to seek redress at home rather than abroad, limiting the transnational activism that took place.

This evidence suggests a distinctive relationship between the level of authoritarian governance and the predominance of local or transnational activism under hybrid or authoritarian regimes. There appeared to be an inverse relationship between the level of authoritarian governance and the level of local activism, with increasing authoritarianism beyond a tipping point resulting in less local activism, but a direct relationship between the level of authoritarian governance and the level of transnational activism, with increasing authoritarianism beyond a tipping point resulting in greater transnational activism (see Table 7.2).

Table 7.2   Local and transnational activism under political regimes

Regime type	Level of authoritarianism	Adverse impact on environmental insecurity	Local activism	Transnational activism
Hybrid (before tipping point)	Low	Low	High	Low
Authoritarian (after tipping point)	High	High	Low	High

This relationship arose because although the desire to express dissent increased as authoritarianism increased so did the personal costs of expressing that dissent. Under hybrid competitive authoritarianism, before the tipping point, there was sufficient authoritarianism to stimulate high levels of local activism but not enough to suppress it. Activists therefore focused on local activism as there were opportunities to mobilise populations and governments were more responsive to public pressure. In Thailand the government under Thaksin became more authoritarian and provoked greater local activism but it did not reach the tipping point. Under military rule in Myanmar, however, local dissent was so brutally crushed that transnational activism in exile was the only significant outlet. This authoritarianism, while suppressing opposition at home, actually appeared to stimulate transnational linkages and activism, indicating that more traditional authoritarian regimes may be fuelling the growth of the global justice movement in the South by expanding transnational networks of activists.

Movement Organisations

Authoritarian governance and environmental insecurity also impacted on the issues focused upon by environment movements and the philosophies and practices that they employed. Activists and organisations within these movements broadly adhered to the four green pillars of environmental politics in their aims and organisational structure: participatory democracy, ecological sustainability, social justice and nonviolence. Both social relations and ecological issues were therefore accorded equal priority. The emancipatory goals that underpinned these movements resonated with the intellectual traditions of critical security studies and critical international political economy. As environmental movements, however, achieving justice for marginalised communities was intimately linked to issues of ecological health and environmental security.

The campaigns against the projects in Myanmar – a country characterised by poverty, marginalisation and authoritarianism – therefore focused more specifically on these issues, linking environmental destruction to increased militarisation and the resultant undermining of democracy, justice and nonviolence. While the formation of a nominally 'civilian' government in 2011 provided some hope of a progressive transition towards more democratic rule the continued dominance of the military in the political and economic spheres of the country meant that progress would be fragmentary at best. While militarisation was considered the prime cause of injustice and insecurity the accompanying large-scale development paradigm aggravated the marginalisation of local communities. Activists argued that small-scale technologies, such as micro-hydro, solar and biomass, could better satisfy the energy needs of local ethnic minority communities than large-scale socially and environmentally destructive projects, even if they were used to improve local energy security.

During protests on the Thai side of the border, under a more competitive regime, issues of forest ecology were prominent but, equally importantly, activists linked business dealings with the Myanmar military to the suppression of human rights in Myanmar and the delaying of democratic reform. Relatively affluent local activists from the Kanchanaburi Conservation Group (KCG) initially focused on preservation of the forest in the campaign against the Yadana Pipeline but through this activism their awareness of linkages to wider social and political issues tended to radicalise their approach. Activists in these campaigns understood that engaging in activism itself was a transformative process, a result also seen in the North whereby NIMBY local environment groups become emancipatory social movement actors (Doherty 2002: 185; Wall 1999: 25).

A similar transformative process also characterised the campaign against the TTM in Thailand. Most local Muslim villagers were less affluent than the activists from Kanchanaburi and their concerns were predominantly associated with potential pollution from the project and the industrialisation of the region, with the resultant impacts on livelihoods, ecosystems and their way of life. As the campaign progressed, however, they began to view the project both as an attack on their religion and as part of a wider pattern of globalised capitalism. The villagers argued that the undemocratic and unjust decisions on the project were consistent with a pattern of marginalisation perpetrated by the predominantly Buddhist Thai government.

Activists in these various movements also perceived, often through their experience of torture and other violent conflict, that nonviolent activism was the only path to achieving their goals. Although nonviolence took a pre-eminent role in the campaigns, the repression faced by ethnic minorities in Myanmar meant that exiled activists often remained networked to ethnic minority insurgent groups, particularly as these groups often provided the governance for 'liberated areas' under their control. Despite adhering to nonviolence themselves, Northern activists based in the South understood that their experiential understanding in this area would always be limited by their relatively secure life experiences in the North.

With a general concern for social justice within the campaigns, women's issues came to the fore. The campaigns highlighted the adverse impacts of the energy projects on women, particularly ethnic minorities in Myanmar but also in Thailand. The introduction of women's issues into the Thai Yadana campaign provided symbols of feminine strength and values, with activists being aligned with the forest against the brute masculinity of the mechanised excavators, evoking the 'ecological insights' of the women of the Chipko movement in India. These inferences of femininity found parallels in women's movements elsewhere in the South, although bringing women's issues into the campaigns did not suggest a feminine essentialism, but simply that as the prime caretakers, women tend to prioritise education, health and human welfare needs

Although activists promoted participatory democracy for Myanmar under military rule democratic governance was non-existent. The more competitive political environment in Thailand provided limited outlets for two types of participation. The first was activism that registered dissent through protest actions while the second was engagement with the state and business through public participation in state-sanctioned fora. The extent of protest in Thailand was related to the political openings available within formal public fora and the two forms of activism sometimes converged due to dubious participation regimes. State-sanctioned fora in this context included consultative processes whereby participation from the public was sought in development decisions, primarily through processes linked to environmental impact assessment (EIA). Despite criticism of some EIA processes as providing cover for 'business as usual' there is little doubt that the existence of EIA processes provided some transparency and accountability while in Myanmar, where EIA requirements were absent, options for participation were more limited. Despite attempts by activists to accommodate government processes the few occurrences of EIA public participation in these case studies illustrated the large gap that existed between theory and practice.

The four green pillars also influenced the organisational structure of movement organisations, generally resulting in more effective emancipatory activist environmental governance. Analysis of the aims and activities and the structure of organisations within these environment movements produced a typology of activist environmental governance groups. This typology drew on a distinction proposed by Doyle and Doherty (2006) between emancipatory groups and organisations of the environmental governance state (EGS) but provided a more complex and nuanced model of what constitutes environmental governance. Emancipatory actors are quite capable of pursuing a positive and constructive form of environmental governance by engaging in sensitive, localised and interactive activities that contrasts with actors within the neoliberal EGS. By comparing the structures and activities of environmental organisations I identified three categories: emancipatory governance groups (EGGs), which adhered to the four green pillars in both their structures and their activities; compromise governance groups (CGGs), which had conservative structures but broadly adhered to the green pillars in their aims and activities; and organisations of the EGS with conservative aims and structures. EGGs were therefore more 'process-oriented' than the 'outcome-oriented' CGGs, with a focus on providing an emancipatory environment within the organisation. As a result of their conservative structures CGGs were less effective in achieving emancipatory outcomes than EGGs. Organisations within the EGS, however, remained entirely counter-productive in relation to emancipatory objectives.

Due to the emancipatory nature of the campaigns discussed in this book most of the participating organisations qualified as EGGs. The ecological component in their organisation and activities was allied to more central emancipatory concerns focused on achieving justice and human rights for marginalised communities through nonviolent means. CGGs did exist but were rare and often criticised within the movements for not practising the emancipatory values that informed the broader campaign goals. Northern organisations from the EGS did not directly participate in the campaigns but the actions of some did, nevertheless, adversely affect the campaigns in Myanmar by cooperating with the elite and oppressive social structures of the military in the pursuit of post-materialist wilderness values.

Throughout the case studies it was also clear that both local and transnational activist environmental governance was characterised by local cultural or religious influences, although they were more prominent in local campaigns. With Buddhism being the dominant religion of both Thailand and Myanmar, the practice of Engaged Buddhism, either by lay activists or activist monks within the *sangha*, featured prominently in environmental activism despite a diversity of other religious and cultural beliefs. Buddhism was used both strategically as a cultural symbol and tactically during street protests, although often the state, including the Thai monarch and Myanmar's military, competed for a hegemonic Buddhist legitimacy. Activists saw the state as appropriating and subverting the true message of Buddhism and in the Yadana campaign this struggle for legitimacy led to activist Buddhist monks ordaining trees in their attempts to stymie the project. Engaged Buddhist activists framed the project as being contrary to Buddhist values, as inappropriate development for a 'Buddhist economics'.

While Buddhist symbols were frequently deployed, Muslim villagers in the south of Thailand fighting the TTM Pipeline increasingly framed their opposition through their Islamic identity, placing the local marginalisation of Muslims in Thailand into the larger context of a global attack on Muslims in the 'War on Terror'. Environmental activism can be a transformative process and this activity contributed to the villagers' increased awareness of their Muslim identity, which then became a campaign tool. In the Salween campaign the indigenous knowledge and animist beliefs of the Kayin were emphasised in opposition to an unresponsive military that consciously favoured Buddhist institutions. In these cases activists and communities, while strategically employing the symbols of their culture and religion, saw the radical essence of their beliefs as safeguarding their communities against the authoritarian tendencies of governments and complicit business interests. Despite the global reach and essential transnationalism of Islam and Buddhism, communities used these religions as a local symbol of differentiation.

While religious belief systems provided these points of differentiation, other culturally specific forms of activism also resulted in novel forms of resistance. Irony is used in environment movements in the North but the Moustache Brothers employed irony as both rhetorical form and philosophical content in their specifically Myanmar *a-nyeint* satire (Szerszynski 2007: 348). In their improvised performances they demonstrated their opposition to the Myanmar military and their support for Aung San Suu Kyi's ban on foreign investment and the Yadana Project. They continued to perform despite the likelihood of arrest leading to over six years of incarceration for two of the troupe. Doherty has identified manufactured vulnerability as a tactic employed during protests in the North (Doherty 1999) but the torture and hard labour that resulted in this case demonstrated the wide gulf in consequences between manufactured vulnerability in the North and the South.

Although culturally specific symbols were central to the local campaigns they were less significant at the transnational scale. Religious and cultural symbols were still essentialised but in the transnational campaigns' appeals to the international community were often couched in more universalised concepts such as human rights and democracy. While activism celebrated, and was influenced by, local cultural factors, these influences were sometimes sidelined in favour of the rationalist bureaucratic framework of a rights-based approach. Despite this perspective, the adoption of the human rights discourse also lent support to the cultural particularism expressed by some groups in the form of rights for the protection of indigenous knowledge and environments.

Critical Energy and Environmental Security

Analysis of the case studies in this book suggests that it was often concerns over environmental security for local communities that led activists to initiate the campaigns against the energy projects. While activists themselves did not often use the term, the focus of the campaigns on ensuring security for marginalised populations in energy, food, water and other environmental services through the removal of oppressive social structures indicated concerns that are central to critical conceptions of environmental security. Although there has been significant work on critical approaches to environmental security since Barnett's (2001) seminal book, critical approaches to energy security are relatively rare. Some of those that do exist, as I argued in Chapter 1, could also benefit from more 'critical' inquiry. Although the very nature of critical analysis tends to militate against developing exclusive definitive frameworks, in this last section I draw on the case study analysis throughout this book to argue that there are four broad criteria that characterise a more 'critical' approach to energy security. As energy security is a key component of environmental security I then argue that these criteria can be generalised to develop a critical environmental security framework, which can be deployed in the exercise of emancipatory activist environmental governance. Due to its wide usage in more traditional and state-centric security studies, energy security provides a more challenging, and therefore robust, basis than, say, food or water security, for developing a critical model of environmental security. Scholars, policy makers and activists can employ the criteria of the critical energy security model below to assess whether particular energy projects or policies promote genuinely critical energy security. Any approach that satisfies the criteria is likely to result in emancipatory outcomes that foster justice and mitigate existing inequities and insecurities.

The first criterion for a critical approach to energy security is that energy security should be considered as a fundamental and integral component of critical environmental security rather than being addressed as either an isolated concern or simply part of national economic or military security. As Booth argues, environmental security will be the 'central battleground for the theory and practice of world politics over the decades to come' (2007: 57) but if energy security is not considered in this context the link that exists between energy consumption, production and distribution and the social and geographical distribution of any resulting environmental degradation is lost.

Traditional realist analysis in particular lacks this embedding of energy within a broader environmental context. Moran and Russell argue that the realist view remains largely unchallenged and has become entrenched within the energy security discourse: '"[e]nergy security" is now deemed so central to "national security" that threats to the former are liable to be reflexively interpreted as threats to the latter' (2009: 2). As a result they contend that energy security is the 'one area of international life' where large-scale conflict among developed states remains a possibility. This linking of energy with traditional notions of national security has resulted in policies and decisions that have caused widespread social and environmental dislocation. The projects in the case studies throughout this book are all justified on the basis of energy security by receiver countries but because the energy security concept employed is conceived as part of a traditional approach to national security rather than as part of a broader critical environmental security the adverse localised implications of the projects are ignored. This shortcoming reflects a broader critique of traditional conceptions of security studies by critical scholars; that 'states are more often part of the problem than the source of the solution' (Bellamy and McDonald 2002: 373).

This analysis contributes to the second criterion, which is that the referent object of energy security shifts away from the state in the North to marginalised individuals and communities in the South. Adopting the state as the referent object of energy security has historically been most clearly associated with realist or mercantilist schools of thought but a national energy security perspective also predominates within most liberal approaches. Traditionally there has been a focus on energy security in the US and Europe but this analysis now also includes Asia, and particularly China and India (see, for example, Cole 2008; Li 2007; Luft and Korin 2009; Wesley 2007; Yergin 1993; 2011). While from a critical perspective any shift away from the North is welcome the context for this shift remains largely centred on the threats to national energy security in the North that a 'rising Asia' is likely to precipitate.

From a security studies perspective Ciuta also argues that 'the story of energy security is relatively straightforward ... energy *in*security is the product of the contradiction between a general trend of increasing energy consumption and a contradictory trend of decreasing energy reserves' (Ciuta 2010: 126). This definition assumes national or global concerns and a focus on the affluent North. In the case of many communities in the South, particularly those in Myanmar associated with the energy projects examined here, 'increasing energy consumption' in their communities is hardly the problem at all.

These traditional discourses of energy security have been focused on securing energy supplies to ensure the industrial development and, particularly, the military security of the nation-state. This quest for national energy security has barely progressed since the nineteenth century and it is still the main reason for many modern imperial interventions and conflicts (Stokes and Raphael 2010). Klare (2012), a prominent realist scholar, argues that diminishing energy resources, and particularly oil, is a likely cause of severe and enduring threats to energy and national security from both regional competition for the control of energy resources and great power rivalry and conflict.

More liberal and optimistic approaches to energy security also tend to remain focused on the nation-state or the North; they assign a central role to the market in driving innovation and allocating energy resources (Yergin 2011). Some liberal approaches do argue for transcending the state as the route to global energy security (Lesage, Van de Graaf and Westphal 2010; Müller-Kraenner 2008) but these approaches allow for little of the normative emphasis on marginalised populations that characterises a critical approach. Throughout the case studies in this book emancipatory activists sought to protect marginalised local communities whose overall environmental security was threatened by the energy projects but without any prospect of improvement in their energy security.

There is also the potential for this criterion to be reinterpreted by undertaking critical energy security analysis with marginalised communities in the North, who can be considered the South in the North (Doherty and Doyle 2006: 706) as the referent object of security. Due to the concentration of academics and grant funding in the North, however, the existing attempts at critical energy security analysis already tend to focus on these areas (see, for example, the energy chapters in Schnurr and Swatuk 2012). A 'decentring' (Gole 2000) of this analysis towards the less affluent countries of the South would therefore do much more to promote a globalised justice.

Another analytical perspective below the level of the state that can be employed to consider lack of adequate access to energy at the household level is the relatively new concept of 'energy poverty' (Bazilian et al. 2010). Energy poverty is generally defined as a lack of household access to electricity and reliance on traditional biomass fuels for cooking (Sovacool and Drupady 2012: 5). As a model it is more closely related to development studies but viewing energy security from a critical perspective results in a more human-centred security approach that could clearly encompass this concept. As might be expected from its development studies lineage, however, energy poverty tends to be far more quantitative than most critical theoretical analysis; it provides specific daily energy consumption quantities, depending on social and environmental requirements, while critical analysis focuses more on challenging social and political structures.

This critical challenge to dominant structures results in the third criterion, which is that a critical energy security analysis focuses on renewable, low carbon and decentralised energy sources rather than centralised fossil fuel, nuclear and large-scale hydro technologies with energy production owned or controlled by local communities. Realist and liberal approaches to energy security offer no normative preferences in relation to technologies but being embedded within existing hierarchical social and political structures they often result in the persistence of the traditionally subsidised and less sustainable energy sectors. Technology is not neutral (Kranzberg 1986) and it often 'reflects and reinforces existing power relations' (Curran 2006: 75). The socio-cultural appeal of sophisticated technologies has been used by elites to garner support for controversial and expensive technologies in the traditional security field (Peoples 2010) and the preference for large and technologically challenging energy projects permeates much state decision making. In addition the 'interests of those designing the megaproject get built into the system, rather than becoming a latent or unintended result' (Sovacool and Cooper 2013: 67). Any critical approach to energy security, therefore, should include some discussion of the intrinsic bias attached to the energy technologies (Mander 1996: 347–48).

Fossil fuel, nuclear and large-scale hydroelectric energy technologies all favour large-scale industrial development and have centralising political and economic implications. Nuclear technology in particular requires a centralisation of political and economic power that frequently results in a lack of democratic oversight and transparency in relation to the corporations that operate power stations (Kaufmann and Penciakova 2012). Nuclear power has never been economic without significant subsidies and pursuing energy security without exacerbating climate change requires investments in other technologies, 'which pose far fewer problems and which yield quicker returns' (Duffy 2011: 683).

Similarly, despite assurances to the contrary, the electricity from large dam projects is often too expensive for displaced local or rural communities in the South who pay the same rates as wealthy urbanites (Liu 2005). Previous studies have demonstrated that the expected returns on dams have been exaggerated and the impacts on communities and their environments understated (Barber and Ryder 1993; Hildyard 1998; Probe International 2006; WCD 2000). Indigenous peoples and marginalised communities, who often live particularly close to the land, are usually the worst affected by large dams as the land impounded for the reservoir is often forested and provides a livelihood for indigenous communities (Goodland 2006: 21; Newson 1997: 202). Various studies have demonstrated that the practice of 'pork-barrelling' is rife in the large-dam industry with widespread corruption both preceding the decision to build the dams and continuing during construction (Goldsmith and Hildyard 1984: 259–62; McCully 1996: 242–62). These corrupt relationships between governments and business are at the core of many of the problems associated with large-scale energy projects.

Electricity from fossil fuel sources in the South may also be too expensive for local communities unless it is heavily subsidised. While natural gas emits less greenhouse gases than other fossil fuels it remains at best a temporary solution to long-term energy security. As countries in the South such as Myanmar have such poor electricity networks it is the decentralised renewable technologies that are likely to provide more reliable and sustainable energy sources for most communities. NGOs such as the Border Green Energy Team provide solar, micro-hydro and biogas technologies to isolated and ethnic minority communities on the Thai–Myanmar border (Border Green Energy Team 2012) and these decentralised technologies offer the communities a measure of independence from reliance on the state and related oppressive governance structures. These technologies also allow easier ownership and control of energy production by local communities, which is particularly beneficial in environments often characterised by corruption and rent seeking.

These technologies also limit adverse impacts on other ecological processes, which leads to the fourth criterion of critical energy security, that the pursuit of energy security should not adversely affect other aspects of environmental security and that total environmental security should therefore be improved. Realism and liberalism have little to say on the impacts that the pursuit of energy security have on other ecological or social processes. A critical approach to energy security takes as its starting point the integral nature of environmental security and therefore the pursuit of energy security is considered in the context of this wider framework. Many large-scale energy projects have adverse environmental outcomes with large hydroelectric dams often having dire ecological impacts through the disruption of water and nutrient flows. Natural gas use may be less damaging in some respects but potential mercury poisoning at the site of extraction, ecosystem disruption due to pipeline construction and greenhouse gas emissions all cause ecological degradation. Large corporations also often control these energy technologies with resultant environmental security impacts felt far from the end point of consumption; producing increased energy security for relatively affluent consumers but reduced environmental security for those living in the vicinity of the energy sources. These impacts are exacerbated when energy projects are transnational with the environmental externalities largely impacting on communities in the producer country while the benefits are enjoyed across the border.

The above four criteria for critical energy security can also be generalised to provide four criteria for a critical approach to environmental security. First, critical environmental security is comprised of its integral components that include, for example, water, food, energy and climate security (Doyle and Risely 2008; Floyd and Matthew 2013). Each constituent is significant and a reduction in the security for any constituent reduces overall environmental security. This model does not offer a definitive list of what actually comprises environmental security; it is up to each individual critical analysis to justify the inclusion of its components. Second, the referent object shifts away from the state to marginalised individuals and communities, predominantly in the global South. Third, the focus is on sustainable, renewable and decentralised solutions to environmental security that are owned or controlled by local communities rather than centralised, large-scale technologies. In relation to food security, for example, industrial agriculture may increase food production in the short term but it also reduces ecological diversity and food nutrition, concentrates wealth and power and destroys the connection to the land that sustains many communities in the South (Shiva 2008). Fourth, each component of environmental security should not be pursued at the expense of another component, such that total environmental security is improved. These models of critical energy and environmental security can be deployed in the analysis of energy or other development policies to promote emancipatory outcomes focused on sustainability and justice. The criteria can therefore be applied to provide a critical security perspective for emancipatory actors engaging in activist environmental governance.

Conclusion

Activist environmental governance of transnational energy projects in the South is often a response to the environmental insecurity faced by marginalised populations resulting from the symbiotic interests of large business interests and the most powerful political actors. Throughout this book unjust social structures played a significant role in determining the nature and extent of the environmental activism against the energy projects at both local and transnational scales. There appeared to be a consistent thread running through these energy projects where a disregard for human rights was linked to environmental insecurity, with large business interests complicit in the perpetuation of authoritarian governance under various political regimes. In both Thailand and Myanmar political and business elites pursued transnational energy projects that, while enriching these elites, caused human and environmental insecurity for local communities in the vicinity of the projects. Many of these communities already faced marginalisation, being indigenous or ethnic minorities. Reciprocity between elites in government and business therefore facilitated both the causes of insecurity and the attempts to silence voices of dissent. The attempt to close down political debate on these energy projects was representative of wider disjunctures in society between the interests of large capital and marginalised communities. These communities received support in their struggles from emancipatory governance groups (EGGs), which adhered to the core green pillars of participatory democracy, social justice, ecological sustainability and nonviolence. These EGGs faced repression for challenging powerful elites and activists in both Thailand and Myanmar risked incarceration, injury and death.

The rationale of energy security was often used to justify the energy projects in this book but the need for these projects remains in doubt. Thailand imported gas from Myanmar in the west via the Yadana Pipeline while it exported gas to Malaysia in the south via the Thai–Malaysian Pipeline. Both these projects caused dislocation to various local communities and their environments while business, government and military elites reaped the dividends. The Salween Dams in Myanmar were designed to supply EGAT and the inefficient Thai electricity market but the affected ethnic communities would likely face displacement and continued energy insecurity without receiving any related revenue. The Yadana Pipeline has been sending gas from Myanmar to Thailand since 1999 while throughout Myanmar electricity access and reliability is something of a regional embarrassment. While the lessons learnt from the Yadana Project could have been used to mitigate the suffering of local communities in other projects, it appeared that collusion between TNCs and illiberal governments resulted in similar outcomes.

The motivation behind these projects, despite the discourse of energy security and development, must be seriously questioned. Nowhere in these case studies do benefits flow to local inhabitants who invariably bore the true costs of the projects through increased insecurity and human rights violations. The top-down hierarchical development model with its attendant over-dependence on centralised and capital-intensive sources of energy is unsustainable both in the North and the South. It is only by increasing public participation and transparency in decision making (Forsyth and Walker 2008: 244–46), together with a 'deconcentration and decentralisation of institutional power' (Bello 2004: 115) at national, regional and global levels that appropriate energy projects are likely to provide sustainable and socially beneficial outcomes for the future. Resource exploitation in the South can either promote or hinder equitable development and embedding democratic processes can thwart the formation of patrimonial states, which hamper effective resource use (Thorp et al. 2012). Gas and micro-hydro projects, in particular, could be beneficial in alleviating energy insecurity for marginalised communities, especially in Myanmar, but until local communities fully participate in decision making and share the benefits of their local resources environmental insecurity is likely to remain widespread. Only the application of a truly critical energy and environmental security framework is likely to consider the needs of the most marginalised communities and provide an analysis founded on justice.

The seemingly inextricable link between major energy projects, authoritarian governments and human rights violations continues to this day in Myanmar, Thailand and across the South. Emancipatory environment movements across the South continue to promote a paradigm shift from the dominant large-scale energy security discourse of nation-states and their elites to a more holistic, small-scale and localised environmental security agenda whereby security, for marginalised communities, becomes a term linked to their well-being rather than their oppression.

Environmental actors throughout these campaigns pursued these perspectives, founded on the philosophical bedrock of radical environmental politics. Activists, groups, networks and coalitions demonstrated a common commitment to the green pillars, resulting in emancipatory activist environmental governance. These green ideals were not only pursued in the strategies and tactics of the campaigns but also in the structure and composition of their organisations with a focus on equity and a lack of hierarchy. Despite often being unable to halt the projects, EGGs in these campaigns succeeded in other important respects by promoting a generation of socially and environmental aware community activists and leaders. Engaging in activism was itself a transformative process that stimulated the pursuit of justice. Improvements in the behaviour of TNCs, particularly from the North, that have mitigated some environmental insecurities over time can also be attributed to the tremendous international pressure brought to bear on the corporations through this activist environmental governance.

Although the campaigns focused upon in Thailand could be considered the highpoint of Thai environmental activism, activists there continue their important work in promoting environmental security. The Map Ta Phut community, for instance, continues its struggle, on the streets and in the courts, to provide its children with a clean and safe environment in which to grow and live. In Myanmar a glimmer of hope has emerged among its long-suffering communities, with emergent democratic openings and a responsive government for the first time in generations. Newly introduced fora such as the EITI Multi-Stakeholder Group have provided a hitherto absent space for government, industry and civil society to openly discuss and collaborate to help build trust for the future. The domestic environment movement is embryonic but with the assistance of the activist diaspora there is the possibility of developing a more robust activist community within Myanmar itself that promotes socially and environmentally appropriate development through effective and emancipatory activist environmental governance.

Affiliation of Interviewees

Acaroglu, L. Coordinator of Salween Dams Protest. Melbourne, Australia.

ADB Myanmar Country Coordinator (name withheld). Asian Development Bank (ADB). Bangkok, Thailand.

Alex Shwe. Activist with Karen Environmental and Social Action Network (KESAN) (aka Ko Shwe). Chiang Mai, Thailand.

Ariyaratne, A. T. Founder and President Sarvodaya Shramadana Movement. Moratuwa, Sri Lanka and Devon, UK.

Aung Aung Naing. Member of 24 Hour Electricity Committee in Rakhine State. Yangon, Myanmar.

Aung Htoo. Burma Lawyer's Council (BLC) General Secretary. Mae Sot, Thailand.

Aung Kyaw Soe. Activist with Thazin Development Foundation. Yangon, Myanmar.

Aung Naing Oo. Associate Program Director, Peace Dialogue Program, Myanmar Peace Centre. Yangon, Myanmar.

Aung Ngyeh (pseudonym). Activist with Karenni Development Research Group (KDRG) and Coordinator of Burma Rivers Network (BRN). (via email from) Mae Hong Song, Thailand.

Aung-Thwin, M. Director of the Burma Project/Southeast Asia Initiative at the Open Society Foundations. New York, USA.

Aye Tha Aung. Translated from Rakhine by Zaw Myat Lin. General Secretary, Arakan League for Democracy (ALD), Secretary of Committee Representing the People's Parliament (CRPP) Yangon, Myanmar.

Banjong Nasae. Director of Southern Coastal Management Project. (via email from) Hat Yai, Songkhla Province, Thailand.

Beng Tuan. Activist with Third World Network. Penang, Malaysia.

Bergoffen, M. Lawyer with KESAN. Chiang Mai, Thailand.

Bobby Maung. CEO, Network Activities Group (NAG). Yangon, Myanmar.

CGG Employees (names withheld), Employees of anonymous Compromise Governance Group (CGG). Thailand.

CGG, Former employee (name withheld), Former employee of anonymous CGG. Thailand.

CGG, External activist (name withheld), Long-term Myanmar activist based in Thailand commenting on CGG. Thailand

Chana Maung. Southeast Asia Office Director, EarthRights International (ERI). Chiang Mai, Thailand.

Choe, M. Executive Director, Korean House for International Solidarity (KHIS). (via email from) Seoul, South Korea.

Cynthia Maung. Founder of Mae Tao Clinic. Mae Sot, Thailand.

Donowitz, P. Campaigns Director. EarthRights International (ERI). Chiang Mai, Thailand.

Eddie Mee Reh. Karenni Development Research Group (KDRG). Ban Nai Soi, Mae Hong Son, Thailand.

ERI Program Coordinator (name withheld). EarthRights International (ERI). Chiang Mai, Thailand.

ERI Assistant Team Leader (name withheld). EarthRights International (ERI). Chiang Mai, Thailand.

Founder of Myanmar-based Environmental NGO (details withheld). Yangon, Myanmar.

Giannini, T. Co-Founder EarthRights International (ERI). Now Clinical Professor of Law and Director of the International Human Rights Clinic, Harvard Law School. Bangkok, Thailand.

Green, S. (pseudonym). Activist with KESAN, formerly with Images Asia E-Desk. On the Thai-Myanmar border (location witheld).

Hsiplopo. Ei Tu Hta Camp Leader/Chairman. Ei Tu Hta Camp, KNU controlled Myanmar on the Salween River.

Ida Aroonwong. Activist with Alternative Energy Project for Sustainability. Bangkok, Thailand.

Jockai Khaing. Executive Director of Arakan Oil Watch (AOW). Chiang Mai, Thailand.

Junatoo. Volunteer nurse from Mae La Refugee Camp. Ei Tu Hta Camp, KNU controlled Myanmar on the Salween River.

K'nyaw Paw. Karen Women's Organization (KWO) and 2006 International Women's Development Agency (IWDA) Ambassador. (via email from) Mae Sariang, Thailand.

Ka Hsaw Wa. Co-Founder/Executive Director EarthRights International (ERI). Chiang Mai, Thailand.

Khin Nanda. Program Coordinator, EarthRights School Myanmar. Chiang Mai, Thailand.

Khor, M. Formerly Director, Third World Network. Penang, Malaysia.

King, D. Asia Legal Director, EarthRights International (ERI). Chiang Mai, Thailand.

Kyaw Han. President, All Arakan Students and Youth Congress (AASYC). New Delhi, India.

Kyaw Thu Aung. Director, Paung Ku. Yangon, Myanmar.

Lazarus, K. Oxfam America (at time of interview). Phnom Penh, Cambodia.

Lohmann, L. Researcher at The Corner House. (via email from) Dorset, UK.

Lwin Lwin Nao. Activist with Palaung Women's Organisation (PWO). Mae Sot, Thailand.

Mai Aung Ko. Ta-aung Student and Youth Organisation (TSYO). Mae Sot, Thailand.

Marshall, S. Liaison Officer, International Labour Organization (ILO). Yangon, Myanmar.

Melanson, H. Teacher and Conflict Transformation Coordinator, EarthRights School Myanmar. Chiang Mai, Thailand.

Min Zar Ni Lin. Research Associate, Centre for Economic and Social Development, Myanmar Development Resource Institute. Yangon, Myanmar.

Morris, R. Assistant Director, Southeast Asia Office of ERI. (via email from) Chiang Mai, Thailand.

Myint Aung. Union Minister, Ministry of Mines. Nay Pyi Taw, Myanmar.

Myint Thein. NLD-Liberated Area General Secretary and Burma Lawyers' Council (BLC) Founding Executive Council Member. Mae Sot, Thailand.

Naing Htoo. Acting Myanmar Project Director, EarthRights International (ERI). Chiang Mai, Thailand.

Nay Tha Blay. Director of Karen Office of Relief and Development (KORD) and activist with Karen Rivers Watch (KRW). Mae Sariang, Thailand.

Nge Reh. Karenni Evergreen (KEG). Ban Nai Soi, Mae Hong Son, Thailand.

Nizam Mahshar. Research Coordinator, Friends of the Earth Malaysia, Sahabat Alam Malaysia. Penang, Malaysia.

Oo Hla Saw. General Secretary, Rakhine Nationalities Development Party (RNDP). Yangon, Myanmar.

Par Par Lay (deceased) and Lu Maw. 'The Moustache Brothers'. Mandalay, Myanmar.

Patrick (pseudonym). Northern activist with 'Myanmar NGO' with Thai and foreign employees. Chiang Mai, Thailand.

Paul Sein Twa. Founder and Director, Karen Environmental and Social Action Network (KESAN). On the Thai-Myanmar border (location witheld).

Paw Wah. Activist with Karen Rivers Watch (KRW). (via email from) Chiang Mai, Thailand.

Penchom Tang. Director of Campaign for Alternative Industry Network. (via email from) Bangkok, Thailand.

Phinan Chotirosseranee. Co-President, Kanchanaburi Conservation Group. Kanchanaburi Province, Thailand.

Phyo Phyo. Activist with Shwe Gas Movement (SGM). Yangon, Myanmar.

Pipob Udomittipong. Activist with Salween Watch (formerly of EarthRights International).

Prasart Meetam. Assistant Professor in Mathematics, Prince of Songkla University. Hat Yai, Thailand.

Premrudee Daoroung. Co-Director Towards Ecological Recovery and Regional Alliance (TERRA). Bangkok, Thailand.

Quigley, J. US Campaign for Burma Staff Member. Washington DC, USA.

Raynaud, M. Activist with Myanmar Egress. Yangon, Myanmar.

Redford, K. Co-Founder/Director EarthRights International (ERI). Chiang Mai, Thailand.

Reungchai Tansakul. Professor, Prince of Songkla University. (via email from) Hat Yai, Thailand.

Ryrie, R. Activist with EarthRights International (ERI). Yangon, Myanmar.

Sai Sai. Founder of Shan Sapawa Environmental Organization (Sapawa) and coordinator of Salween Watch and Burma Rivers Network (BRN). Chiang Mai, Thailand.

San Ray Kyaw. Central Executive Committe Member, All Arakan Students and Youth Congress (AASYC). Chiang Mai, Thailand.

Sanchai. Activist and wife of village deputy chairman. Ban Ta Tar Fung on the Salween River, Thailand.

Shiva, V. Indian environmental activist. Dehra Dun, India.

Simons, M. Legal Director, EarthRights International (ERI). Washington DC.

Smith, M. Senior Consultant to EarthRights International (ERI) (and later Human Rights Watch consultant). Chiang Mai, Thailand.

Soe Myint. Editor and Founder, Mizzima News. New Delhi, India.

Sulak Sivaraksa. Thai social activist and Co-Founder International Network of Engaged Buddhists. Bangkok, Thailand.

Supara Janchitfah. Journalist at *Bangkok Post*. Bangkok, Thailand.

Tenzin Gyatso. His Holiness the Fourteenth Dalai Lama. Dharamsala, India.

Terraz, N. General Manager, Total E&P Myanmar. Yangon, Myanmar.

Thein Oo. Chairman, Burma Lawyers' Council (BLC). Mae Sot, Thailand.

Thu Rein. Campaign Coordinator with Arakan Oil Watch (AOW). Chiang Mai, Thailand.

Tun Lwin. Former Director General of Bureau of Meteorology. Yangon, Myanmar.

Varaporn Chamsanit. Former human rights activist with Thailand Research Fund. Bangkok, Thailand.

Wandee Suntivutimetee. Editor, Salween News Network. Chiang Mai, Thailand.

Win Myo Thu. Founder and Managing Director, ECODEV. Yangon, Myanmar.

Winston Set Aung. Deputy Minister, Ministry for National Planning and Economic Development. Nay Pyi Taw, Myanmar.

Wolsak, R. Alumni Program Coordinator, EarthRights School Myanmar. Chiang Mai, Thailand.

Wong Aung. Global Coordinator, Shwe Gas Movement (SGM). Chiang Mai, Thailand.

Woo, J. Vice-President of Buddhist TV Network (BTN). Seoul, South Korea.

World Bank, Myanmar Country Program Coordinator (name withheld). Washington DC, USA.

Zaw Oo. Director of Research and Programs, Centre for Economic and Social Development, Myanmar Development Resource Institute. Yangon, Myanmar.

References

ACE 2003. *Proposed TAGP in AEEMTRC 1996 Masterplan.* ASEAN Centre for Energy (ACE). http://www.aseanenergy.org/energy_sector/natural_gas/ proposed_tagp_aeemtrc_1996_mplan.htm. Updated: 14 January. Accessed: 12 June 2005.

Achara Ashayagachat 15 February 2008. 'Jon says future's bleak for human rights here.' *Bangkok Post.* http://www.bangkokpost.net/News/15Feb2008_news03. php. Accessed: 15 February 2008.

Acuto, M. 2008. 'Edges of the conflict: A three-fold conceptualization of national borders.' *Borderlands,* 7(1). http://www.borderlands.net.au/vol7no1_2008/ acuto_edges.htm. Accessed:1 December 2011.

Ad Hoc Commission on the Depayin Massacre 2003. 'Preliminary report of the Ad Hoc Commission on the Depayin Massacre (Burma).' *Legal Issues on Burma.* August(15): 1–62.

AFP 4 July 2005a. 'France rejects rights groups' calls for Total to leave Myanmar.' *Agence France Presse (AFP).*

—— 29 November 2005b. 'Total settles out of court in Myanmar forced labour case.' *Agence France Presse (AFP).*

—— 16 January 2007a. 'China's CNPC inks exploration deal for Myanmar oil, gas.' *Agence France Presse (AFP).*

—— 23 August 2007b. 'Protesters continue to defy Burmese junta.' *ABC News.* http://www.abc.net.au/news/stories/2007/08/23/2013678.htm.

—— 26 July 2013. 'Myanmar revises Monywa copper mine deal.' *Myanmar Times.* http://www.mmtimes.com/index.php/national-news/7581-myanmar- revises-monywa-copper-mine-deal.html. Accessed: 10 September 2013.

AHRC 1998. *Sulak charged but Yadana gas pipeline continues.* Asian Human Rights Commission (AHRC). 25 March. http://www.ahrchk.net/ua/mainfile. php/1998/23/.

AITPN 2005. *Written statement submitted by the Asian Indigenous and Tribal Peoples Network (AITPN) a NGO with special consultative status.* UN Commission on Human Rights, Promotion and Protection of Human Rights: Human Rights Defenders. 27 January. E/CN.4/2005/NGO/15. http://ap.ohchr. org/documents/dpage_e.aspx?c=182&su=180.

Al Jazeera. 31 January 2008. 'Myanmar resources offer few riches'. http://english. aljazeera.net/NR/exeres/9AB69756-8B71-470B-B30D-2996DE70D360.htm. Accessed:3 March 2008.

Alamgir, J. 2008. 'Myanmar's foreign trade and its political consequences.' *Asian Survey,* 48(6): 977–96.

Albritton, R. B. and Thawilwadee Bureekul 2007. 'Public opinion and political power: Sources of support for the coup in Thailand.' *Crossroads: An Interdisciplinary Journal of Southeast Asian Studies*, 19(1): 20–49.

Alexander's Gas and Oil Connections. 3 June 2003. 'Thai-Malaysian gas pipeline project expects further opposition'. http://www.gasandoil.com/goc/news/nts32305.htm. Accessed: 5 July 2003.

Allison, T. 5 August 2000. 'The Trans Thai-Malaysia natural gas project.' *Asia Times Online.* http://www.atimes.com/se-asia/BH05Ae02.html.

Amnesty International 2001. *The Moustache Brothers Are Released!* Amnesty International (AI). http://web.amnesty.org/library/Index/ENGASA160182001. Updated: 13 July. Accessed: 2007.

—— 2004. *Thailand: Memorandum on Human Rights Concerns.* Amnesty International.   27   October.   http://web.amnesty.org/library/Index/ENGASA390132004?open&of=ENG-THA.

Anchalee Kongrut 30 June 2003. 'Children call on police to return home.' *Bangkok Post.*

Anderson, M. R. 1996. 'Human rights approaches to environmental protection: An overview.' in *Human Rights Approaches to Environmental Protection* A. E. Boyle and M. R. Anderson eds. Oxford, Clarendon Press.

Anthias, F. 1998. 'Evaluating "diaspora": Beyond ethnicity.' *Sociology*, 32(3): 557–80.

AP 4 April 2006. 'Myanmar, Thailand sign US$6 billion hydropower plant agreement.' *Associated Press (AP).*

—— 15 November 2007a. '14 S Koreans convicted of exporting weapons technology to Myanmar.' *The Times of India (Associated Press).* http://timesofindia.indiatimes.com/World/14_S_Koreans_convicted_of_exporting_weapons_technology_to_Myanmar/articleshow/2543381.cms.

—— 4 September 2007b. 'Thai hydro project workers evacuated from Myanmar after shelling kills 1.' *International Herald Tribune (Associated Press).*

Apple, B. 1998. *School for Rape: The Burmese Military and Sexual Violence.* EarthRights International (ERI), Washington DC. http://www.earthrights.org/files/Reports/schoolforrape.pdf. Accessed: 18 February 2008.

Apple, B. and Martin, V. 2003. *No Safe Place: Burma's Army and the Rape of Ethnic Women.* Refugees International. April. http://www.refugeesinternational.org/content/publication/detail/3023/.

Arakan Oil Watch 2008. *Blocking Freedom: A Case Study of China's Oil and Gas Investments in Burma.* Arakan Oil Watch, Chiang Mai. October. http://www.shwe.org/wp-content/uploads/2011/03/Blocking-Freedom-English.pdf. Accessed:4 February 2012.

Ardeth Maung Thawnghmung 2004. *Behind the Teak Curtain: Authoritarianism, Agricultural Policies and Political Legitimacy in Rural Burma/Myanmar.* London: Kegan Paul.

Arnold, W. 7 October 2000. 'Power plant in Thailand is criticized.' *New York Times.* http://query.nytimes.com/gst/fullpage.html?res=950CE3DE153CF934

A35753C1A9669C8B63&n=Top/Reference/Times Topics/Organizations/W/ World Bank.

Arora-Jonsson, S. 2013. *Gender, Development and Environmental Governance.* London and New York: Routledge.

Ashoka 2006. *Ashoka Fellows: Katie Redford.* http://www.ashoka.org/node/3710. Accessed: 2006.

Asian Development Bank 2003. *Indonesia-Malaysia-Thailand Growth Triangle (IMT-GT).* Asian Development Bank (ADB). http://www.adb.org/IMT/default. asp. Updated: 15 September. Accessed: 2007.

—— 2012. *Myanmar: Energy Sector Initial Assessment.* Asian Development Bank (ADB), Manila. October. http://www.adb.org/sites/default/files/myanmar-energy-sector-assessment.pdf. Accessed: 1 December 2012.

Askew, M. 2008. 'Thailand's intractable southern war: Policy, insurgency and discourse.' *Contemporary Southeast Asia.* August, 30(2): 186–214.

Atkinson, J. and Scurrah, M. 2009. *Globalizing Social Justice: The Role of Non-Government Organizations in Bringing About Social Change.* Basingstoke: Palgrave Macmillan.

Aung Hla 12 February 1998. 'Thailand pipeline controversy.' *Voice of America (VOA).* Washington.

Aung Htoo 2003. 'Depayin massacre, crime against humanity, national reconciliation and democratic transition.' *Legal Issues on Burma.* August(15): 63–69.

Aung Lwin Oo 30 November 2005. 'Total settles out of court over pipeline.' *The Irrawaddy.* http://www.irrawaddy.org/article.php?art_id=5232. Accessed: 13 February 2008.

Aung San Suu Kyi 1997. *Letters from Burma.* London: Penguin Books.

Aye Mi San 2007. *Arrest of Suu Suu Nway and Protesters.* http://vids.myspace. com/index.cfm?fuseaction=vids.individual&VideoID=16956884. Updated: 29 August. Accessed: 25 April 2012.

Ba Kaung March 2010. 'Selling off the state silver.' *The Irrawaddy.* http://www. irrawaddymedia.com/article.php?art_id=17928. Accessed: 5 March 2011.

Baker, C. and Pasuk Phongpaichit 2005. *A History of Thailand.* Cambridge: Cambridge University Press.

Bandy, J. and Smith, J. eds. 2005. *Coalitions across Borders: Transnational Protest and the Neoliberal Order.* Lanham, Maryland and Oxford: Rowman and Littlefield.

Bangkok Post. 9 November 1997. 'Environmentalists turn to religion to save doomed forest'.

—— 17 July 2001. 'Rangoon free to spend gas money on anything it wants'. http://www.bangkokpost.net/. Accessed: 20 August 2005.

—— 22 October 2007. 'Two-thirds willing to sell their vote for cash'. http://www. bangkokpost.net/News/22Oct2007_news02.php. Accessed: 22 October 2007.

—— 6 May 2012. 'Map Ta Phut toll: 12 dead, 129 injured'. http://www.bangkokpost.com/news/local/292091/map-ta-phut-toll-12-killed129-wounded. Accessed: 9 May 2012.

Barber, M. and Ryder, G. eds. 1993. *Damming The Three Gorges: What Dam Builders Don't Want You To Know.* Second edition. Probe International, Earthscan Publications. http://www.threegorgesprobe.org/pi/documents/three_gorges/Damming3G/index.html.

Barnett, J. 2001. *The Meaning of Environmental Security: Ecological Politics and Policy in the New Security Era.* New York: Zed Books.

Bartlett, R. V. 2005. 'Ecological reason in administration: Environmental impact assessment and green politics.' in *Managing Leviathan.* Second edition R. Paehlke and D. Torgerson eds. Peterborough, Ontario: Broadview Press.

Bazilian, M., Sagar, A., Detchon, R. and Yumkella, K. 2010. 'More heat and light.' *Energy Policy,* 38(10): 5409–12.

BBC News. 23 August 2007. 'Burma activists protest over fuel'. http://news.bbc.co.uk/2/hi/asia-pacific/6959724.stm. Accessed: 4 June 2008.

BCN 2006. *Worldwide Protests Will Mark 1st Anniversary of Total Campaign.* Burma Centre Netherlands (BCM), Amsterdam. 31 January.

BCUK 2005a. *Burma to Dominate TOTAL Oil AGM.* Burma Campaign UK (BCUK). 16 May. http://www.burmacampaign.org.uk/pm/more.php?id=163_0_1_0_M.

—— 2005b. *TOTALitarian Oil: Fuelling the Oppression in Burma.* Burma Campaign UK (BCUK). February. http://www.burmacampaign.org.uk/PDFs/total report.pdf.

Beck, U. 1999. *World Risk Society.* Cambridge: Polity Press.

Beder, S. 1996. *The Nature of Sustainable Development.* Second edition. Carlton: Scribe.

Beeson, M. 2010. 'The coming of environmental authoritarianism.' *Environmental Politics,* 19(2): 276–94.

Beeson, M. and Bellamy, A. J. 2003. 'Globalisation, security and international order after 11 September.' *Australian Journal of Politics and History,* 49(3): 339–54.

—— 2008. *Securing Southeast Asia: The Politics of Security Sector Reform.* Oxford and New York: Routledge.

Beetham, D. 1992. 'Liberal democracy and the limits of democratization.' *Political Studies,* XL Special Issue: 40–53.

Bellamy, A. J. and McDonald, M. 2002. 'The utility of human security: Which humans? What security? A reply to Thomas and Tow.' *Security Dialogue,* 33(2): 373–77.

Bello, W. 2004. *Deglobalization: Ideas for a New World Economy.* Updated edition. London and New York: Zed Books.

Bello, W., Cunningham, S. and Poh, L. K. 1998. *A Siamese Tragedy: Development and Disintegration in Modern Thailand.* London and New York: Zed Books.

Biel, E., Hicks, N. and McClintock, M. eds. 2006. *Losing Ground: Human Rights Defenders and Counterterrorism in Thailand*. New York and Washington DC: Human Rights First (HRF).

Birtwell, P. 2005. *Letter to Ponglert Pongwanan, Alternative Energy Project for Sustainability (AEPS)*. Barclays Bank, London. 21 September.

Blair, A. and Hitchcock, D. 2000. *Environment and Business*. London and New York: Routledge.

Blondet, C. 2002. 'The "devil's deal": Women's political participation and authoritarianism in Peru.' in *Gender, Justice, Development and Rights* M. Molyneux and S. Razavi eds. Oxford: Oxford University Press.

Blowfield, M. 2007. 'Reasons to be cheerful? What we know about CSR's impact.' *Third World Quarterly*. June, 28(4): 683–95.

Bo Kong 2010. 'The geopolitics of the Myanmar-China oil and gas pipelines.' in *Pipeline Politics in Asia: The Intersection of Demand, Energy Markets, and Supply Routes* E. Chow, L. E. Hendrix, M. E. Herberg et al. eds. Seattle: The National Bureau of Asian Research.

Bollard, R. 1999. 'Environmental justice challenges at home and abroad.' in *Global Ethics and Environment* N. Low ed. London: Routledge.

Bomberg, E. and Schlosberg, D. eds. 2008. *Environmentalism in the United States: Changing Patterns of Activism and Advocacy*. London and New York: Routledge.

Bond, G. D. 2003. 'Sarvodaya Shramadana's quest for peace.' in *Action Dharma: New Studies in Engaged Buddhism* C. Queen, C. Prebish and D. Keown eds. London: RoutledgeCurzon.

Booth, K. 2007. *Theory of World Security*. Cambridge: Cambridge University Press.

Border Green Energy Team 2012. *Border Green Energy Team (BGET)*. Mae Sot. http://www.bget.org. Accessed: 10 December 2012.

Boyle, A. E. and Anderson, M. R. eds. 1996. *Human Rights Approaches to Environmental Protection*. Oxford: Clarendon Press.

Brecher, J., Costello, T. and Smith, B. 2000. *Globalization from Below: The Power of Solidarity*. Cambridge, Massachusetts: South End Press.

Bretherton, C. 2003. 'Movements, networks, hierarchies: A gender perspective on global environmental governance.' *Global Environmental Politics*. May, 3(2): 103–19.

Brewer, N. 2008. *The New Great Walls: A Guide to China's Overseas Dam Industry*. International Rivers, Berkeley. http://www.internationalrivers.org/files/New Great Walls low res.pdf. Accessed: 18 August 2008.

Brown, A. 2004. *Labour, Politics and the State in Industrializing Thailand*. London and New York: Routledge.

—— 2007. 'Labour and modes of participation in Thailand.' *Democratization*. December, 14(5): 816–33.

Browning, C. S. and McDonald, M. 2011. 'The future of critical security studies: Ethics and the politics of security.' *European Journal of International Relations* OnlineFirst(27 October 2011).

Brownlee, J. 2007. *Authoritarianism in an Age of Democratization*. New York: Cambridge University Press.

Brubaker, R. 2005. 'The "diaspora" diaspora.' *Ethnic and Racial Studies*, 28(1): 1–19.

Bruff, I. and Tepe, D. 2011. 'What is critical IPE?' *Journal of International Relations and Development*, 14(3): 354–58.

Brunner, J., Talbott, K. and Elkin, C. 1998. *Logging Burma's Frontier Forests: Resources and the Regime*. World Resources Institute, Washington DC. August. http://pdf.wri.org/loggingburmasfrontierforests_bw.pdf. Accessed: 22 May 2012.

Burma Rivers Network 2007. *Press Release: Dam-Affected Communities Unite to form Burma Rivers Network*. Burma Rivers Network (BRN). 5 June. http://www.ksdf.org/read.asp?title=Dam-affected communities unite to form Burma Rivers Network&CatId=Publications&id=24. Accessed: 14 April 2008.

—— 2008. *Press Release: Nature's Wake-Up Call to Burma*. Burma Rivers Network, Chiang Mai. 7 May 2008.

—— 2012. *Burma Rivers Network*. Chiang Mai. http://www.burmariversnetwork. org/. Updated: 10 April. Accessed: 21 May 2012.

Butler, K. D. 2001. 'Defining diaspora, refining a discourse.' *Diaspora*, 10(2): 189–219.

Cabezon, J. I. 1996. 'Buddhist principles in the Tibetan liberation movement.' in *Engaged Buddhism: Buddhist Liberation Movements in Asia* C. S. Queen and S. B. King eds. Albany: State University of New York Press.

Callahan, M. 2009. 'Myanmar's perpetual junta.' *New Left Review*, 60: 27–63.

Callahan, M. P. 2003. *Making Enemies: War and State Building in Burma*. Ithaca: Cornell University Press.

Calvert, P. and Calvert, S. 1999. *The South, the North and the Environment*. London and New York: Pinter.

Caniglia, B. 2002. 'Elite alliances and transnational environmental movement organizations.' in *Globalization and Resistance: Transnational Dimensions of Social Movements* J. Smith and H. Johnston eds. Lanham, MD: Rowman and Littlefield Publishers.

Carmin, J. and Agyeman, J. eds. 2011. *Environmental Inequalities beyond Borders: Local Perspectives on Global Injustices*. Cambridge, MA: MIT Press.

Carmin, J. and Fagan, A. 2010. 'Environmental mobilisation and organisations in post-socialist Europe and the former Soviet Union.' *Environmental Politics*, 19(5): 689–707.

Carroll, T. and Sovacool, B. 2010. 'Pipelines, crisis and capital: Understanding the contested regionalism of Southeast Asia.' *The Pacific Review*. December, 23(5): 625–47.

Carter, N. 2007. *The Politics of the Environment: Ideas, Activism, Policy*. Second edition. Cambridge: Cambridge University Press.

Case, W. 2007. 'Democracy's quality and breakdown: New lessons from Thailand.' *Democratization*. September, 14(4): 622–42.

—— 2009. 'Low-quality democracy and varied authoritarianism: Elites and regimes in Southeast Asia today.' *The Pacific Review*, 22(3): 255–69.

Castells, M. 2000. 'Materials for an exploratory theory of the network society.' *British Journal of Sociology*. January–March, 51(1): 5–24.

—— 2003. *The Power of Identity*. Second edition. Oxford: Blackwell Publishing.

Catney, P. and Doyle, T. 2011. 'The welfare of now and the green (post) politics of the future.' *Critical Social Policy*, 31(2): 174–93.

CCR and ERI 2009. *Settlement Reached in Human Rights Cases Against Royal Dutch/Shell*. The Center for Constitutional Rights (CCR) and EarthRights International (ERI), New York. 8 June. http://www.ccrjustice.org/newsroom/press-releases/settlement-reached-human-rights-cases-against-royal-dutch/shell. Accessed: 25 June 2009.

Cerletti, F. 2013. *Engaging with the Total Oil Corporation in Myanmar: The Impact of Dialogue as a Tool for Change Towards Greater Conflict Sensitivity*. Coventry, Unpublished PhD Thesis, Coventry University.

Cerny, P. G. 2000. 'Restructuring the political arena: Globalization and the paradoxes of the competition state.' in *Globalization and its Critics: Perspectives from Political Economy* R. D. Germain ed. London: Macmillan Press.

Chaiwat Satha-Anand 1999. 'Thailand: The layers of a strategic culutre.' in *Strategic Cultures in the Asia-Pacific Region* K. Booth and R. Trood eds. London: Macmillan.

—— 2006a. 'Fostering "authoritarian democracy": The effect of violent solutions in Southern Thailand.' in *Empire and Neoliberalism in Asia* V. R. Hadiz ed. London and New York: Routledge.

—— 2006b. 'Transforming terrorism with Muslims' nonviolent alternatives?' in *Contemporary Islam: Dynamic, not Static* A. A. Said, M. Abu-Nimer and M. Sharify-Funk eds. London and New York: Routledge.

—— 2007. 'Reflections on October 6, 1976: Time and violence.' *Crossroads: An Interdisciplinary Journal of Southeast Asian Studies*, 19(1): 185–97.

Chang Noi 2009. *Jungle Book: Thailand's Politics, Moral Panic and Plunder, 1996-2008*. Chiang Mai: Silkworm.

Charney, M. W. 2009. *A History of Modern Burma*. Cambridge: Cambridge University Press.

Chatchai Ratanachai 2000a. *Environmental Impact Assessment Trans Thailand Malaysia (TTM) Project: Gas Pipeline*. Faculty of Environmental Management, Prince of Songkla University, Had Yai, Thailand. March. http://www.envi.psu.ac.th/ptt/ttm_eng.html. Accessed: 6 December 2007.

—— 2000b. *Environmental Impact Assessment Trans Thailand Malaysia (TTM) Project: GSP*. Faculty of Environmental Management, Prince of Songkla

University, Had Yai, Thailand. March. http://www.envi.psu.ac.th/ptt/ttm_eng. html. Accessed: 6 December 2007.

Chatterjee, P. and Finger, M. 1994. *The Earth Brokers: Power, Politics, and World Development*. London and New York: Routledge.

Chaturvedi, S. 1998. 'Common security? Geopolitics, development, South Asia and the Indian Ocean.' *Third World Quarterly*, 19(4): 701–24.

—— 2003. *Geopolitics of India's Cultural Diversity: Conceptualization and Contestations*. GERM (Groupe d'Etudes et de Recherches sur les Mondialisations). http://www.mondialisations.org/php/public/art. php?id=11963&lan=EN. Updated: 31 May. Accessed: 2007.

—— 2005. 'India's quest for energy security: The geopolitics and geoeconomics of pipelines.' in *Energy Security in the Indian Ocean Region* D. Rumley and S. Chaturvedi eds. New Delhi: South Asian Publishers.

Cheesman, N., Skidmore, M. and Wilson, T. eds. 2012. *Myanmar's Transition: Openings, Obstacles and Opportunities*. Singapore, ISEAS.

Chenyang, L. 2010. 'The policies of China and India toward Myanmar.' in *Myanmar/ Burma: Inside Challenges, Outside Interests* L. Rieffel ed. Washington DC: Brookings Institution Press and the Konrad Adenauer Foundation: 113–33.

Chevron 2005. *The Chevron Way*. http://www.chevron.com/about/chevtex_way/ values.asp. Accessed: 2005.

—— 2007. *Chevron Statement on Myanmar*. Chevron. http://www.chevron. com/news/press/Release/?id=2007-10-02. Updated: 2 October. Accessed: 14 February 2008.

Chimprabha, M. 28 May 2003. 'UN envoy cites climate of fear.' *The Nation*. http://www.nationmultimedia.com/specials/humanrights/index_may28.php.

Cho, V. 14 March 2008. 'Ban the dam, say activists.' *The Irrawaddy*. http://www. irrawaddy.org/article.php?art_id=10888. Accessed: 17 March 2008.

Chomsky, N. 2004. *Hegemony or Survival: America's Quest for Global Dominance*. Crows Nest, NSW: Allen & Unwin.

Christiansen, T., Petito, F. and Tonra, B. 2000. 'Fuzzy politics around fuzzy borders: The European Union's "near abroad".' *Cooperation and Conflict*, 35(4): 389–415.

Christmann, T. 2000. 'The "Unocal case": potential liability of multinational companies for investment activities in foreign countries.' *Southern Cross University Law Review*, 4: 206–24.

Christoff, P. 2005. 'Green governance and the green state: Capacity building as a political project.' in *Managing Leviathan: Environmental Politics and the Administrative State*. Second edition. R. Paehlke and D. Torgerson eds. Peterborough, Canada: Broadview Press.

Chuenchom Sangarasri Greacen and Greacen, C. 2004. 'Thailand's electricity reforms: Privatization of benefits and socialization of costs and risks.' *Pacific Affairs*. Fall, 77(3): 517–41.

Ciuta, F. 2010. 'Conceptual notes on energy security: Total or banal security?' *Security Dialogue*. April, 41(2): 123–44.

Clark, J. N. 2012. 'Fieldwork and its ethical challenges: Reflections from research in Bosnia.' *Human Rights Quarterly*, 34(3): 823–39.

Clarke, R. and Dalliwall, S. 4 September 2008. 'Sino-Indian competition for Burmese oil and natural gas.' *Harvard International Review*. http://www. harvardir.org/articles/1751/. Accessed: 15 September 2008.

Cohen, I. 1997. *Green Fire*. Sydney: Angus and Robertson.

Cohen, R. and Rai, S. M. eds. 2000. *Global Social Movements*. London and New Brunswick: The Athlone Press.

Cole, B. D. 2008. *Sea Lanes and Pipelines: Energy Security in Asia*. Westport: Praeger Security International.

Committee to Protect Journalists 2010. *Attacks on the Press 2009: Burma*. http:// cpj.org/2010/02/attacks-on-the-press-2009-burma.php. Accessed: 1 November 2010.

Connelly, J., Smith, G., Benson, D. and Saunders, C. 2012. *Politics and the Environment: From Theory to Practice*. Third edition. London and New York: Routledge.

Connors, M. K. 2004. *Thaksin's Thailand – to have and to hold: Thai politics in 2003–2004*. Thai Update Conference, Macquarie University, Sydney. 20–21 April. http://www.latrobe.edu.au/socsci/staff/connors/connors-pub.htm.

—— 2005. 'Democracy and the mainstreaming of localism in Thailand.' in *Southeast Asian Responses to Globalization: Restructuring Governance and Deepening Democracy* F. L. K. Wah and J. Ojendal eds. Copenhagen: Nordic Institute of Asian Studies.

—— 2007. *Democracy and National Identity in Thailand*. Revised edition. Copenhagen: Nordic Institute of Asian Studies.

—— 2008. 'Thailand – Four elections and a coup.' *Australian Journal of International Affairs*, 62(4): 478–96.

—— 2009. 'Liberalism, authoritarianism and the politics of decisionism in Thailand.' *The Pacific Review*, 22(3): 355–73.

Connors, M. K. 2011. 'Thailand's emergency state: Transformation and struggle.' *Southeast Asian Affairs*: 287–305.

Connors, M. K. and Hewison, K. 2008. 'Introduction: Thailand and the "good coup".' *Journal of Contemporary Asia*. February, 38(1): 1–10.

Corben, R. 11 July 2006. 'Chinese-Thai-Burmese dam projects raise humanitarian, environmental concerns.' *VOA.com (Voice of America)*. http://www.voanews. com/english/archive/2006-07/2006-07-11-voa10.cfm?CFID=29441064&CFT OKEN=71678385.

CORE 2005. *A Big Deal? Corporate Social Responsibility and the Finance Sector in Europe*. Corporate Responsibility (CORE), London. December. http://www. corporate-responsibility.org/module_images/A_Big_Deal_CSR_&_the_ Finance_Sector_in_Europe_Report_(Dec2005).pdf.

Cornwall, A. and Molyneux, M. 2006. 'The politics of rights – dilemmas for feminist praxis: An introduction.' *Third World Quarterly*, 27(7): 1175–91.

Cox, R. 1981. 'Social forces, states and world orders: Beyond International Relations theory.' *Millennium: Journal of International Studies*, 10(2): 126–55.

Cox, R. W. 1993. 'Gramsci, hegemony and international relations: An essay in method.' in *Gramsci, Historical Materialism and International Relations* S. Gill ed. Cambridge: Cambridge University Press.

—— 1999. 'Civil society at the turn of the millennium: Prospects for an alternative world order.' *Review of International Studies*, 25(1): 3–28.

Cumming-Bruce, N. 4 August 1988. 'Burma's new leader imposes martial law.' *The Guardian*. http://www.guardian.co.uk/burma/story/0,13373,1275024,00. html.

Curran, G. 2006. *21st Century Dissent: Anarchism, Anti-Globalization and Environmentalism*. Basingstoke, Hampshire: Palgrave Macmillan.

Daewoo International 2006. *Daewoo International Corporation Annual Report 2005*. Daewoo International, Seoul. http://www.daewoo.com/korean/investor/2005_annual_report_en.pdf.

Dalby, S. 2009. *Security and Environmental Change*. Cambridge: Polity Press.

Dale, J. G. 2011. *Free Burma: Transnational Legal Action and Corporate Accountability*. Minneapolis, MN: University of Minnesota Press.

Darlington, S. M. 1998. 'The ordination of a tree: The Buddhist ecology movement in Thailand.' *Ethnology*. Winter, 37(1): 1–15.

—— 2003. 'Buddhism and development: The ecology monks of Thailand.' in *Action Dharma: New Studies in Engaged Buddhism* C. Queen, C. Prebish and D. Keown eds. London: RoutledgeCurzon.

Davis, M. and Kumar, C. R. 2003. 'An opinion on the Depayin Massacre as a crime against humanity.' *Article 2 (Special Edition: The Depayin Massacre, Burma)*. December, 2(6). http://www.article2.org/pdf/v02n06.pdf. Accessed: 1 May 2012.

della Porta, D., Andretta, M., Mosca, L. and Reiter, H. 2006. *Globalization from Below: Transnational Activists and Protest Networks*. Minneapolis and London: University of Minnesota Press.

della Porta, D. and Diani, M. 2006. *Social Movements: An Introduction*. Second edition. Malden, MA: Blackwell.

della Porta, D. and Mosca, L. 2007. 'In movimento: "Contamination" in action and the Italian Global Justice Movement.' *Global Networks*. January, 7(1): 1–27.

Desai, M. 2002. 'Transnational solidarity: Women's agency, structural adjustment, and globalization.' in *Women's Activism and Globalization: Linking Local Struggles and Transnational Politics* N. A. Naples and M. Desai eds. New York and London: Routledge.

Di Gregorio, M. 2012. 'Networking in environmental movement organisation coalitions: Interest, values or discourse?' *Environmental Politics*, 21(1): 1–25.

Diamond, L. 2002. 'Elections without democracy: Thinking about hybrid regimes.' *Journal of Democracy*. April, 13(2): 21–35.

Diamond, L., Linz, J. and Lipset, S. 1988. 'Democracy in developing countries: Facilitating and obstructing factors.' in *Freedom in the World: Political Rights and Civil Liberties 1987-88* R. Gastil ed. New York: Freedom House.

Diamond, L. and Plattner, M. F. eds. 2012. *Liberation Technology: Social Media and the Struggle for Democracy*. Baltimore: Johns Hopkins University Press.

Dittmer, L. ed. 2010. *Burma or Myanmar? The Struggle for National Identity*. Singapore: World Scientific.

Dobson, A. 1998. *Justice and the Environment: Conceptions of Environmental Sustainability and Theories of Distributive Justice*. New York: Oxford University Press.

—— 2007. *Green Political Thought*. Fourth edition. London and New York: Routledge.

Dodds, F. and Pippard, T. 2005. *Human and Environmental Security: An Agenda for Change*. Sterling, VA: Earthscan.

Doherty, B. 1999. 'Manufactured vulnerability: Eco-activist tactics in Britain.' *Mobilization*, 4(1): 75–89.

—— 2002. *Ideas and Actions in the Green Movement*. London and New York: Routledge.

—— 2006. 'Friends of the Earth International: Negotiating a transnational identity.' *Environmental Politics*. November, 15(5): 860–80.

—— 2007. 'Environmental movements.' in *The Politics of the Environment: A Survey* C. Okereke ed. London and New York: Routledge.

Doherty, B. and de Geus, M. eds. 1996. *Democracy and Green Political Thought: Sustainability, Rights, and Citizenship*. London and New York: Routledge.

Doherty, B. and Doyle, T. 2006. 'Beyond borders: Transnational politics, social movements and modern environmentalisms.' *Environmental Politics*. November, 15(5): 697–712.

—— 2008. *Beyond Borders: Environmental Movements and Transnational Politics*. London and New York: Routledge.

Douthwaite, R. 1996. *Short Circuit: Strengthening Local Economies for Security in an Unstable World*. Totnes, Devon: Green Books.

Dove, M. R. 2006. 'Indigenous people and environmental politics.' *Annual Review of Anthropology*. October, 35: 191–208.

Doyle, T. 1998. 'Sustainable development and Agenda 21: The secular bible of global free markets and pluralist democracy.' *Third World Quarterly*, 19(4): 771–86.

—— 2000. *Green Power: The Environment Movement in Australia*. Sydney: UNSW Press.

—— 2004. 'Dam disputes in Australia and India: Appreciating differences in struggles for sustainable development.' in *India and Australia: Issues and Opportunities* D. Gopal and D. Rumley eds. New Delhi: Authorspress.

—— 2005. *Environmental Movements in Majority and Minority Worlds*. New Brunswick, New Jersey and London: Rutgers University Press.

Doyle, T. and Doherty, B. 2006. 'Green public spheres and the green governance state: The politics of emancipation and ecological conditionality.' *Environmental Politics*. November, 15(5): 881–92.

Doyle, T. and Kellow, A. 1995. *Environmental Politics and Policy Making in Australia*. South Melbourne: Macmillan.

Doyle, T. and McEachern, D. 2008. *Environment and Politics*. Third edition. London & New York: Routledge.

Doyle, T. and Risely, M. eds. 2008. *Crucible for Survival: Environmental Security and Justice in the Indian Ocean Region*. New Brunswick, New Jersey and London, Rutgers University Press.

Doyle, T. and Simpson, A. 2006. 'Traversing more than speed bumps: Green politics under authoritarian regimes in Burma and Iran.' *Environmental Politics*. November, 15(5): 750–67.

DPA 28 February 2006. 'Villagers want end of Salween dams.' *Bangkok Post (Deutsche Presse-Agentur GmbH)*. http://www.bangkokpost.com/breaking_news/breakingnews.php?id=117124.

Dressel, B. 2009. 'Thailand's elusive quest for a workable constitution, 1997–2007.' *Contemporary Southeast Asia*, 31(2): 296–325.

—— 2010. 'Judicialization of politics or politicization of the judiciary? Considerations from recent events in Thailand.' *The Pacific Review*, 23(5): 671–91.

Dryzek, J. S. 1987. *Rational Ecology: Environment and Political Economy*. Oxford and New York: Basil Blackwell.

—— 1999. 'Global ecological democracy.' in *Global Ethics and Environment* N. Low ed. London: Routledge.

—— 2005a. 'Designs for environmental discourse revisited: A greener administrative state?' in *Managing Leviathan*. Second edition. R. Paehlke and D. Torgerson eds. Peterborough, Ontario: Broadview Press.

—— 2005b. *The Politics of the Earth: Environmental Discourses*. Second edition. Oxford and New York: Oxford University Press.

—— 2006. *Deliberative Global Politics: Discourse and Democracy in a Divided World*. Cambridge: Polity.

—— 2010. *Green Democracy, Global Governance*. Canberra: Academy of the Social Sciences in Australia.

Dryzek, J. S., Downes, D., Hunold, C., Schlosberg, D. and Hernes, H.-K. 2003. *Green States and Social Movements: Environmentalism in the United States, United Kingdom, Germany and Norway*. New York: Oxford University Press.

Dryzek, J. S. and Schlosberg, D. eds. 1998. *Debating the Earth: The Environmental Politics Reader*. Oxford: Oxford University Press.

Duffy, R. 2006. 'Non-governmental organisations and governance states: The impact of transnational environmental management networks in Madagasacar.' *Environmental Politics*. November, 15(5): 731–49.

—— 2011. 'Déjà vu all over again: Climate change and the prospects for a nuclear power renaissance.' *Environmental Politics*, 20(5): 668–86.

DVB 13 February 2007. 'Daewoo arms trial delayed until March.' *Democratic Voice of Burma (DVB)*.

Dwivedi, R. 1997. *People's Movements in Environmental Politics: A Critical Analysis of the Narmada Bachao Andolan in India*. The Hague: Institute of Social Studies.

—— 2001. 'Environmental movements in the global South: Outline of a critique of the "livelihood" approach.' in *Globalization and Social Movements* P. Hamel, H. Lustiger-Thaler, J. N. Pieterse and S. Roseneil eds. Basingstoke and New York: Palgrave.

Dyer, H. 2011. 'Eco-imperialism: Governance, resistance, hierarchy.' *Journal of International Relations and Development*, 14(2): 186–212.

EarthRights School of Burma 2008. *Gaining Ground: Earth Rights Abuses in Burma Exposed*. EarthRights International, Chiang Mai, Thailand. July. http://www.earthrights.org/files/Reports/Gaining Ground – ERSB 2008.pdf. Accessed: 10 September 2008.

Echoing Green 2007. *Echoing Green Fellowship Program*. Echoing Green. http://www.echoinggreen.org/index.cfm?fuseaction=Page.viewPage&pageID=41. Accessed: 2007.

Eckersley, R. 1992. *Environmentalism and Political Theory: Toward an Ecocentric Approach*. London: UCL Press.

—— 1996. 'Greening liberal democracy: The rights discourse revisited.' in *Democracy and Green Political Thought: Sustainability, Rights and Citizenship* B. Doherty and M. de_Geus eds. London and New York: Routledge.

Eckl, J. and Weber, R. 2007. 'North–South? Pitfalls of dividing the world by words.' *Third World Quarterly*. February, 28(1): 3–23.

The Economic Times. 30 June 2008. 'ONGC, GAIL share in Myanmar blocks fall'. India. http://economictimes.indiatimes.com/News/News_By_Industry/Energy/Oil__Gas/ONGC_GAIL_share_in_Myanmar_blocks_fall/rssarticleshow/3179239.cms. Accessed: 22 July 2008.

The Economist. June 19 2003. 'The alien problem'.

Edwards, M. and Gaventa, J. eds. 2001. *Global Citizen Action*. Boulder: Lynne Rienner.

EGAT and DHP 2005. *Memorandum of Agreement*. EGAT, Department of Hydroelectric Power, Union of Myanmar. December. http://timesonline.typepad.com/times_tokyo_weblog/2006/03/documents_on_th.html - more.

Egreteau, R. 2008. 'India's ambitions in Burma: More frustration than success?' *Asian Survey*, 48(6): 936–57.

Egreteau, R. and Jagan, L. 2013. *Soldiers and Diplomacy in Burma: Understanding the Foreign Relations of the Burmese Praetorian State*. Singapore: NUS Press.

EIA 2011. *India: Analysis*. US Energy Information Administration (EIA), Department of Energy, Washington DC. 21 November. http://www.eia.gov/countries/cab.cfm?fips=IN. Accessed: 10 January 2013.

el-Ojeili, C. and Hayden, P. 2006. *Critical Theories of Globalization*. Basingstoke and New York: Palgrave Macmillan.

Elliott, L. 2004. *The Global Politics of the Environment*. Second edition. New York: New York University Press.

—— 2009. 'Australian scholarship, International Relations and the environment: Commitment, critique and contestation.' *Australian Journal of Politics and History*, 55(3): 394–414.

Epstein, B. 2001. 'Anarchism and the Anti-Globalization Movement.' *Monthly Review*, 53(4): 1.

Equator Principles Secretariat 2008. *The Equator Principles*. http://www.equator-principles.com/index.html. Updated: 3 March. Accessed: 17 April 2008.

ERI 2004a. *Another Yadana: The Shwe Natural Gas Pipeline Project*. EarthRights International (ERI). http://www.earthrights.org/burma/shwepipeline.shtml. Updated: 27 August. Accessed: 2005.

—— 2004b. *In Our Court: ATCA, Sosa, and the Triumph of Human Rights*. EarthRights International (ERI), Washington DC and Chiang Mai. July. http://www.earthrights.org/files/Reports/inourcourt.pdf. Accessed:11 February 2008.

—— 2005. *Historic Advance for Universal Human Rights: Unocal to Compensate Burmese Villagers*. EarthRights International (ERI). http://www.earthrights.org/legalfeature/historic_advance_for_universal_human_rights_unocal_to_compensate_burmese_villagers.html. Updated: 2 April. Accessed: 2006.

—— 2006a. *Annual Report 2005*. EarthRights International (ERI), Washington DC and Chiang Mai, Thailand. http://www.earthrights.org/files/Documents/Annual Report 2005 latest.pdf. Accessed: 4 February 2008.

—— 2006b. *Documentary about Doe v. Unocal Wins Vaclav Havel Award at One World Film Festival*. EarthRights International. http://www.earthrights.org/content/view/288/25/. Updated: 5 March. Accessed: 2006.

—— 2006c. *The Situation of Women in Burma*. EarthRights International (ERI). http://www.earthrights.org/misc/the_situation_of_women_in_burma.html. Updated: 27 February. Accessed: 1 February 2008.

—— 2007a. *Activists Working for Burma Meet with Senior Government Officials at the ADB*. EarthRights International (ERI). http://www.earthrights.org/images/stories/Media/adb_agm_burma_press_release_may_4.pdf. Updated: 4 May. Accessed: 2007.

—— 2007b. *Annual Report 2006*. EarthRights International (ERI), Washington DC and Chiang Mai, Thailand. http://www.earthrights.org/files/Documents/ERI_annual06_spreads latest.pdf. Accessed: 4 February 2008.

—— 2007c. *Inaugural Training for Judges and Lawyers*. EarthRights International (ERI). http://www.earthrights.org/misc/inaugural_training_for_judges_and_lawyers.html. Updated: 30 March. Accessed: 14 February 2008.

—— 2007d. *Long-Term Volunteer Teacher, EarthRights Burma School*. EarthRights International (ERI). Chiang Mai. http://www.earthrights.org/misc/ERSB_cuso_volunteer.html. Updated: 30 March. Accessed: 2007.

—— 2008a. *Campaigns*. EarthRights International (ERI). http://www.earthrights.org/campaigns/. Accessed: 18 February 2008.

—— 2008b. *China in Burma: The Increasing Investment of Chinese Multinational Corporations in Burma's Hydropower, Oil and Natural Gas, and Mining Sectors*. EarthRights International (ERI), Chiang Mai and Washington DC. September. http://www.earthrights.org/files/Reports/China in Burma – BACKGROUNDER – 2008 Update – FINAL.pdf. Accessed: 30 September 2010.

—— 2008c. *The Human Cost of Energy: Chevron's Continuing Role in Financing Oppression and Profiting From Human Rights Abuses in Military-Ruled Burma (Myanmar)*. EarthRights International (ERI), Washington DC and Chiang Mai. April. http://www.earthrights.org/files/Burma Project/Yadana/HCoE_pages. pdf. Accessed: 1 May 2008.

—— 2008d. *Training*. EarthRights International (ERI). http://www.earthrights. org/training/. Accessed: 14 February 2008.

—— 2010. *Energy Insecurity: How Total, Chevron, and PTTEP Contribute to Human Rights Violations, Financial Secrecy, and Nuclear Proliferation in Burma (Myanmar)*. EarthRights International (ERI), Washington DC, USA. July. http://www.earthrights.org/sites/default/files/documents/energy-insecurity.pdf. Accessed: 20 May 2012.

—— 2011. *The True Cost of Chevron: An Alternative Annual Report*. EarthRights International (ERI), Washington DC, USA. May. http://www.earthrights.org/ sites/default/files/documents/true-cost-of-chevron-may-2011.pdf. Accessed: 20 May 2012.

ERI and SAIN 1996. *Total Denial*. EarthRights International (ERI) & Southeast Asian Information Network (SAIN), Bangkok. 10 July.

Eschle, C. 2005. 'Constructing "the anti-globalisation movement".' in *Critical Theories, International Relations and 'the Anti-Globalisation Movement': The Politics of Global Resistance* C. Eschle and B. Maiguashca eds. London and New York: Routledge.

Eschle, C. and Maiguashca, B. eds. 2005. *Critical Theories, International Relations and 'the Anti-Globalisation Movement': The Politics of Global Resistance*. London and New York: Routledge.

—— 2006. 'Bridging the academic/activist divide: Feminist activism and the teaching of global politics.' *Millennium: Journal of International Studies*. December, 35(1): 119–37.

Escobar, A. 2004. 'Other worlds are (already) possible: Self-organisation, complexity, and post-capitalist cultures.' in *World Social Forum: Challenging Empires* J. Sen, A. Anand, A. Escobar and P. Waterman eds. New Delhi: The Viveka Foundation.

ESCR-Net 2007. *International Network for Economic, Social and Cultural Rights*. ESCR-Net. http://www.escr-net.org/. Updated: 5 March. Accessed: 2007.

Evans, J. P. 2012. *Environmental Governance*. London and New York: Routledge.

Eviatar, D. 30 June 2003. 'Profits at gunpoint.' *The Nation (US)*. http://www. thenation.com/doc/20030630/eviatar.

—— 9 May 2005. 'A big win for human rights.' *The Nation (US)*. http://www. thenation.com/doc.mhtml?i=20050509&s=eviatar.

Faber, D. 2005. 'Building a transnational environmental justice movement: Obstacles and opportunities in the age of globalization.' in *Coalitions across Borders: Transnational Protest and the Neoliberal Order* J. Bandy and J. Smith eds. Lanham, Maryland and Oxford: Rowman and Littlefield.

Fagan, A. 2006. 'Neither "North" nor "South": The environment and civil society in post-conflict Bosnia-Herzegovina.' *Environmental Politics*. November, 15(5): 787–802.

Fahn, J. 2003. *A Land on Fire: The Environmental Consequences of the Southeast Asian Boom*. Chiang Mai: Silkworm.

Falk, R. 2000. 'Resisting "globalization-from-above" through "globalization-from-below".' in *Globalization and the Politics of Resistance* B. K. Gills ed. Basingstoke and New York: Macmillan.

—— 2009. *Achieving Human Rights*. New York and London: Routledge.

Farrelly, N. 2013a. 'Discipline without democracy: Military dominance in post-colonial Burma.' *Australian Journal of International Affairs*, 67(3): 312–26.

—— 2013b. 'Why democracy struggles: Thailand's elite coup culture.' *Australian Journal of International Affairs*, 67(3): 281–96.

Financial Express. 11 January 2006. 'Myanmar pact with Petro China for sale of gas: India's efforts to import gas hit'. http://www.financialexpress.com/ fe_full_story.php?content_id=114117. Accessed: 20 January 2006.

Fink, C. 2008. 'Militarization in Burma's ethnic states: Causes and consequences.' *Contemporary Politics*. December, 14(4): 447–62.

—— 2009. *Living Silence in Burma: Surviving under Military Rule*. Second edition. London and New York: Zed Books.

Florini, A. and Sovacool, B. K. 2011. 'Bridging the gaps in global energy governance.' *Global Governance*. January–March, 17(1): 57–74.

Floyd, R. 2010. *Security and the Environment: Securitisation Theory and US Environmental Security Policy*. Cambridge: Cambridge University Press.

Floyd, R. and Matthew, R. eds. 2013. *Environmental Security: Approaches and Issues*. London and New York: Routledge.

FoE 2005. *Barclays, Human Rights and the Trans Thai-Malaysia Gas Pipeline*. Friends of the Earth (FoE), London. September. http://www.foe.co.uk/ resource/briefings/barclays_thai_malaysia.pdf. Accessed: 16 April 2008.

Ford, L. H. 2003. 'Challenging global environmental governance: Social movement agency and global civil society.' *Global Environmental Politics*. May, 3(2): 120–34.

Ford, M. ed. 2013. *Social Activism in Southeast Asia*. London and New York: Routledge.

Foreman, D. 1991. *Confessions of an Eco-Warrior*. New York: Harmony Books.

Foreman, D. and Haywood, B. eds. 1993. *Ecodefense: A Field Guide to Monkeywrenching*. Chico, California: Abbzug Press.

Forest Peoples Programme and Tebtebba Foundation 2006. *Indigenous Peoples' Rights, Extractive Industries and Transnational and Other Business Enterprises: A Submission to the Special Representative of the Secretary-General on Human Rights and Transnational Corporations and Other Business Enterprises.* Forest Peoples Programme and Tebtebba Foundation, UK and Philippines. 29 December. http://www.business-humanrights.org/ Documents/Forest-Peoples-Tebtebba-submission-to-SRSG-re-indigenous-rights-29-Dec-2006.pdf.

Forsyth, T. 1997. 'Industrial pollution and government policy in Thailand: Rhetoric versus reality.' in *Seeing Forests for the Trees* P. Hirsch ed. Chiang Mai: Silkworm Books.

—— 2001. 'Environmental social movements in Thailand: How important is class?' *Asian Journal of Social Sciences*, 29(1): 35–51.

—— 2010. 'Thailand's red shirt protests: Popular movement or dangerous street theatre?' *Social Movement Studies*. November, 9(4): 461–67.

Forsyth, T. and Walker, A. 2008. *Forest Guardians, Forest Destroyers: The Politics of Environmental Knowledge in Northern Thailand.* Seattle and London: University of Washington Press.

Fraser, N. and Honneth, A. 2003. *Redistribution or Recognition: A Political-Philosophical Exchange.* London: Verso.

Fredriksson, P. G. and Wollscheid, J. R. 2007. 'Democratic institutions versus autocratic regimes: The case of environmental policy.' *Public Choice.* March, 130(3–4): 381–93.

Freedman, A. L. 2006. *Political Change and Consolidation: Democracy's Rocky Road in Thailand, Indonesia, South Korea, and Malaysia.* New York: Palgrave Macmillan.

FTUB 2005. *Federation of Trade Unions - Burma (FTUB).* FTUB. http://www. tradeunions-burma.org/. Updated: 21 October. Accessed: 2006.

Funston, J. 2006. 'Thailand.' in *Voices of Islam in Southeast Asia: A Contemporary Sourcebook* G. Fealy and V. Hooker eds. Singapore: Institute for Southeast Asian Studies.

—— ed. 2009. *Divided Over Thaksin: Thailand's Coup and Problematic Transition.* Chiang Mai: Silkworm.

Gainsborough, M. ed. 2009. *On The Borders of State Power: Frontiers in the Greater Mekong Sub-Region.* London and New York: Routledge.

Galbreath, D. J. ed. 2010. *Contemporary Environmentalism in the Baltic States: From Phosphate Springs to 'Nordstream'.* London and New York: Routledge.

Garner, R. 2011. *Environmental Politics: The Age of Climate Change.* Third edition. Basingstoke and New York: Palgrave Macmillan.

Geddes, B. 1999. 'What do we know about democratization after twenty years.' *Annual Review of Political Science*, 2: 115–44.

Giannini, T. 1994. *Reforming the Environmental Policies of the World Bank.* Sierra Club Legal Defense Fund [now EarthJustice]. 20 December.

—— 1999. *Destructive Engagement: A Decade of Foreign Investment in Burma.* EarthRights International (ERI), Bangkok. http://www.earthrights.org/pubs. shtml - burma.

Giannini, T. and Friedman, A. 2005. *A Report on Forced Labour in Burma.* EarthRights International. March. http://earthrights.org/docs/ILO_ ForcedLaborReportinBurma2005.pdf. Accessed:15 October 2009

Giannini, T. and Redford, K. 1994. *Echoing Green: Grant Proposal.* Proposal submitted to Echoing Green, New York. 5 December.

Giannini, T., Redford, K., Apple, B., Greer, J. and Simons, M. 2003. *Total Denial Continues.* Second edition. Washington and Chiang Mai: EarthRights International (ERI).

Giddens, A. 1987. *The Nation-State and Violence.* Berkeley and Los Angeles: University of California Press.

—— 1998. *The Third Way: The Renewal of Social Democracy.* Cambridge: Polity Press.

Gill, S. 2008. *Power and Resistance in the New World Order.* Second edition. Basingstoke: Palgrave Macmillan.

Gillan, K., Pickerill, J. and Webster, F. 2008. *Anti-War Activism: New Media and Protest in the Information Age.* Basingstoke: Palgrave Macmillan.

Gilley, B. 2012. 'Authoritarian environmentalism and China's response to climate change.' *Environmental Politics*, 21(2): 287–307.

Gilquin, M. 2005. *The Muslims of Thailand.* Translated by Michael Smithies, Chiang Mai: Silkworm.

Girion, L. 30 June 2004. 'Court OKs foreign-abuse suits.' *Los Angeles Times.* http://www.latimes.com/.

Glassman, J. 2004. *Thailand at the Margins: Internationalization of the State and the Transformation of Labour.* Oxford: Oxford University Press.

—— 2010. 'Cracking hegemony in Thailand: Gramsci, Bourdieu and the dialectics of rebellion.' *Journal of Contemporary Asia*, 41(1): 25–46.

Glassman, J., Park, B.-G. and Choi, Y.-J. 2008. 'Failed internationalism and social movement decline: The cases of South Korea and Thailand.' *Critical Asian Studies*, 40(3): 339–72.

Gleditsch, N. P., Furlong, K., Hegre, H., Lacina, B. and Owen, T. 2006. 'Conflicts over shared rivers: Resource scarcity or fuzzy boundaries?' *Political Geography.* May, 25(4): 361–82.

Global Exchange 2005. *'Most Wanted' Corporate Human Rights Violators of 2005.* http://www.globalexchange.org/getInvolved/corporateHRviolators. html. Updated: 12 December. Accessed: 2005.

Goldsmith, E. and Hildyard, N. 1984. *The Social and Environmental Effects of Large Dams, Volume 1: Overview.* Camelford, Cornwall: Wadebridge Ecological Centre.

Gole, N. 2000. 'Snapshots of Islamic modernities.' *Daedalus.* Winter, 29(1).

Goodland, R. 2006. *Suriname: Environmental and Social Reconnaissance – the Bakhuys Bauxite Mine Project.* The Association of Indigenous Village Leaders in Suriname (VIDS) and The North-South Institute (NSI). http://www.nsi-ins. ca/english/pdf/Robert_Goodland_Suriname_ESA_Report.pdf.

Goodman, A. 2 October 2007. 'Chevron's pipeline is the Burmese regime's lifeline.' *Truthdig*. http://www.truthdig.com/report/item/20071002_chevrons_ pipeline_is_the_burmese_regimes_lifeline/. Accessed: 8 April 2008.

Gottlieb, R. 2005. *Forcing the Spring: The Transformation of the American Environmental Movement*. Revised and updated edition. Washington, DC: Island Press.

Gramsci, A. 1971. *Selections from the Prison Notebooks of Antonio Gramsci*. London: Lawrence & Wishart.

Greer, J. and Giannini, T. 1999. *Earth Rights: Linking the Quests for Human Rights and the Environment*. Washington DC: EarthRights International.

Gret, M. and Sintomer, Y. 2005. *The Porto Alegre Experiment: Learning Lessons for Better Democracy*. London and New York: Zed Books.

Haacke, J. 2006. *Myanmar's Foreign Policy: Domestic Influences and International Implications*. London and New York: Routledge.

—— 2010a. 'China's role in the pursuit of security by Myanmar's State Peace and Development Council: Boon and bane?' *The Pacific Review*, 23(1): 113–37.

—— 2010b. 'The Myanmar imbroglio and ASEAN: Heading towards the 2010 elections.' *International Affairs*, 86(1): 153–74.

Haas, M. 2008. *International Human Rights: A Comprehensive Introduction*. London and New York: Routledge.

Hadenius, A. and Teorell, J. 2007. 'Pathways from authoritarianism.' *Journal of Democracy*. January, 18(1): 143–56.

Hadiz, V. R. ed. 2006. *Empire and Neoliberalism in Asia*. London and New York: Routledge.

Hajer, M. A. 1995. *The Politics of Environmental Discourse: Ecological Modernization and the Policy Process*. Oxford: Oxford University Press.

Hale, A. and Wills, J. 2007. 'Women Working Worldwide: Transnational networks, corporate social responsibility and action research.' *Global Networks*. October, 7(4): 453–76.

Hamburg, C. G. 2008. 'Prohibited spaces: Barriers and strategies in women's NGO work in Isaan, north-eastern Thailand.' in *Women and Politics in Thailand: Continuity and Change* K. Iwanaga ed. Copenhagen: NIAS Press.

Hancock, J. 2003. *Environmental Human Rights: Power, Ethics and Law*. Burlington, Vermont: Ashgate.

Handley, P. M. 2006. *The King Never Smiles: A Biography of Thailand's Bhumibol Adulyadej*. New Haven, CT: Yale University Press.

Hardt, M. and Negri, A. 2004. *Multitude: War and Democracy in the Age of Empire*. New York: Penguin Press.

Hares, M. 2006. *Community forestry and environmental literacy in northern Thailand: Towards collaborative natural resource management and conservation*. Unpublished thesis. Faculty of Agriculture and Forestry.

Helsinki, University of Helsinki. http://ethesis.helsinki.fi/julkaisut/maa/mekol/vk/hares/communit.pdf. Accessed: 20 December 2008.

Harish, S. P. and Liow, J. C. 2007. 'The coup and the conflict in southern Thailand.' *Crossroads: An Interdisciplinary Journal of Southeast Asian Studies*, 19(1): 161–84.

Harris, M. 1979. *Cultural Materialism: The Struggle for a Science of Culture*. New York: Random House.

Harris, P. G. 2010. *World Ethics and Climate Change: From International to Global Justice*. Edinburgh: Edinburgh University Press.

Harrison, G. 2004. *The World Bank and Africa: The Construction of Governance States*. London: Routledge.

Harvey, D. 1989. *The Condition of Postmodernity: An Enquiry into the Origins of Cultural Change*. Oxford: Blackwell.

—— 2005. *A History of Neoliberalism*. Oxford: Oxford University Press.

Harvey, N. 1998. *Environmental Impact Assessment: Procedures, Practice and Prospects in Australia*. Melbourne: Oxford University Press.

Haufler, V. 2010. 'Disclosure as governance: The Extractive Industries Transparency Initiative and resource management in the developing world.' *Global Environmental Politics*. August, 10(3): 53–73.

Hay, P. 2002. *Main Currents in Western Environmental Thought*. Sydney: UNSW Press.

Haynes, J. 1999. 'Power, politics and environmental movements in the Third World.' *Environmental Politics*, 8(1): 222–42.

Hayward, T. 2005. *Constitutional Environmental Rights*. Oxford: Oxford University Press.

Heiduk, F. 2011. 'From guardians to democrats? Attempts to explain change and continuity in the civil–military relations of post-authoritarian Indonesia, Thailand and the Philippines.' *The Pacific Review*. May, 24(2): 249–71.

Hewison, K. 2002. 'Thailand: Boom, bust and recovery.' *Perspectives on Global Development and Technology*, 1(3–4): 225–49.

—— 2004. 'Crafting Thailand's new social contract.' *The Pacific Review*, 17(4): 503–22.

—— 2005. 'Neo-liberalism and domestic capital: The political outcomes of the economic crisis in Thailand.' *Journal of Development Studies*. February, 41(2): 310–30.

—— 2007. 'Constitutions, regimes and power in Thailand.' *Democratization*. December, 14(5): 928–45.

—— 2008. 'A book, the king and the 2006 coup.' *Journal of Contemporary Asia*. February, 38(1): 190–211.

—— 2010. 'Thaksin Shinawatra and the reshaping of Thai politics.' *Contemporary Politics*. June, 16(2): 119–33.

Hicken, A. 2007. 'The 2007 Thai Constitution: A return to politics past.' *Crossroads: An Interdisciplinary Journal of Southeast Asian Studies*, 19(1): 128–60.

Hildyard, N. 1998. *Dams on the Rocks: The Flawed Economics of Large Hydroelectric Dams*. The Corner House, London. August. http://www. thecornerhouse.org.uk/item.shtml?x=51963 - index-01-00-00-00.

Hirsch, P. 1997. 'Introduction: Seeing forests for trees.' in *Seeing Forests for Trees: Environment and Environmentalism in Thailand* P. Hirsch ed. Chiang Mai: Silkworm Books.

—— 1998. 'Dams resources and the politics of environment in mainland Southeast Asia.' in *The Politics of Environment in Southeast Asia: Resources and Resistance* P. Hirsch and C. Warren eds. London: Routledge.

—— 2001. 'Globalisation, regionalisation and local voices: The Asian Development Bank and re-scaled politics of environment in the Mekong Region.' *Singapore Journal of Tropical Geography*, 22(3): 237–51.

—— 2002. 'Global norms, local compliance and the human rights-environment nexus: A case study of the Nam Theun II Dam in Laos.' in *Human Rights and the Environment: Conflicts and Norms in a Globalizing World* L. Zarsky ed. London: Earthscan.

Holland, I. 2002. 'Consultation, constraints and norms: The case of nuclear waste.' *Australian Journal of Public Administration*. March, 61(1): 76–86.

Holliday, I. 2005a. 'Doing business with rights violating regimes: Corporate social responsibility and Myanmar's military junta.' *Journal of Business Ethics*, 61(4): 329–42.

—— 2005b. 'The Yadana syndrome? Big oil and principles of corporate engagement in Myanmar.' *Asian Journal of Political Science*, 13(2): 29–51.

—— 2008. 'Voting and violence in Myanmar: Nation building for a transition to democracy.' *Asian Survey*, 48(6): 1038–58.

—— 2011. *Burma Redux: Global Justice and the Quest for Political Reform in Myanmar*. Hong Kong: Hong Kong University Press.

Horkheimer, M. 1972. *Critical Theory: Selected Essays*. New York: Herder and Herder.

Horsey, R. 2011. *Ending Forced Labour in Myanmar: Engaging a Pariah Regime*. London and New York: Routledge.

Howes, M. 2005. *Politics and the Environment: Risk and the Role of Government and Industry*. St Leonards, NSW: Allen & Unwin.

Hughes, C. 2011. 'Soldiers, monks, morders: Violence and contestation in the Greater Mekong Sub-region.' *Journal of Contemporary Asia*, 41(2): 181–205.

Human Rights Watch 2003. *US: Ashcroft Attacks Human Rights Law*. Human Rights Watch, New York. 15 May. http://hrw.org/english/docs/2003/05/15/usdom6050.htm.

—— 2005. *'They Came and Destroyed Our Village Again': The Plight of Internally Displaced Persons in Karen State*. Human Rights Watch, New York. June. http://hrw.org/reports/2005/burma0605/6.htm - _ftn105.

—— 2007a. *No One Is Safe: Insurgent Violence Against Civilians in Thailand's Southern Border Provinces*. Human Rights Watch, New York. August. http://hrw.org/reports/2007/thailand0807/thailand0807webwcover.pdf.

—— 2007b. *Sold to be Soldiers: The Recruitment and Use of Child Soldiers in Burma.* Human Rights Watch, New York. October. http://hrw.org/reports/2007/burma1007/burma1007webwcover.pdf. Accessed: 30 October 2008.

—— 2013. *'All You Can Do is Pray': Crimes Against Humanity and Ethnic Cleansing of Rohingya Muslims in Burma's Arakan State.* Human Rights Watch, New York. April. http://www.hrw.org/sites/default/files/reports/burma0413webwcover_0.pdf. Accessed: 1 September 2013.

Humphries, B., Mertens, D. M. and Truman, C. 2000. 'Arguments for an "emancipatory" research paradigm.' in *Research and Inequality* C. Truman, D. M. Mertens and B. Humphries eds. London: UCL Press.

Huntington, S. 1991. *The Third Wave: Democratization in the Late Twentieth Century.* London: University of Oklahoma Press.

Hurrell, A. and Sengupta, S. 2012. 'Emerging powers, North–South relations and global climate politics.' *International Affairs*, 88(3): 463–84.

Hutton, D. and Connors, L. 1999. *A History of the Australian Environment Movement.* Cambridge and Melbourne: Cambridge University Press.

ICFTU 2005. *Doing Business in or with Burma.* International Confederation of Free Trade Unions (ICFTU), Brussels. January. http://www.icftu.org/www/PDF/Burma-ICFTUReport-January.pdf.

ICG 2005. *Thailand's Emergency Decree: No Solution.* International Crisis Group (ICG), Jakarta and Brussels. 18 November. Asia Report No.105. http://www.crisisgroup.org/library/documents/asia/south_east_asia/105_thailand_s_emergency_decree_no_solution_web.pdf.

—— 2007. *Southern Thailand: The Impact of the Coup.* International Crisis Group (ICG), Jakarta and Brussels. 15 March. Asia Report 129. http://www.crisisgroup.org/library/documents/asia/south_east_asia/129_southern_thailand___the_impact_of_the_coup_web.pdf.

—— 2008a. *Burma/Myanmar: After the Crackdown.* International Crisis Group (ICG), Brussels. 31 January. Asia Report No 144. http://www.crisisgroup.org/library/documents/asia/burma_myanmar/144_burma_myanmar___after_the_crackdown.pdf. Accessed: 1 February 2008.

—— 2008b. *China's Thirst for Oil.* International Crisis Group (ICG), Brussels. 9 June. Asia Report No. 153. http://www.crisisgroup.org/library/documents/asia/153_china_s_thirst_for_oil.pdf. Accessed: 10 June 2008.

IISS 2005. *The Military Balance 2005/2006.* International Institute for Strategic Studies (IISS), London. October.

ILO 2004. *92nd Annual Conference of the ILO Concludes its Work.* International Labour Organization (ILO). 17 June. http://www.ilo.org/public/english/bureau/inf/pr/2004/32.htm.

—— 2006. *ILO Governing Body concludes 297th Session.* International Labour Organization (ILO). 17 November. http://www.ilo.org/public/english/bureau/inf/pr/2006/53.htm.

Imhof, A. 2005. 'Making smart choices for the Mekong.' *World Rivers Review.* October/December, 20(5/6): 8–9. http://www.irn.org/pubs/wrr/issues/WRR. V20.N5.pdf.

Independent Bangladesh. 12 April 2005. 'Myanmar's mangrove eco-system faces devastation'. http://www.independent-bangladesh.com/. Accessed: 15 June 2006.

India Daily. 10 January 2006. 'Myanmar has refused to supply natural gas to New Delhi and instead preferred doing business with China'. http://www.indiadaily. com/default.asp. Accessed: 20 January 2006.

International Commission of Jurists 2008. *Corporate Complicity and Legal Accountability: Volume 3 – Civil Remedies.* International Commission of Jurists, Geneva. http://icj.org/IMG/Volume_3.pdf. Accessed: 18 September 2008.

International Energy Agency 2010a. *Electricity/Heat in Thailand in 2009.* International Energy Agency (IEA). Paris. http://www.iea.org/stats/ electricitydata.asp?COUNTRY_CODE=TH. Accessed: 25 April 2012.

—— 2010b. *World Energy Outlook 2010: The Electricity Access Database.* International Energy Agency (IEA). Paris. http://www.iea.org/weo/database_ electricity10/electricity_database_web_2010.htm. Accessed: 19 January 2012.

International Rivers 2007. *Burma's Salween Dams Threaten Over Half a Million Lives Downstream.* International Rivers (formerly International Rivers Network (IRN)). http://internationalrivers.org/en/southeast-asia/burmas-salween-dams-threaten-over-half-million-lives-downstream. Accessed: 20 November 2011.

—— 2008. *International Day of Action for Rivers.* International Rivers (formerly International Rivers Network (IRN)). http://internationalrivers.org/en/day-of-action. Accessed: 28 March 2008.

The Irrawaddy. November 2004. 'Corporate responsibility and despotism'. http:// www.irrawaddy.org/aviewer.asp?a=4192&z=108. Accessed: 7 February 2006.

—— 31 October 2006. 'Popular outrage sparked by "Wedding of the Year" video'. http://www.irrawaddy.org/aviewer.asp?a=6324&z=154. Accessed: 5 December 2006.

ITUC 2008. *Rich Pickings: How Trade and Investment keep the Burmese Junta Alive and Kicking.* International Trade Union Confederation (ITUC), Brussels, Belgium. April. http://www.ituc-csi.org/IMG/pdf/BirmanieEN.pdf. Accessed: 5 May 2008.

Iwanaga, K. ed. 2008. *Women and Politics in Thailand: Continuity and Change.* Copenhagen, NIAS Press.

Jamal, A. A. 2007. *Barriers to Democracy: The Other Side of Social Capital in Palestine and the Arab World.* Princeton, New Jersey: Princeton University Press.

James, H. 2006. *Security and Sustainable Development in Myanmar.* Abingdon, Oxon: Routledge.

Jayasuriya, K. and Hewison, K. 2004. 'The antipolitics of good governance: From global social policy to a global populism?' *Critical Asian Studies*, 36(4): 571–90.

Jayasuriya, K. and Rodan, G. 2007. 'Beyond hybrid regimes: More participation, less contestation in Southeast Asia.' *Democratization*. December, 14(5): 773–94.

Jilani, H. 2004. *Report by the Special Representative of the Secretary-General on the Situation of Human Rights Defenders: Mission to Thailand.* UN Commission on Human Rights, Promotion and Protection of Human Rights: Human Rights Defenders. 12 March. E/CN.4/2004/94/Add.1. http://ap.ohchr.org/documents/dpage_e.aspx?c=182&su=180. Accessed: 8 May 2008.

Jones, L. 2010. 'ASEAN's unchanged melody? The theory and practice of "non-interference" in Southeast Asia.' *The Pacific Review*. September, 23(4): 479–502.

—— 2012. *ASEAN, Sovereignty and Intervention in Southeast Asia.* Basingstoke: Palgrave Macmillan.

—— 2013. 'The political economy of Myanmar's transition.' *Journal of Contemporary Asia.* DOI:10.1080/00472336.2013.764143.

Kaiser, R. and Nikiforova, E. 2006. 'Borderland spaces of identification and dis/location: Multiscalar narratives and enactments of Seto identity and place in the Estonian-Russian borderlands.' *Ethnic and Racial Studies*. September, 29(5): 928–58.

Kalayanamitra Council 1998. *Thailand: Forest Defenders Arrested.* World Rainforest Movement. 6 March. http://www.wrm.org.uy/bulletin/10/Thailand.html. Accessed: 23 January 2008.

Kalland, A. and Persoon, G. 1997. 'An anthropological perspective on environmental movements.' in *Environmental Movements in Asia* A. Kalland and G. Persoon eds. , Richmond, Surrey: Curzon, 1–43.

Kamol Sukin 22 July 2007. 'Up in arms over pipeline.' *The Nation*. Bangkok. http://www.nationmultimedia.com/2007/07/22/headlines/headlines_30041891.php.

Kaneva, M. 2006. 'Total Denial'. MK Production. Rome, Italy. http://www.totaldenialfilm.com/. 2006.

Karlsson-Vinkhuyzen, S. I. 2010. 'The United Nations and global energy governance: past challenges, future choices.' *Global Change, Peace & Security*, 22(2): 175–95.

Kate, D. T. 25 December 2005. 'Analysis: Business with Myanmar thrives amid diplomatic stress.' *IHT Thai Day*. http://www.manager.co.th/IHT/ViewNews.aspx?NewsID=9480000176528.

Kaufmann, D. and Penciakova, V. 2012. *Japan's Triple Disaster: Governance and the Earthquake, Tsunami and Nuclear Crises.* Brookings Institution, Washington DC. 16 March 2012. http://www.brookings.edu/opinions/2011/0316_japan_disaster_kaufmann.aspx. Accessed: 28 March 2012.

KDNG 2007. *Valley of Darkness: Gold Mining and Militarization in Burma's Hugawng Valley.* Kachin Development Networking Group (KDNG). http://

kdng.org/images/stories/publication/ValleyofDarkness_english.pdf. Accessed: 20 May 2012.

—— 2010. *Tyrants, Tycoons and Tigers: Yuzana Company Ravages Burma's Hugawng Valley*. Kachin Development Networking Group (KDNG). http://www.kdng.org/images/stories/publication/TyrantsTycoonsandTigers.pdf. Accessed: 20 May 2012.

KDRG 2006. *Dammed by Burma's Generals*. Karenni Development Research Group (KDRG). March. http://www.salweenwatch.org/images/stories/downloads/publications/dammed-eng.pdf. Accessed:10 December 2011.

Keck, M. E. and Sikkink, K. 1998. *Activists Beyond Borders: Advocacy Networks in International Politics*. Ithaca and London: Cornell University Press.

Kelly, J. and Etling, B. 2008. *Mapping Iran's Online Public: Politics and Culture in the Persian Blogosphere*. Internet and Democracy Project, Berkman Center for Internet and Society, Harvard University, Cambridge, Massachusetts. 6 April. http://cyber.law.harvard.edu/sites/cyber.law.harvard.edu/files/Kelly&Etling_Mapping_Irans_Online_Public_2008.pdf. Accessed: 8 October 2008.

Kengkij Kitirianglarp and Hewison, K. 2009. 'Social movements and political opposition in contemporary Thailand.' *The Pacific Review*, 22(4): 451–77.

Kerényi, S. and Szabó, M. 2006. 'Transnational influences on patterns of mobilisation within environmental movements in Hungary.' *Environmental Politics*. November, 15(5): 803–20.

Kerr, P. 2007. 'Human security.' in *Contemporary Security Studies* A. Collins ed. Oxford: Oxford University Press.

KESAN 2008. *Khoe Kay: Biodiversity in Peril*. Karen Environmental and Social Action Network (KESAN), Wanida Press, Chiang Mai. July. http://www.salweenwatch.org/images/stories/downloads/brn/2008_009_24_khoekay.pdf. Accessed: 1 October 2011.

Khagram, S. and Ali, S. H. 2008. 'Transnational transformations: From government-centric interstate regimes to cross-sectoral multi-level networks of global governance.' in *The Crisis of Global Environmental Governance: Towards a New Political Economy of Sustainability* J. Park, K. Conca and M. Finger eds. London and New York: Routledge.

Khagram, S., Riker, J. V. and Sikkink, K. eds. 2002. *Restructuring World Politics: Transnational Social Movements, Networks and Norms*. Minneapolis and London: University of Minnesota Press.

KHRG 2007. *Development by Decree: The Politics of Poverty and Control in Karen State*. Karen Human Rights Group (KHRG). April. http://www.khrg.org/khrg2007/khrg0701.pdf. Accessed: 15 December 2011.

Kinver, M. 6 May 2008. 'Mangrove loss "left Burma exposed".' *BBC News*. http://news.bbc.co.uk/2/hi/science/nature/7385315.stm. Accessed: 7 May 2008.

Klare, M. T. 2012. *The Race for What's Left: The Global Scramble for the World's Last Resources*. New York: Metropolitan Books.

Klein, N. 2001. 'Farewell to "The End of History": Organization and vision in anti-corporate movements.' in *Socialist Register 2002: A World of Contradictions* L. Panitch and C. Leys eds. London: Merlin Press.

Korean House for International Solidarity 2006. *Joint Statement on Daewoo International's Suspicious Illegal Exports of Defense Industry Materials.* Seoul. 18 September.

Kranzberg, M. 1986. 'Technology and history: "Kranzberg's Laws".' *Technology and Culture*, 27(3): 544–60.

KRW 2004. *Damming at Gunpoint: Burma Army Atrocities Pave the Way for Salween Dams in Karen State.* Karen Rivers Watch (KRW), Kawthoolei, Burma. November. http://www.burmariversnetwork.org/images/stories/publications/english/dammingatgunpointenglish.pdf. Accessed:12 May 2012.

—— 2006. *Joint Statement on the International Day of Action Against Dams for Water and Life.* Karen Rivers Watch (KRW). http://www.salweenwatch.org/news/march282006-30.php - 1. Updated: 16 March. Accessed: 2006.

—— 2007a. *Grassroots Calling for the Withdrawal of the Salween Dam Projects.* EarthRights International (ERI) for Karen Rivers Watch (KRW). Chiang Mai, Thailand. http://www.earthrights.org/burmafeature/grassroots_calling_for_the_withdrawal_of_the_salween_dam_projects.html. Updated: 20 March. Accessed: 2007.

—— 2007b. *Submission to ASEAN Inter-Parliamentary Myanmar Caucus (AIPMC).* Karen Rivers Watch (KRW), Mae Sot, Thailand. 24 March. http://www.karenriverwatch.net/statement2007/Karen_Rivers_WatchLetterAIPMC%5B1%5D.doc.

KRW and SEARIN 2006. *Joint Statement on the International Day of Action Against Dams for Water and Life.* Karen Rivers Watch (KRW) and Southeast Asia Rivers Network (SEARIN). 16 March. http://www.salweenwatch.org/news/march282006-30.php - 1.

Ksentini, F. Z. 1994. *Human Rights and the Environment: Final Report.* Sub-Commission on Prevention of Discrimination and Protection of Minorities, Commission of Human Rights, United Nations Economic and Social Council, Geneva. 6 July. E/CN.4/Sub.2/1994/9. http://www.unhchr.ch/Huridocda/Huridoca.nsf/0/eeab2b6937bccaa18025675c005779c3?Opendocument.

Kudo, T. 2008. 'The impact of US sanctions on the Myanmar garment industry.' *Asian Survey*, 48(6): 997–1017.

Kudo, T. and Mieno, F. 2009. 'Trade, foreign investment and Myanmar's economic develiopment.' in *The Economic Transition in Myanmar after 1988: Market Economy versus State Control* K. Fujita, F. Mieno and I. Okamot eds. Singapore: NUS Press.

Kultida Samabuddhi 18 September 2002. 'Pollution slowly killing Songkhla lake.' *Bangkok Post.* http://www.ecologyasia.com/news-archives/2002/sep-02/bangkokpost_180902_news24.htm.

—— 5 May 2006. 'New call to scrap dams after geologist loses leg.' *Bangkok Post.* http://www.bangkokpost.net/News/05May2006_news10.php.

Kultida Samabuddhi and Yuthana Praiwan December 18 2003. 'China plans 13 dams on Salween.' *Bangkok Post.* 1.

Kusnetz, N. 8 June 2008. 'Burma dams would flood rebel territories.' *San Francisco Chronicle.* A-4. http://www.sfgate.com/cgi-bin/article.cgi?f=/c/a/2008/06/08/MNIM10GJUC.DTL. Accessed: 10 June 2008.

Kutting, G. 2000. *Environment, Society and International Relations: Towards More Effective International Environmental Agreements.* London and New York: Routledge.

—— ed. 2011. *Global Environmental Politics: Concepts, Theories and Case Studies.* London and New York: Routledge.

Kutting, G. and Lipschutz, R. D. eds. 2009. *Environmental Governance: Power and Knowledge in a Local-Global World.* London and New York: Routledge.

KWO 2007. *State of Terror: The Ongoing Rape, Murder, Torture and Forced Labour Suffered by Women Living Under the Burmese Military Regime in Karen State.* Karen Women's Organisation (KWO), Mae Sariang and Mae Sot, Thailand. February. http://www.womenofburma.org/Statement&Release/state_of_terror_report.pdf. Accessed: 20 June 2011.

Kyaw Hsu Mon 13–19 February 2012. 'Groups back expansion of small-scale power sources.' *Myanmar Times.* http://www.mmtimes.com/2012/news/614/news61417.html. Accessed: 1 June 2012.

Kyaw Yin Hlaing 2009. 'Setting the rules for survival: Why the Burmese military regime survives in an age of democratization.' *The Pacific Review,* 22(3): 271–91.

—— 2012. 'Understanding recent political changes in Myanmar.' *Contemporary Southeast Asia,* 34(2): 197–216.

Laird, J. 2000. *Money Politics, Globalisation, and Crisis: The Case of Thailand.* Singapore: Graham Brash.

Lamb, V. 2007. 'Between tigers and triggers: Conservation and conflict in Burma's Hugawng Valley.' *Watershed.* March–October, 12(2): 43–49. http://www.terraper.org/pic_water/WS12_2_web.pdf. Accessed: 17 March 2008.

Lambrecht, C. W. 2004. 'Oxymoronic development: The military as benefactor in the border regions of Burma.' in *Civilizing the Margins: Southeast Asian Government Policies for the Development of Minorities* C. R. Duncan ed. Ithaca, New York: Cornell University Press.

Larkin, E. 2010. *Everything is Broken: A Tale of Catastrophe in Burma.* New York: Penguin Press.

Lassalle, J.-F. 1 December 2005. 'Why Total agrees to compensation in forced labor suit.' *The Irrawaddy.* http://www.irrawaddy.org/article.php?art_id=5235. Accessed: 15 April 2008.

Lee, S.-H. 1999. 'Environmental movements in South Korea.' in *Asia's Environmental Movements: Comparative Perspectives* Y.-s. F. Lee and A. Y. So eds. Armonk, New York: M.E. Sharpe.

Lee-Treweek, G. and Linkogle, S. eds. 2000. *Danger in the Field: Risk and Ethics in Social Research.* London and New York: Routledge.

Leone, F. and Giannini, T. 2005. *Traditions of Conflict Resolution in Burma.* EarthRights International (ERI), Chiang Mai and Washington DC. December. http://www.earthrights.org/files/Reports/ctwp_paper.pdf. Accessed: 15 February 2008.

Lesage, D., Van de Graaf, T. and Westphal, K. 2010. *Global Energy Governance in a Multipolar World.* Farnham: Ashgate.

Levi, M. and Murphy, G. H. 2006. 'Coalitions of contention: The case of the WTO protests in Seattle.' *Political Studies*, 54(4): 651–70.

Levitsky, S. and Way, L. A. 2002. 'The rise of competitive authoritarianism.' *Journal of Democracy*, 13(2): 51–65.

—— 2006. 'Linkage versus leverage: Rethinking the international dimension of regime change.' *Comparative Politics*, 38(4): 379–400.

—— 2010. *Competitive Authoritarianism: Hybrid Regimes After the Cold War.* Cambridge: Cambridge University Press.

Levy, A., Scott-Clark, C. and Harrison, D. 23 March 1997. 'Save the rhino, kill the people.' *The Observer.* 9.

Lewis, G. 2006. *Virtual Thailand: The Media and Cultural Politics in Thailand, Malaysia and Singapore.* London and New York: Routledge.

Li, M. 2007. 'Peak oil, the rise of China and India, and the global energy crisis.' *Journal of Contemporary Asia*, 37(4): 449–71.

Lintner, B. 1990. *Outrage: Burma's Struggle for Democracy.* Second edition. London and Bangkok: White Lotus.

—— 1999. *Burma in Revolt: Opium and Insurgency since 1948.* Second edition. Chiang Mai: Silkworm.

Liotta, P. H., Mouat, D. A., Kepner, W. G. and Lancaster, J. M. eds. 2008. *Environmental Change and Human Security: Recognizing and Acting on Hazard Impacts.* Dordrecht: Springer.

Lipschutz, R. D. and Mayer, J. 1996. *Global Civil Society and Global Environmental Governance: The Politics of Nature from Place to Planet.* Albany: SUNY.

Litfin, K. 2009. 'Reinventing the future: The global ecovillage movement as a holistic knowledge community.' in *Environmental Governance: Power and Knowledge in a Local-Global World* G. Kutting and R. D. Lipschutz eds. London and New York: Routledge.

Liu, Y. 2005. *Missing Voices on the Nu River Dam Project.* Worldwatch Institute. 29 November. http://www.worldwatch.org/features/chinawatch/stories/20051129-2.

Liverani, A. 2008. *Civil Society in Algeria: The Political Functions of Associational Life.* London and New York: Routledge.

Lohmann, L. ed. 2007. *The Struggle of Villagers in Chana District, Southern Thailand in Defence of Community, Land and Religion against the Trans Thai-Malaysia Pipeline and Industrial Project (TTM), 2002–2007.* Dorset, UK: The Corner House.

—— 2008. 'Gas, waqf and Barclays Capital: A decade of resistance in Southern Thailand.' *Race and Class*, 50(2): 89–100.

Luft, G. and Korin, A. eds. 2009. *Energy Security Challenges for the 21st Century.* Santa Barbara: Praeger Security International.

Macdonald, A. P. 2013. 'From military rule to electoral authoritarianism: The reconfiguration of power in Myanmar and its future.' *Asian Affairs: An American Review*, 40(1): 20–36.

MacLean, K. 2003. *Capitalizing on Conflict: How Logging and Mining Contribute to Environmental Destruction in Burma.* Thailand: EarthRights International & Karen Environmental & Social Action Network.

Maddison, S. and Scalmer, S. 2006. *Activist Wisdom: Practical Knowledge and Creative Tension in Social Movements.* Sydney: UNSW Press.

Mahn Nay Myo, Imamura, M., Foley, J., Robinson, N., Shwe Maung, Naing Htoo and Giannini, T. 2003a. *Entrenched: An Investigative Report on the Systematic Use of Forced Labor by the Burmese Army in a Rural Area.* EarthRights International, Chiang Mai.

—— 2003b. *Entrenched: Supplemental Report.* EarthRights International, Chiang Mai.

Mander, J. 1996. 'Technologies of globalization.' in *The Case Against the Global Economy and For a Turn Toward the Local* J. Mander and E. Goldsmith eds. San Francisco: Sierra Club Books.

Mariner, J. 2003. 'Ashcroft's justice, Burma's crimes and Bork's revenge.' *FindLaw.* 26 May. http://writ.news.findlaw.com/mariner/20030526.html.

Markar, M. M. 29 September 2006. 'Dams on Salween: Test for Burmese, Thai juntas.' *IPS/IFEJ.* http://www.ipsnews.net/news.asp?idnews=34935.

Markels, A. 15 June 2003. 'Showdown for a tool in human rights lawsuits.' *New York Times.*

Marks, D. 2011. 'Climate change and Thailand: Impact and response.' *Contemporary Southeast Asia*, 33(2): 229–58.

Martinez-Alier, J. 2002. *Environmentalism of the Poor: A Study of Ecological Conflicts and Valuation.* Cheltenham, UK and Northampton, MA: Edward Elgar.

Mason, M. 1999. *Environmental Democracy.* London: Earthscan.

Matisoff, A. 2012. *Crude Beginnings: An Assessment of China National Petroleum Corporation's Environmental and Social Performance Abroad.* Friends of the Earth, San Francisco. February. http://www.foe.org/news/news-releases/2012-02-new-investor-brief-exposes-china-national-petroleum. Accessed: 20 May 2012.

Matthews, B. 1999. 'The legacy of tradition and authority: Buddhism and the nation in Myanmar.' in *Buddhism and Politics in Twentieth Century Asia* I. Harris ed. London and New York: Pinter.

Maung Aung Myoe 2009. *Building the Tatmadaw: Myanmar Armed Forces Since 1948.* Singapore: ISEAS.

Mavroudi, E. 2008. 'Palestinians in diaspora, empowerment and informal political space.' *Political Geography.* January, 27(1): 1–138.

McCargo, D. 2000. *Politics and the Press in Thailand: Media Machinations.* London and New York: RoutledgeCurzon.

—— 2005. 'Network monarchy and legitimacy crises in Thailand.' *The Pacific Review.* December, 18(4): 499–519.

—— ed. 2007. *Rethinking Thailand's Southern Violence.* Singapore, NUS Press.

—— 2009. *Tearing Apart the Land: Islam and Legitimacy in Southern Thailand.* Singapore: NUS Press.

—— 2012. *Mapping National Anxieties: Thailand's Southern Conflict.* Copenhagen: NIAS.

McCargo, D. and Ukrist Pathmanand 2005. *The Thaksinization of Thailand.* Copenhagen: NIAS Press.

McCarthy, S. 2000. 'Ten years of chaos in Burma: Foreign investment and economic liberalization under the SLORC-SPDC, 1988 to 1998.' *Pacific Affairs*, 73(2): 233–62.

—— 2008. 'Overturning the alms bowl: the price of survival and the consequences for political legitimacy in Burma.' *Australian Journal of International Affairs*, 62(3): 298–314.

McCully, P. 1996. *Silenced Rivers: The Ecology and Politics of Large Dams.* London: Zed Books.

McDonald, M. 2012. *Security, the Environment and Emancipation: Contestation over Environmental Change.* London and New York: Routledge.

McGready, R. 2003. 'How a refugee community deals with adoption.' *British Medical Journal.* 19 April, 326(7394): 873. http://www.pubmedcentral.nih. gov/articlerender.fcgi?artid=1125775.

McLeod, G. 8 July 2007. 'Atrocity before the deluge.' *Bangkok Post.* http://www. bangkokpost.com/topstories/topstories.php?id=119986.

MDF 2006. *Metta Development Foundation.* Metta Development Foundation (MDF). http://www.metta-myanmar.org/. Updated: 3 April. Accessed: 2006.

Melucci, A. 1989. *Nomads of the Present: Social Movements and Individual Needs in Contemporary Society.* Philadelphia: Temple University Press.

—— 1996. *Challenging Codes: Collective Action in the Information Age.* Cambridge and New York: Cambridge University Press.

Mendez, J. B. 2002. 'Creating alternatives from a gender perspective: Transnational organizing for maquila workers' rights in Central America.' in *Women's Activism and Globalization: Linking Local Struggles and Transnational Politics* N. A. Naples and M. Desai eds. New York and London: Routledge.

Mertha, A. 2009. '"Fragmented authoritarianism 2.0": Political pluralization in the Chinese policy process.' *The China Quarterly*, 200: 995–1012.

Midnight University 2002. *Academics Appeal to the Thai Society to urge the Government to Review the Thai-Malaysia Gas Pipeline Project.* Midnight University. 24 November. http://www.geocities.com/miduniv888/newpage19. html. Accessed: 2007.

Mies, M. and Shiva, V. 1993. *Ecofeminism.* North Melbourne: Spinifex Press.

Miller, C. 2002. *Environmental Rights: Critical Perspectives*. London and New York: Routledge.

Miller, K. 2010. 'Coping with China's financial power: Beijing's financial foreign policy.' *Foreign Affairs*. July–August, 89(4): 96–109.

Miller, M. A. L. 1995. *The Third World in Global Environmental Politics*. Buckingham: Open University Press.

Misra, A. 2007. 'Contours of India's energy security: Harmonising domestic and external options.' in *Energy Security in Asia* M. Wesley ed. New York: Routledge.

Missingham, B. D. 2003. *The Assembly of the Poor in Thailand*. Chiang Mai: Silkworm Books.

Mitchell, R. E. 2006. 'Green politics or environmental blues? Analyzing ecological democracy.' *Public Understanding of Science*. October, 15(4): 459–80.

Mizzima. 8 February 2012. 'PTT ready to expand its Burma oil business'. http://www.mizzima.com/business/6558-ptt-ready-to-expand-its-burma-oil-business.html. Accessed: 12 February 2012.

Modins.net 2004. *State and Division*. Modins.net. http://modins.net/MyanmarInfo/state_division/. Accessed: 2006.

Mol, A. P. J. and Carter, N. T. 2006. 'China's environmental governance in transition.' *Environmental Politics*. April, 15(2): 149–70.

Montesano, M. J., Pavin Chachavalpongpun and Aekapol Chongvilaivan eds. 2012. *Bangkok May 2010: Perspectives on a Divided Thailand*. Singapore: Institute of Southeast Asian Studies (ISEAS).

Moran, D. and Russell, J. A. 2009. 'Introduction: The militarization of energy security.' in *Energy Security and Global Politics: The Militarization of Resource Management* D. Moran and J. A. Russell eds. London and New York: Routledge.

Moses, B. 9 July 2003. 'Thaksin: Create strategic alliances.' *New Straits Times*. http://www.nst.com.my/.

Mounier, A. and Voravidh Charoenloet 2010. 'New challenges for Thailand: Labour and growth after the crisis.' *Journal of Contemporary Asia*, 40(1): 123–43.

Mukherjee, S. and Chakraborty, D. eds. 2014. *Environmental Challenges and Governance*. London and New York: Routledge.

Mukul, J. 6 December 2007. 'GAIL, ONGC lose out on Myanmar gas contract.' *Daily News & Analysis (DNA)*. Mumbai. http://www.dnaindia.com/report.asp?newsid=1137525.

Müller-Kraenner, S. 2008. *Energy Security: Re-measuring the World*. London: Earthscan.

Mulligan, S. 2010. 'Energy, environment, and security: Critical links in a post-peak world.' *Global Environmental Politics*, 10(4): 79–100.

—— 2011. 'Energy and human ecology: A critical security approach.' *Environmental Politics*, 20(5): 633–49.

The Myanmar Times. July 2008. 'Energy: A special Myanmar Times feature'. http://www.mmtimes.com/feature/energy08/feamain.htm.   Accessed: 15 September 2008.

Myat Thein 2004. *Economic Development of Myanmar.* Singapore: Institute of Southeast Asian Studies.

MYPO 2007. *In the Balance: Salween Dams Threaten Downstream Communities in Burma.* Mon Youth Progressive Organization (MYPO). http://www. salweenwatch.org/downloads/IntheBalance.pdf.

Naing Htoo, Shwe Maung, Oum Kher, Mahn Nay Myo, MacLean, K., Imamaura, M. and Giannini, T. 2002. *We Are Not Free to Work for Ourselves: Forced Labor and other Human Rights Abuses in Burma.* EarthRights International, Chiang Mai.

Naing Htoo, Smith, M. F., Khun Ko Wein, Donowitz, P. and Redford, K. 2009. *Getting it Wrong: Flawed 'Corporate Social Responsibility' and Misrepresentations Surrounding Total and Chevron's Yadana Gas Pipeline in Military-Ruled Burma (Myanmar).* EarthRights International, Chiang Mai, Thailand. September.

Nantiya Tangwisutijit 21 January 1998. 'Rival groups square off over gas pipeline.' *The Nation.* Bangkok.

The Nation. 29 December 2002. 'The week that was'. http://www.nationmultimedia. com/search/page.arcview.php?clid=11&id=71559&usrsess=.   Accessed: 28 March 2006.

—— 4 January 2004. 'Editorial: The silencing of the wolves'. http://www. nationmultimedia.com/page.arcview.php3?clid=11&id=91324&usrsess=1. Accessed: 6 April 2004.

—— 4 September 2007. 'Appeals Court acquits 19 pipeline protesters'. Bangkok. http://www.nationmultimedia.com/breakingnews/read.php?newsid-30047283. Accessed: 4 September 2007.

Naw, A. 2001. *Aung San and the Struggle for Burmese Independence.* Chiang Mai: Silkworm.

NCGUB 2006. *NCGUB Thanks Pro-Democracy, Support Groups for Burma Protesting Total.* National Coalition Government of the Union of Burma (NCGUB). 2 February.

Nelson, M. H. 2005. 'Thailand and Thaksin Shinawatra: from election triumph to political decline.' *Eastasia.at.* December, 4(2). http://www.eastasia.at/vol4_2/article01.htm.

Neuman, W. L. 2000. *Social Research Methods: Qualitative and Quantitative Approaches.* Fourth edition. Boston: Allyn and Bacon.

The New Light of Myanmar. 30 July 2013. 'Shwe natural gas project starts delivering natural gas through Myanmar-China gas pipeline': 16.

New York Times. 14 September 2004. 'Judge OKs human rights suit against Unocal'.   http://www.nytimes.com/aponline/business/AP-Unocal-Myanmar. html. Accessed: 14 September 2004.

Newell, P. 2001. 'Campaigning for corporate change: Global citizen action and the environment.' in *Global Citizen Action* M. Edwards and J. Gaventa eds. Boulder: Lynne Rienner.

—— 2005. 'Race, class and the global politics of environmental inequality.' *Global Environmental Politics*. August, 5(3): 70–94.

Newmyer, J. 2008. 'Chinese energy security and the Chinese regime.' in *Energy Security and Global Politics: The Militarization of Resource Management* D. Moran and J. A. Russell eds. London and New York: Routledge.

Newson, M. 1997. *Land, Water and Development: Sustainable Management of River Basin Systems*. Second edition. London and New York: Routledge.

NGO-COD-North and Salween Watch Coalition 2007. *Letter to Thai Prime Minister: Request withdrawal from cooperation with the Burmese military regime for the construction of hydropower dams on the Salween River*. NGO Coordinating Committee on Development (NGO-COD-North) and Salween Watch Coalition, Chiang Mai. 28 February. http://salweenwatch.org/downloads/PetitionLetterFinal.pdf. Accessed: 31 August 2007.

Noam, Z. 14 March 2008. 'Damming Salween needs proper study first.' *Bangkok Post*. http://www.bangkokpost.com/140308_News/14Mar2008_news20.php. Accessed: 17 March 2008.

Nunes, J. 2012. 'Reclaiming the political: Emancipation and critique in security studies.' *Security Dialogue*, 43(4): 345–61.

O'Kane, M. 2005. *Borderlands and Women: Transversal Political Agency on the Burma-Thailand Border*. Victoria, Australia: Monash University Press.

O'Neill, K. 2009. *The Environment and International Relations*. Cambridge: Cambridge University Press.

O'Neill, K. and VanDeveer, S. D. 2005. 'Transnational environmental activism after Seattle: Between emancipation and arrogance.' in *Charting Transnational Democracy: Beyond Global Arrogance* J. Leatherman and J. Webber eds. New York and Basingstoke: Palgrave Macmillan.

Ockey, J. 2007. 'Thailand's "professional soldiers" and coup-making: The coup of 2006.' *Crossroads: An Interdisciplinary Journal of Southeast Asian Studies*, 19(1): 95–127.

Oehlers, A. 2005. 'Public health in Burma: Anatomy of a crisis.' *Journal of Contemporary Asia*, 35(2): 195–206.

—— 2006. 'A critique of ADB policies towards the Greater Mekong Sub-Region.' *Journal of Contemporary Asia*. October, 36(4): 464–78.

Ogunlana, S., Yotsinsak, T. and Yisa, S. 2001. 'An assessment of people's satisfaction with the public hearing on the Yadana natural gas pipeline project.' *Environmental Monitoring and Assessment*. November, 72(2): 207–25.

Oilwatch SEA 2004. *Barclays Bank warned to withdraw from controversial Thai-Malaysian Gas Project*. Oilwatch SEA. 11 June. http://oilwatch-sea.org/index.php.

—— 2005a. *Pipeline opponents insist 'State has no right to force Muslims to commit a sin': Rally at Chana Land Office to demand return of common land.* Oilwatch SEA. 27 July. http://oilwatch-sea.org/index.php.

—— 2005b. *Stop destroying Islam, pipeline opponents demand.* Oilwatch SEA. 7 April. http://oilwatch-sea.org/index.php.

Okereke, C. 2008. 'Equity norms in global environmental governance.' *Global Environmental Politics.* August, 8(3): 25–50.

One World 2006. *8th International Human Rights Documentary Film Festival.* Prague, Czech Republic. 1–9 March. http://www.jedensvet.cz/ow/2006/index_en.php. Accessed: 7 March 2006.

Osborne, M. 2007. *The Water Politics of China and Southeast Asia II: Rivers, Dams, Cargo Boats and the Environment.* Lowy Institute for International Policy, Sydney. May. http://www.lowyinstitute.org/.

Ottaway, M. 2003. *Democracy Challenged: The Rise of Semi-Authoritarianism.* Washington DC: Carnegie Endowment for International Peace.

Paehlke, R. 2005. 'Democracy and environmentalism: Opening the door to the administrative state.' in *Managing Leviathan.* Second edition. R. Paehlke and D. Torgerson eds. Peterborough, Ontario: Broadview Press.

Paehlke, R. and Torgerson, D. eds. 2005. *Managing Leviathan: Environmental Politics and the Administrative State.* Second edition. Peterborough, Ontario: Broadview Press.

Paik, W. 2011. 'Authoritarianism and humanitarian aid: Regime stability and external relief in China and Myanmar.' *The Pacific Review,* 24(4): 439–62.

Painter, M. 2006. 'Thaksinisation or managerialism? Reforming the Thai bureaucracy.' *Journal of Contemporary Asia.* March, 36(1): 26–47.

Parker, C. 17 August 2005. 'Wrangle prolongs allocation of Unocal payout.' *The Irrawaddy.* http://www.irrawaddy.org/article.php?art_id=4915. Accessed: 13 February 2008.

Parry, R. L. 22 March 2006. 'Sold down the river: Tribe's home to be a valley of the dammed.' *Times Online.* http://www.timesonline.co.uk/tol/news/world/asia/article743880.ece. Accessed: 18 February 2008.

Pasuk Phongpaichit and Baker, C. 2000. *Thailand's Crisis.* Chiang Mai: Silkworm Books.

—— 2004. *Thaksin: The Business of Politics in Thailand.* Chiang Mai: Silkworm Books.

—— 2005. '"Business Populism" in Thailand.' *Journal of Democracy,* 16(2): 58–72.

—— eds. 2008. *Thai Capital after the 1997 Crisis.* Chiang Mai: Silkworm.

—— 2009. *Thaksin.* Second edition. Chiang Mai: Silkworm.

Paterson, M., Humphreys, D. and Pettiford, L. 2003. 'Conceptualizing global environmental governance: From interstate regimes to counter-hegemonic struggles.' *Global Environmental Politics.* May, 3(2): 1–10.

Paterson, M., Doherty, B. and Seel, B eds. 2000. *Direct Action in British Environmentalism.* London and New York: Routledge.

Pavin Chachavalpongpun 2005. *A Plastic Nation: The Curse of Thainess in Thai-Burmese Relations.* Lanham, Maryland: University Press of America.

—— 2009. 'Diplomacy under siege: Thailand's political crisis and the impact on foreign policy.' *Contemporary Southeast Asia*, 31(3): 447–67.

—— 2010. *Reinventing Thailand: Thaksin and His Foreign Policy.* Singapore: Institute for Southeast Asian Studies (ISEAS).

Pax Romana 2003. *Pax Romana ICMICA/MIIC.* Pax Romana ICMICA/MIIC. Geneva. http://www.icmica-miic.org/node/32. Updated: 23 May. Accessed: 1 June 2012.

Pearce, F. 2006. 'Mega-dams back on the agenda.' *New Scientist.* 16 September, 191(2569).

Pedersen, M. B. 2008. 'Burma's ethnic minorities.' *Critical Asian Studies*, 40(1): 45–66.

—— 2011. 'The politics of Burma's "democratic" transition.' *Critical Asian Studies*, 43(1): 49–68.

Penchom Tang and Pipob Udomittipong 2003. *The Resistance Against the Thai-Malaysian Gas Pipeline.* Oilwatch Assembly. September. Cartagena, Columbia.

Peoples, C. 2010. *Justifying Ballistic Missile Defence: Technology, Security and Culture.* Cambridge: Cambridge University Press.

Pepper, D. 1996. *Modern Environmentalism: An Introduction.* Abingdon: Routledge.

Perlez, J. 17 November 2006. 'Myanmar is left in dark, an energy-rich orphan.' *The New York Times.* http://www.nytimes.com/2006/11/17/world/asia/17myanmar. html.

Petronas 2000. *PETRONAS and PTT Conclude JDA Gas Purchase, Sign Bilateral Agreements for Related Infrastructure Projects.* http://www.petronas.com.my/internet/corp/news.nsf/24f01a9e1040218948256abf00275c65/b3947edbacf92 51a48256adf0049858d?OpenDocument. Accessed: 25 June 2009.

—— 2004. *Petronas signs Production Sharing Contract for Block B-17-01 in Malaysia-Thailand Joint Development Area.* http://www.petronas.com.my/internet/corp/news.nsf/24F01A9E1040218948256ABF00275C65/026C3DB C6EE8FD6048256F20002B49AF?OpenDocument. Updated: 30 September. Accessed: 6 December 2005.

Philp, J. and Mercer, D. 2002. 'Politicised pagodas and veiled resistance: Contested urban space in Burma.' *Urban Studies*, 39(9): 1587–610.

Pianporn Deetes 28 February 2007. 'The invisible costs of the Salween dam project.' *The Nation.* Bangkok. http://nationmultimedia.com/2007/02/28/opinion/opinion_30028089.php.

Pibhop Dhongchai, Boonthan Verawongse, Bhinand Jotiroseranee, Sulak Sivaraksa, Wanida Tantiwittayapitak and Penchom Tang 2005. *Response to Initial Report of Thailand: Submission to the 83rd Session of the United Nations Human Rights Committee.* Various NGOs, Bangkok. 14 March.

Pieck, S. K. 2006. 'Opportunities for transnational indigenous eco-politics: The changing landscape in the new millennium.' *Global Networks*. July, 6(3): 309–29.

Pipeline and Gas Journal 2007. 'CNPC, Myanmar launch feasibility study on gas pipeline.' March, 234(3): 16.

Piper, N. and Uhlin, A. eds. 2004. *Transnational Activism in Asia: Problems of Power and Democracy*. London and New York: Routledge.

Piya Pangsapa and Smith, M. J. 2008. 'Political economy of Southeast Asian borderlands: Migration, environment, and developing country firms.' *Journal of Contemporary Asia*, 38(4): 485–514.

Piyaporn Wongruang 4 September 2006. 'Salween dam project "likely to go ahead without study".' *Bangkok Post*.

Porter, E. 2003. 'Women, political decision-making, and peace-building.' *Global Change, Peace and Security*. October, 15(3): 245–62.

Princen, T. and Finger, M. 1994. *Environmental NGOs in World Politics: Linking the Local and the Global*. London and New York: Routledge.

Probe International 2006. *Three Gorges Probe*. Probe International. http://www. threegorgesprobe.org/tgp/index.cfm. Updated: 7 December. Accessed: 2006.

Provincial Court of Songkhla, Thailand. 30 December 2004. 'Suppression of demonstration of protestors against the Thai-Malaysian Gas Pipeline Project'. [Translated by Pipob Udomittipong].

PTTEP 2004. *PTTEP Signed Production Sharing Contract for Block B-17-01 in the MTJDA*. PTT Exploration and Production Public Company Ltd (PTTEP). http://www.pttep.com/en/news/index.asp?id=428. Updated: 30 September. Accessed: 6 December 2005.

Pulido, L. 1997. *Environmentalism and Economic Justice*. Tucsan: University of Arizona.

Pye, O. and Schaffar, W. 2008. 'The 2006 anti-Thaksin movement in Thailand: An analysis.' *Journal of Contemporary Asia*. February, 38(1): 38–61.

Quintana, T. O. 2011. *Situation of Human Rights in Myanmar*. United Nations, New York. 16 September. http://www.ohchr.org/Documents/Countries/MM/A-66-365.pdf. Accessed: 1 May 2012.

Rabinowitz, A. 2007. *Life in the Valley of Death: The Fight to Save Tigers in a Land of Guns, Gold, and Greed*. Washington DC: Island Press.

Razavi, S. 2006. 'Islamic politics, human rights and women's claims for equality in Iran.' *Third World Quarterly*. October, 27(7): 1223–37.

Reddel, T. and Woolcock, G. 2004. 'From consultation to participatory governance? A critical review of citizen engagement strategies in Queensland.' *Australian Journal of Public Administration*. September, 63(3): 75–87.

Reilly, B. 2004. 'The global spread of preferential voting: Australian institutional imperialism?' *Australian Journal of Political Science*. July, 39(2): 253–66.

Reitan, R. 2007. *Global Activism*. London and New York: Routledge.

Reuters 22 April 2007a. 'China-Myanmar oil pipe work to begin this year.' *Reuters UK*. http://uk.reuters.com/article/oilRpt/idUKPEK7637220070422.

—— 13 January 2007b. 'China, Russia veto UN resolution on Burma.' *ABC News Online*. http://www.abc.net.au/news/newsitems/200701/s1826277.htm.

—— 16 August 2007c. 'Myanmar confirms agreement to sell gas to China.' *Yahoo! Asia News*. http://asia.news.yahoo.com/070816/3/36hf0.html.

Richards, G. 1991. *The Philosophy of Gandhi: A Study of His Basic Ideas*. Richmond: Curzon Press.

Rigg, J. 1991. 'Thailand's Nam Choan Dam project: A case study in the "greening" of South-East Asia.' *Global Ecology and Biogeography Letters*, 1: 42–54.

Roberts, C. 2010. *ASEAN'S Myanmar Crisis: Challenges to the Pursuit of a Security Community*. Singapore: Institute of Southeast Asian Studies.

Rodan, G. 2004. *Transparency and Authoritarian Rule in Southeast Asia: Singapore and Malaysia*. London and New York: RoutledgeCurzon.

Rodan, G. and Hewison, K. 2006. 'Neoliberal globalization, conflict and security: New life for authoritarianism in Asia?' in *Empire and Neoliberalism in Asia* V. R. Hadiz ed. London and New York: Routledge.

Rodger, A. 1 April 2013. 'Myanmar: Dawning of a new era?' *E&P*. http://www. epmag.com/item/Myanmar-dawning-a-era_114364. Accessed: 1 September 2013.

Rootes, C. 2005. 'Globalisation, environmentalism and the Global Justice Movement.' *Environmental Politics*, 14(5): 692–96.

—— 2006. 'Facing south? British environmental movment organisations and the challenge of globalisation.' *Environmental Politics*. November, 15(5): 768–86.

—— ed. 2007. *Environmental Protest in Western Europe*. Oxford: Oxford University Press.

—— ed. 2008. *Acting Locally: Local Environmental Mobilizations and Campaigns*. London: Routledge.

Rosenbaum, W. A. 2010. *Environmental Politics and Policy*. Eighth edition. Warriewood, NSW: Footprint.

Routledge, P. 2003. 'Convergence space: process geographies of grassroots globalization networks.' *Transactions of the Institute of British Geographers*, 28(3): 333–49.

Routledge, P., Nativel, C. and Cumbers, A. 2006. 'Entangled logics and grassroots imaginaries of Global Justice Networks.' *Environmental Politics*. November, 15(5): 839–59.

Royal Thai Government 2004. *Initial Report: Thailand*. UN Human Rights Committte, Report No: CCPR/C/THA/2004/1. 2 August. http://www2.ohchr. org/english/bodies/hrc/hrcs84.htm. Accessed: 16 April 2008.

RSF 2010. *Worldwide Press Feedom Index 2010*. Reporters Sans Frontières (Reporters Without Borders) (RSF). http://en.rsf.org/press-freedom-index-2010,1034.html. Accessed: 1 November 2010.

Rucht, D. 1999. 'The transnationalization of social movements: Trends, causes, problems.' in *Social Movements in a Globalizing World* D. della Porta, H. Kriesi and D. Rucht eds. Basingstoke and New York: Palgrave Macmillan.

Rupert, M. 2000. *Ideologies of Globalization: Contending Visions of a New World Order*. London: Routledge.

Sai Silp 18 August 2006. 'Yadana pipeline protest case dismissed.' *The Irrawaddy*. http://www.irrawaddy.org/article.php?art_id=6076.

Salween Watch 2012. *Salween Watch: Threatened Peoples, Threatened River*. Salween Watch. Chiang Mai, Thailand. http://www.salweenwatch.org/index. html. Updated: 14 March. Accessed: 25 May 2012.

Salween Watch and SEARIN 2004. *The Salween Under Threat*. Salween Watch and Southeast Asia Rivers Network (SEARIN). September. http://www. salweenwatch.org/downloads/UnderThreat.pdf.

Sandler, R. D. and Pezzullo, P. C. 2007. *Environmental Justice and Environmentalism: The Social Justice Challenge to the Environmental Movement*. Cambridge, Mass: MIT Press.

Sanger, D. E. and Myers, S. L. 28 September 2007. 'US steps up confrontation with Myanmar.' *New York Times*. http://www.nytimes.com/2007/09/29/ washington/29policy.html?_r=1&scp=1&sq=chevron+burma&st=nyt&oref=s login. Accessed: 8 April 2008.

Sanitsuda Ekachai 8 January 2003. 'Neutral on which side?' *Bangkok Post*. 1.

Sapawa 2006. *Warning Signs: An Update on Plans to Dam the Salween in Burma's Shan State*. Shan Sapawa Environmental Organization (Sapawa), Chiang Mai, Thailand.   http://www.burmariversnetwork.org/images/stories/publications/ english/warningsign.pdf. Accessed:10 December 2011.

—— 2007. *Press Release: 400 Villagers Forced to Attend 'Celebration' by Thai MDX to Launch Construction of Tasang Dam on the Salween River in Shan State*. Shan Sapawa Environmental Organization (Sapawa), Chiang Mai, Thailand.   29   March.   http://www.salweenwatch.org/March29_SAPAWA_ PressRelease_Eng.html. Accessed: 14 December 2007.

Sater, J. 2007. *Civil Society and Political Change in Morocco*. London and New York: Routledge.

Sathirakoses-Nagapradeepa Foundation 2008. http://www.sulak-sivaraksa.org. Sathirakoses-Nagapradeepa Foundation (SNF). Thailand. http://www.sulak-sivaraksa.org/web/index.php. Accessed: 15 April 2008.

Saw Karen 2007. 'Karen communities keep watch.' *Salween Watch Newsletter*. August, 1(1): 5. http://www.salweenwatch.org/downloads/Salween Watch_ Vol 1 August 2007-1 FINAL.pdf. Accessed: 22 October 2007.

Saw Yan Naing 29 August 2007. 'KNU allows EGAT to survey Salween River dam site.' *The Irrawaddy*. http://www.irrawaddy.org/article.php?art_id=8416.

—— 4 March 2008a. 'Mae Hong Son businessmen unhappy with border closure.' *The Irrawaddy*. http://www.irrawaddy.org/article.php?art_id=10695. Accessed: 27 March 2008.

—— 10 November 2008b. 'Young Burmese blogger sentenced to more than 20 years in jail.' *The Irrawaddy*. http://www.irrawaddy.org/article.php?art_ id=14604. Accessed: 11 November 2008.

Sawyer, S. and Gomez, E. T. 2012. 'On indigenous identity and a language of rights.' in *The Politics of Resource Extraction: Indigenous Peoples, Multinational Corporations and the State* S. Sawyer and E. T. Gomez eds. Basingstoke: Palgrave Macmillan.

Schlosberg, D. 1999. *Environmental Justice and the New Pluralism: The Challenge of Difference for Environmentalism.* Oxford: Oxford University Press.

—— 2004. 'Reconceiving environmental justice: Global movements and political theories.' *Environmental Politics,* 13(3): 517–40.

—— 2007. *Defining Environmental Justice: Theories, Movements and Nature.* Oxford: Oxford University Press.

Schnurr, M. A. and Swatuk, L. A. eds. 2012. *Natural Resources and Social Conflict: Towards Critical Environmental Security.* Basingstoke: Palgrave Macmillan.

Scott, J. C. 2009. *The Art of Not Being Governed: An Anarchist History of Upland Southeast Asia.* New Haven: Yale University Press.

Seabrooke, L. and Elias, J. 2010. 'From multilateralism to microcosms in the world economy: The sociological turn in Australian international political economy scholarship.' *Australian Journal of International Affairs,* 64(1): 1–12.

SEAPA 21 October 2005. 'PM Thaksin should take a bow for Thailand's dismal showing in 2005 World Press Freedom Index.' *Southeast Asian Press Alliance (SEPA).* http://www.seapabkk.org/.

Selth, A. 1996. *Transforming the Tatmadaw: The Burmese Armed Forces Since 1988.* Canberra: Strategic and Defence Studies Centre, Australian National University.

—— 1998. 'The armed forces and military rule in Burma.' in *Burma: Prospects for a Democratic Future* R. I. Rotberg ed. Washington, DC and Cambridge, MA: Brookings Institution Press, World Peace Foundation and Harvard Institute for International Development.

—— 2000. 'The future of the Burmese armed forces.' in *Burma-Myanmar: Strong Regime, Weak State?* M. B. Pedersen, E. Rudland and R. J. May eds. Adelaide: Crawford House.

—— 2001. *Burma's Armed Forces: Power Without Glory.* Norwalk: EastBridge.

—— 2008. 'Burma's "saffron revolution" and the limits of international influence.' *Australian Journal of International Affairs,* 62(3): 281–97.

Selway, J. S. 2011. 'Electoral reform and public policy outcomes in Thailand: The politics of the 30-Baht Health Scheme.' *World Politics.* January, 63(1): 165–202.

Shabecoff, P. 1993. *A Fierce Green Fire: The American Environmental Movement.* New York: Hill and Wang.

Shin-who, K. 6 December 2006. 'Daewoo head indicted over illegal exports.' *The Korea Times.* http://times.hankooki.com/lpage/nation/200612/kt2006120617293110510.htm.

Shiva, V. 1989. *Staying Alive: Women, Ecology and Development.* London and New Jersey: Zed Books.

—— 1999. 'Ecological balance in an era of globalization.' in *Global Ethics and Environment* N. Low ed. London: Routledge.

—— 2008. *Soil Not Oil: Climate Change, Peak Oil, and Food Insecurity*. London: Zed Books.

SHRF and SWAN 2002. *Licence to Rape*. Shan Human Rights Foundation (SHRF) & Shan Women's Action Network (SWAN). http://www.shanland.org/resources/bookspub/humanrights/LtoR/.

Shwe Gas Movement 2006. *Supply and Command: Natural Gas in Western Burma set to Entrench Military Rule*. Shwe Gas Movement, Mae Sot, Thailand. July. http://www.shwe.org/SUPPLYANDCOMMAND.pdf.

—— 2007. *Activists Outraged at Lenient Sentencing of Daewoo International Executives for Arms Export to Burma*. Shwe Gas Movement, Chiang Mai and New Delhi. 15 November. http://www.shwe.org/media-releases/statements/Shwe Gas Movement press release, Nov 15-2007.doc.

—— 2009. *Corridor of Power: China's Trans-Burma Oil and Gas Pipelines*. Shwe as Movement (SGM), Chiang Mai, Thailand. September. http://www.shwe.org/wp-content/uploads/2011/03/CorridorofPower.pdf. Accessed: 10 February 2012.

—— 2011. *Sold Out: Launch of China Pipeline Project Unleashes Abuse across Burma*. Shwe Gas Movement (SGM), Chiang Mai, Thailand. September. http://www.shwe.org/wp-content/uploads/2011/09/Sold-Out-English-Web-Version.pdf. Accessed: 11 February 2012.

—— 2012a. *Letter to President U Thein Sein*. Shwe Gas Movement (SGM), Chiang Mai. 1 March.

—— 2012b. *The Shwe Gas Movement*. http://www.shwe.org/. Updated: 9 February. Accessed: 11 February 2012.

—— 2013. *Drawing the Line: The Case Against China's Shwe Gas Project, for Better Extractive Industries in Burma*. Shwe Gas Movement (SGM). September.

Silverman, D. 2001. *Interpreting Qualitative Data: Methods for Analysing Talk, Text and Interaction*. Second edition. London: Sage.

Silverstein, J. 2001. 'Burma and the world.' in *Burma: Political Economy under Military Rule* R. H. Taylor` ed. London: Hurst & Company.

Sim, S.-F. 2006. 'Hegemonic authoritarianism and Singapore: Economics, ideology and the Asian economic crisis.' *Journal of Contemporary Asia*, 36(2): 143–59.

Simons, M. 2006. *Transnational Litigation Manual for Human Rights and Environmental Cases in United States Courts: A Resource for Non-Lawyers*. EarthRights International, Washington DC. October. http://www.earthrights.org/files/Reports/lit_manual_2nd_edition_2007.pdf. Accessed: 14 February 2008.

Simpson, A. 1998. 'Politics plays a part: The case of Shoalwater Bay and Graham Richardson.' *Social Alternatives*. January, 17(1): 29–34.

—— 1999. 'Buddhist responses to globalisation: Thailand and ecology.' in *Socially Engaged Buddhism for the New Millennium* Sulak Sivaraksa ed. Bangkok: The Sathirakoses-Nagapradipa Foundation and Foundation for Children.

—— 2004. 'Gas pipelines and green politics in South and Southeast Asia.' *Social Alternatives*. December, 23(4): 29–36.

—— 2007. 'The environment-energy security nexus: Critical analysis of an energy "love triangle" in Southeast Asia.' *Third World Quarterly*. April, 28(3): 539–54.

—— 2008. 'Gas pipelines and security in South and Southeast Asia: A critical perspective.' in *Crucible for Survival: Environmental Security and Justice in the Indian Ocean Region* T. Doyle and M. Risely eds. New Brunswick, New Jersey and London: Rutgers University Press.

—— 2012. 'Prospects for a policy of engagement with Myanmar: A multilateral development bank perspective.' in *Myanmar's Transition: Openings, Obstacles and Opportunities* N. Cheesman, M. Skidmore and T. Wilson eds. Singapore: ISEAS.

—— 2013a. 'An "activist diaspora" as a response to authoritarianism in Myanmar: The role of transnational activism in promoting political reform.' in *Civil Society Activism under Authoritarian Rule: A Comparative Perspective* F. Cavatorta ed. London and New York: Routledge/ECPR Studies in European Political Science.

—— 2013b. 'Challenging hydropower development in Myanmar (Burma): Cross-border activism under a regime in transition.' *The Pacific Review*. May, 26(2): 129–52.

—— 2013c. 'Challenging inequality and injustice: A critical approach to energy security.' in *Environmental Security: Approaches and Issues* R. Floyd and R. Matthew eds. London and New York: Routledge.

—— 2014a. 'Starting from year zero: Environmental governance in Myanmar.' in *Environmental Challenges and Governance: Diverse Perspectives from Asia* S. Mukherjee and D. Chakraborty eds. London and New York: Routledge.

—— 2014b. 'Democracy and environmental governance in Thailand.' in *Environmental Challenges and Governance: Diverse Perspectives from Asia* S. Mukherjee and D. Chakraborty eds. London and New York: Routledge.

—— 2014c. 'Market building and risk under a regime in transition: The Asian Development Bank in Myanmar (Burma).' in *The Politics of Marketising Asia* T. Carroll and D. S. L. Jarvis eds. Basingstoke and New York: Palgrave Macmillan.

Simpson, A. and Park, S. 2013. 'The Asian Development Bank as a global risk regulator in Myanmar.' *Third World Quarterly*, 34(10): 1858–71.

Skidmore, M. and Wilson, T. 2007. *Myanmar: The State, Community and the Environment*. Canberra: ACT Asia Pacific Press, ANU.

—— eds. 2008. *Dictatorship, Disorder and Decline in Myanmar*. Canberra: ANU E Press. http://epress.anu.edu.au/myanmar02/pdf/whole_book.pdf. Accessed: 3 November 2009.

Smith, Christen A. 2009. 'Strategies of confinement: Environmental injustice and police violence in Brazil.' in *Environmental Justice in the New Millennium: Global Perspectives on Race, Ethnicity, and Human Rights* F. C. Steady ed. New York: Palgrave Macmillan.

Smith, Jackie and Bandy, J. 2005. 'Introduction: Cooperation and conflict in transnational protest.' in *Coalitions across Borders: Transnational Protest and the Neoliberal Order* J. Bandy and J. Smith eds. Lanham, Maryland and Oxford: Rowman and Littlefield.

Smith, Martin. 1994. *Paradise Lost? The Suppression of Environmental Rights and Freedom of Expression in Burma.* UK: Article 19.

—— 1999. *Burma: Insurgency and the Politics of Ethnicity.* Second edition. London: Zed Books.

Smith, Matthew. 2007. 'Environmental governance of mining in Burma.' in *Myanmar: The State, Community and the Environment* M. Skidmore and T. Wilson eds. Canberra: ACT Asia Pacific Press, ANU.

Smith, Matthew and Naing Htoo 2005. 'Another snake in the jungle? Shwe gas development in western Burma.' *Watershed.* July-October, 11(1): 31–38. http://www.earthrights.org/files/Reports/another_snake_in_the_jungle.pdf. Accessed: 1 February 2008.

—— 2006. 'Gas Politics: Shwe Gas Development in Burma.' *Watershed.* June, 11(2). http://www.earthrights.org/files/Burma Project/Shwe/shwe_gas_ watershed_article_sep_06.pdf. Accessed: 1 February 2008.

Sökefeld, M. 2006. 'Mobilizing in transnational space: A social movement approach to the formation of diaspora.' *Global Networks*, 6(3): 265–84.

Somchai Phatharathananunth 2006. *Civil Society and Democratization: Social Movements in Northeast Thailand.* Copenhagen: NIAS Press.

—— 2008. 'The Thai Rak Thai Party and elections in Northeastern Thailand.' *Journal of Contemporary Asia*, 38(1): 106–23.

Somroutai Sapsomboon 31 December 2004. 'Gas pipeline protest: 20 acquitted of all charges.' *The Nation.* http://www.nationmultimedia.com/search/page. arcview.php?clid=3&id=110590&usrsess=.

South, A. 2004. 'Political transition in Myanmar: A new model for democratization.' *Contemporary Southeast Asia*, 26(2): 233–55.

—— 2009. *Ethnic Politics in Burma: States of Conflict.* London and New York: Routledge.

—— 2011. *Burma's Longest War: Anatomy of the Karen Conflict.* Transnational Institute (TNI) and Burma Center Netherlands (BCN), Amsterdam. March.

—— 2012. 'The politics of protection in Burma: Beyond the humanitarian mainstream.' *Critical Asian Studies*, 44(2): 175–204.

Sovacool, B. K. 2009. 'Reassessing energy security and the Trans-ASEAN natural gas pipeline network in Southeast Asia.' *Pacific Affairs*, 82(3): 467–86.

—— 2013. 'Confronting energy poverty behind the bamboo curtain: A review of challenges and solutions for Myanmar (Burma).' *Energy for Sustainable Development*, 17: 305–14.

Sovacool, B. K. and Cooper, C. J. 2013. *The Governance of Energy Megaprojects: Politics, Hubris and Energy Security*. Cheltenham, UK: Edward Elgar.

Sovacool, B. K. and Drupady, I. M. 2012. *Energy Access, Poverty, and Development: The Governance of Small-Scale Renewable Energy in Developing Asia*. Farnham: Ashgate.

Srisompob Jitpiromrsi and McCargo, D. 2008. 'A ministry for the south: New governance proposals for Thailand's southern region.' *Contemporary Southeast Asia*. December, 30(3): 403–28.

Steans, J. 2007. 'Debating women's human rights as a universal feminist project: Defending women's human rights as a political tool.' *Review of International Studies*. January, 33(1): 11–27.

Stein, G. 2004. 'Krue Se Mosque Massacre'. 16 June. SBS. Dateline. http://news.sbs.com.au/dateline/index.php?page=archive&daysum=2004-06-16. Accessed: 23 March 2006.

Steinberg, D. I. 1998. 'The road to political recovery.' in *Burma: Prospects for a Democratic Future* R. I. Rotberg ed. Washington, DC and Cambridge, MA: Brookings Institution Press, World Peace Foundation and Harvard Institute for International Development.

—— 2005a. 'Burma/Myanmar: The role of the military in the economy.' *Burma Economic Watch*, 1/2005: 51–78. http://www.econ.mq.edu.au/__data/assets/pdf_file/17411/1BEW2005.pdf.

—— 2005b. 'Myanmar: The roots of economic malaise.' in *Myanmar: Beyond Politics to Social Imperatives* Kyaw Yin Hlaing, R. H. Taylor and Tin Maung Maung Than eds. Singapore: Institute of Southeast Asian Studies.

—— 2007. 'Legitimacy in Burma/Myanmar: Concepts and implications.' in *Myanmar: State, Society and Ethnicity* N. Ganesan and K. Y. Hlaing eds. Singapore, Institute for Southeast Asian Studies (ISEAS).

—— 2010. *Burma/Myanmar: What Everyone Needs to Know*. Oxford: Oxford University Press.

Steinberg, D. I. and Fan, H. 2012. *Modern China-Myanmar Relations: Dilemmas of Mutual Dependence*. Copenhagen: NIAS.

Stephan, M. J. and Chenoweth, E. 2008. 'Why civil resistance works: The strategic logic of nonviolent conflict.' *International Security*. Summer, 33(1): 7–44.

Stokes, D. and Raphael, S. 2010. *Global Energy Security and American Hegemony*. Washington DC: Johns Hopkins University Press.

Streckfuss, D. 2011. *Truth on Trial in Thailand: Defamation, Treason, and Lèse-Majesté*. New York and London: Routledge.

Subhatra Bhumiprabhas 28 May 2003. 'Special 1: Law, order and a climate of fear.' *The Nation*. http://www.nationmultimedia.com/specials/humanrights/index_special1.php.

Sulak Sivaraksa 1988. *A Socially Engaged Buddhism*. Bangkok: Thai Inter-Religious Commission for Development.

—— 1992. *Seeds of Peace: A Buddhist Vision for Renewing Society*. Berkeley and Bangkok: Parallax Press and International Network of Engaged Buddhists, Sathirakoses-Nagapradipa Foundation.

Supalak Ganjanakhundee 26 April 2005. 'Krue Se, Tak Bai Reports: Muslims hail "first good step" by NRC.' *The Nation*. http://www.nationmultimedia. com/2005/04/26/headlines/index.php?news=headlines_17160271.html.

Supara Janchitfah 15 September 1998. 'Learning to see the wood and the trees.' *Bangkok Post*.

—— 2004. *The Nets of Resistance*. Bangkok: Campaign for Alternative Industry Network.

Swearer, D. K. 1999. 'Centre and periphery: Buddhism and politics in modern Thailand.' in *Buddhism and Politics in Twentieth Century Asia* I. Harris ed. London and New York: Pinter.

Szep, J. 27 January 2012. 'Myanmar has no plans to export new gas finds.' *Reuters*. http://af.reuters.com/article/commoditiesNews/ idAFL4E8CR3SO20120127?sp=true. Accessed: 1 February 2012.

Szerszynski, B. 2007. 'The post-ecologist condition: Irony as symptom and cure.' *Environmental Politics*. April, 16(2): 337–55.

Tadros, M. 2005. 'The Internet and the Arab Middle East.' in *Charting Transnational Democracy: Beyond Global Arrogance* J. Leatherman and J. Webber eds. New York and Basingstoke: Palgrave Macmillan.

Tang, S.-Y. and Zhan, X. 2008. 'Civic environmental NGOs, civil society, and democratisation in China.' *Journal of Development Studies*. March, 44(3): 425–48.

Tangseefa, D. 2006. 'Taking flight in condemned grounds: Forcibly displaced Karens and the Thai-Burmese in-between spaces.' *Alternatives*. October–December, 31(4): 405–29.

Tarrow, S. 2005. *The New Transnational Activism*. New York: Cambridge University Press.

Taylor, J. 1991. 'Living on the rim: Ecology and forest monks in Northeast Thailand.' *Sojourn*. February, 6(1): 106–25.

—— 1993. 'Social activism and resistance on the Thai frontier: The case of Phra Prajak Khuttajitto.' *Critical Asian Studies*. March, 25(2): 3–16.

—— 1997. '"Thamma-chaat": Activist monks and competing discourses of nature and nation in Northeastern Thailand.' in *Seeing Forests for the Trees: Environment and Environmentalism in Thailand* P. Hirsch ed. Chiang Mai: Silkworm Books.

Taylor, R. H. 2005. 'Pathways to the present.' in *Myanmar: Beyond Politics to Social Imperatives* Kyaw Yin Hlaing, R. H. Taylor and Tin Maung Maung Than eds. Singapore: Institute of Southeast Asian Studies.

—— 2009. *The State in Myanmar*. Singapore: National University of Singapore Press.

TERRA 2011. *Key Issues: Salween River*. Towards Ecological Recovery and Regional Alliance (TERRA) and Foundation for Ecological

Recovery (FER). http://www.terraper.org/mainpage/key_issues_detail_en.php?kid=11&langs=en. Updated: March. Accessed: 25 April 2012.

Thak Chaloemtiarana 2007a. 'Distinctions with a difference: The despotic paternalism of Sarit Thanarat and the demagogic authoritarianism of Thaksin Shinawatra.' *Crossroads: An Interdisciplinary Journal of Southeast Asian Studies*, 19(1): 50–94.

—— 2007b. *Thailand: The Politics of Despotic Paternalism*. Chiang Mai: Silkworm.

Thant Myint-U 2006. *The River of Lost Footsteps: A Presonal History of Burma*. New York: Farrar, Straus and Giroux.

—— 2011. *Where China Meets India: Burma and the New Crossroads of Asia*. London: Faber and Faber.

Thiha Aung 2005a. 'Hailing the 58th Anniversary Union Day: Rakhine State marching to a new golden land of unity and amity.' *New Light of Myanmar*. 6 February. http://www.ibiblio.org/obl/docs2/RakhineDvm.pdf. Accessed: 22 July 2008.

—— 2005b. 'Hailing the 58th Anniversary Union Day: Taninthayi Division marching to a new golden land of unity and amity.' *New Light of Myanmar*. 9 February. http://www.ibiblio.org/obl/docs2/TaninthayiDvm.pdf. Accessed: 22 July 2008.

—— 2005c. 'Honouring the 58th Anniversary Union Day: Yangon Division marching to a new golden land of unity and amity.' *New Light of Myanmar*. 14 February. http://www.ibiblio.org/obl/docs2/YGNDvm.pdf. Accessed: 22 July 2008.

Thitinan Pongsudhirak 2012. 'Thailand's uneasy passage.' *Journal of Democracy*. April, 23(2): 47–61.

Thomas, C. 2000. *Global Governance, Development and Human Security: The Challenge of Poverty and Inequality*. London: Pluto Press.

Thongchai Winichakul 2008. 'Nationalism and the radical intelligentsia in Thailand.' *Third World Quarterly*, 29(3): 575–91.

Thorp, R., Battistelli, S., Guichaoua, Y., Orihuela, J. C. and Paredes, M. eds. 2012. *The Developmental Challenges of Mining and Oil: Lessons from Africa and Latin America*. Basingstoke: Palgrave Macmillan.

Tilly, C. 2004. *Social Movements, 1768–2004*. Boulder and London: Paradigm Publishers.

Tin Maung Maung Than 2005a. 'Dreams and nightmares: State building and ethnic conflict in Myanmar (Burma).' in *Ethnic Conflicts in Southeast Asia* K. Snitwongse and W. S. Thompson eds. Bangkok and Singapore: Institute of Security and International Studies and Institute of Southeast Asian Studies.

—— 2005b. 'Myanmar's energy sector: Banking on natural gas.' in *Southeast Asian Affairs* C. K. Wah and D. Singh eds. Singapore: Institute of Southeast Asian Studies.

—— 2007. *State Dominance in Myanmar: The Political Economy of Industrialization*. Singapore: ISEAS.

—— 2011. 'Myanmar's 2010 elections: Continuity and change.' *Southeast Asian Affairs*: 190–207.

Tint Lwin Thaung 2007. 'Identifying conservation issues in Kachin State.' in *Myanmar: The State, Community and the Environment* M. Skidmore and T. Wilson eds. Canberra, ACT: Asia Pacific Press, ANU.

Tisdall, S. 17 January 2007. 'UN vetoes prolong Burma agony.' *Guardian Unlimited.* http://www.guardian.co.uk/commentisfree/story/0,,1992055,00. html.

TNA 5 September 2007. 'Thai dam project in Burma suspended.' *Bangkok Post (Thai News Agency).* http://www.bangkokpost.net/topstories/topstories. php?id=121362.

Torgerson, D. 1999. *The Promise of Green Politics: Environmentalism and the Public Sphere.* Durham and London: Duke University Press.

—— 2006. 'Expanding the green public sphere: Post-colonial connections.' *Environmental Politics.* November, 15(5): 713–30.

Total 29 November 2005. 'Myanmar: Total and the Sherpa Association reach agreement for the creation of a solidarity fund for humanitarian actions.' *Total.* http://www.total.com/en/press/press_releases/pr_2005/051129_myanmar_ agreement_sherpa_8205.htm.

Transparency International 2011. *Corruption Perceptions Index.* Transparency International. http://cpi.transparency.org/cpi2011/results/ – CountryResults. Accessed: 1 May 2012.

Tsikata, D. 2007. 'Announcing a new dawn prematurely? Human rights feminists and the rights-based approaches to development.' in *Feminisms in Development: Contradictions, Contestations and Challenges* A. Cornwall, E. Harrison and A. Whitehead eds. London and New York: Zed Books.

TTM 2012. *About TTM.* Trans Thai-Malaysia Limited (TTM). Bangkok. http:// www.ttm-jda.com/ENG/about_ttm.aspx. Updated: 2 May. Accessed: 16 May 2012.

Tun Myint 2007. 'Environmental governance in the SPDC's Myanmar.' in *Myanmar: The State, Community and the Environment* M. Skidmore and T. Wilson eds. Canberra, ACT: Asia Pacific Press, ANU.

Tunya Sukpanich 11 November 2007. 'Salween on a precipice.' *Bangkok Post.* http://www.bangkokpost.net/.

Turnell, S. 2007. 'Myanmar's economy in 2006.' in *Myanmar: The State, Community and the Environment* M. Skidmore and T. Wilson eds. Canberra, ACT: Asia Pacific Press, ANU.

—— 2008. 'Burma's insatiable state.' *Asian Survey*, 48(6): 958–76.

—— 2009. *Fiery Dragons: Banks, Moneylenders and Microfinance in Burma.* Copenhagen: NIAS Press.

—— 2010. 'Finding dollars and sense: Burma's economy in 2010.' in *Finding Dollars, Sense and Legitimacy in Burma* S. L. Levenstein ed. Washington DC: Woodrow Wilson Centre.

—— 2012. 'Reform and its limits in Myanmar's fiscal state.' in *Myanmar's Transition: Openings, Obstacles and Opportunities* N. Cheesman, M. Skidmore and T. Wilson eds. Singapore: ISEAS.

Ufen, A. 2008. 'Political party and party system institutionalization in Southeast Asia: Lessons for democratic consolidation in Indonesia, the Philippines and Thailand.' *The Pacific Review*, 21(3): 327–50.

UN 2008. *United Nations Declaration on the Rights of Indigenous Peoples*. United Nations, New York. March. http://www.un.org/esa/socdev/unpfii/documents/DRIPS_en.pdf. Accessed: 25 April 2012.

UNDP 2006. *Public Health Expenditure*. United Nations Development Programme (UNDP). November. http://hdr.undp.org/hdr2006/statistics/indicators/50.html.

UNEP 2005. *Sri Lanka: Post-Tsunami Environmental Assessment*. United Nations Environment Programme and Ministry of Environment & Natural Resources of Sri Lanka, Geneva. http://www.unep.org/tsunami/reports/Sri_Lanka_Report_2005.pdf. Accessed: 21 May 2012.

Ungera, D. H. and Patcharee Sirorosb 2011. 'Trying to make decisions stick: Natural resource policy making in Thailand.' *Journal of Contemporary Asia*, 41(2): 206–28.

UNSC 2008. *Security Council demands immediate and complete halt to acts of sexual violence against civilians in conflict zones, unanimously adopting resolution 1820 (2008)*. 19 June. http://www.un.org/News/Press/docs//2008/sc9364.doc.htm. Accessed: 20 June 2008.

USCB 2005. *Protests to Condemn France's Protection of Total Oil Deal with Burma's Dictator*. US Campaign for Burma (USCB). 16 May.

Vasana Chinvarakorn 16 March 2000. 'Learning from experience.' *Bangkok Post*. Outlook.

—— 19 September 2002. 'Fighting to keep the light on.' *Bangkok Post*. Outlook.

Verma, N. 14 August 2007. 'Myanmar to sell gas to China – India minister.' *Reuters India*. http://in.reuters.com/article/businessNews/idINIndia-28977320070814?sp=true.

Vicary, A. 2007. 'Revisiting the financing of health in Burma: A comparison with the other ASEAN countries.' *Burma Economic Watch*, 1: 4–10. http://www.econ.mq.edu.au/Econ_docs/bew/BEW2007.pdf. Accessed: 20 August 2008.

—— 2010. 'The relief and reconstruction programme following Cyclone Nargis: A review of SPDC policy.' in *Ruling Myanmar: From Cyclone Nargis to National Elections* N. Cheesman, M. Skidmore and T. Wilson eds. Singapore: Institute of Southeast Asian Studies: 208–35.

Walker, A. 2012a. *Thailand's Political Peasants: Power in the Modern Rural Economy*. Madison: The University of Wisconsin Press.

Walker, G. 2012b. *Environmental Justice: Concepts, Evidence and Politics*. London and New York: Routledge.

Walker, G. and Bulkeley, H. 2006. 'Geographies of environmental justice.' *Geoforum*. September, 37(5): 655–59.

Wall, D. 1999. *Earth First! and the Anti-Roads Movement: Radical Environmentalism and Comparative Social Movements*. London and New York: Routledge.

Walter, P. 2007. 'Activist forest monks, adult learning and the Buddhist environmental movement in Thailand.' *International Journal of Lifelong Education*, 26(3): 329–45.

Walton, M. J. 2008. 'Ethnicity, conflict, and history in Burma: The myths of Panglong.' *Asian Survey*, 48(6): 889–910.

Wandee Suntivutimetee 1 October 2003. 'Burma in the Thai Press.' *The Irrawaddy*. http://www.irrawaddy.org/article.php?art_id=3132. Accessed: 25 September 2008.

Wapner, P. 1996. *Environmental Activism and World Civic Politics*. Albany, NY: State University of New York Press.

—— 2010. *Living Through the End of Nature: The Future of American Environmentalism*. Cambridge: MIT Press.

Warasak Phuangcharoen 2005. 'The failure of public participation in developing countries: Examples from the Yadana and JDA pipeline projects in Thailand.' *Thai Khadi Journal*, 2(1).

Warr, P. ed. 2005. *Thailand Beyond the Crisis*. London and New York: RoutledgeCurzon.

Watershed. March–October 2007. 'Thailand's Human Rights Commission recommends Hutgyi dam be shelved'. http://www.terraper.org/pic_water/WS12_2_web.pdf. Accessed: 17 March 2008.

WCD 2000. *Dams and Development: A New Framework for Decision Making*. World Commission on Dams (WCD). Earthscan, London. 16 November. http://www.dams.org/report/.

WCS 2003. *Science and Exploration Program: Hukaung Valley Wildlife Sanctuary, Myanmar*. Wildlife Conservation Society (WCS), New York. http://www.savingwildplaces.com/media/file/hukaung2.pdf. Accessed: 14 March 2008.

—— 2006. *Hukaung Valley Tiger Reserve: Myanmar*. Wildlife Conservation Society (WCS). New York. http://www.savingwildplaces.com/swp-home/swp-protectedareas/239764. Accessed: 14 March 2008.

Wellner, P. 1994. 'A pipeline killing field: exploitation of Burma's natural gas.' *The Ecologist*. September–October, 24(5): 189–193.

Welzel, C. and Deutsch, F. 2012. 'Emancipative values and non-violent protest: The importance of "ecological" effects.' *British Journal of Political Science*, 42(2): 465–79.

Wesley, M. ed. 2007. *Energy Security in Asia*. New York: Routledge.

Williams, D. C. 2011. 'Cracks in the firmament of Burma's military government: From unity through coercion to buying support.' *Third World Quarterly*. August, 32(7): 1199–215.

Williams, G. and Mawdsley, E. 2006. 'Postcolonial environmental justice: Government and governance in India.' *Geoforum*. September, 37(5): 660–70.

Wong, W. 2012. *Internal Affairs: How the Structure of NGOs Transforms Human Rights*. Ithaca and London: Cornell University Press.

Woods, K. 2006. 'What does the language of human rights bring to campaigns for environmental justice?' *Environmental Politics*. August, 15(4): 572–91.

—— 2010. *Human Rights and Environmental Sustainability*. Cheltenham: Edward Elgar.

Wright, J. and Escriba-Folch, A. 2012. 'Authoritarian institutions and regime survival: Transitions to democracy and subsequent autocracy.' *British Journal of Political Science*, 42(2): 283–309.

Wyatt, D. K. 2004. *Thailand: A Short History*. Second edition. Chiang Mai: Silkworm.

Xie, L. 2009. *Environmental Activism in China*. London and New York: Routledge.

Xinhua 30 January 2007. 'China's CNPC, Myanmar Oil & Gas launch feasibility study on gas pipeline.' *Xinhua*.

Yanacopulos, H. 2005a. 'Patterns of governance: The rise of transnational coalitions of NGOs.' *Global Society*. July, 19(3): 247–66.

—— 2005b. 'The strategies that bind: NGO coalitions and their influence.' *Global Networks*, 5(1): 93–110.

Ye Lwin 2008. 'Oil and gas ranks second largest FDI at $3.24 billion.' *The Myanmar Times*. 21–27 July, 22(428). http://www.mmtimes.com/feature/energy08/eng002.htm. Accessed: 15 September 2008.

Yeni 6 December 2006. 'S. Korean companies "exported arms technology to Burma".' *The Irrawaddy*. http://www.irrawaddy.org/aviewer.asp?a=6467&z=154.

Yergin, D. 1993. *The Prize: The Epic Quest for Oil, Money, and Power*. New York: Pocket Books.

—— 2011. *The Quest: Energy, Security, and the Remaking of the Modern World*. New York: Penguin.

Yi-chong, X. 2007. 'China's energy security.' in *Energy Security in Asia* M. Wesley ed. New York: Routledge.

Yin, R. K. 2009. 'How to do better case studies.' in *The SAGE Handbook of Applied Social Research Methods*. Second edition. L. Bickman and D. J. Rog eds. Thousand Oaks, CA: SAGE.

Zarsky, L. 2002. 'Global reach: Human rights and environment in the framework of corporate accountability.' in *Human Rights and the Environment: Conflicts and Norms in a Globalizing World* L. Zarsky ed. London: Earthscan.

Index